Mobile Application Development, Usability, and Security

Sougata Mukherjea
IBM, India

A volume in the Advances in Multimedia and Interactive Technologies (AMIT) Book Series

IGI GLOBAL
DISSEMINATOR OF KNOWLEDGE

www.igi-global.com

Published in the United States of America by
IGI Global
Information Science Reference (an imprint of IGI Global)
701 E. Chocolate Avenue
Hershey PA, USA 17033
Tel: 717-533-8845
Fax: 717-533-8661
E-mail: cust@igi-global.com
Web site: http://www.igi-global.com

Library of Congress Cataloging-in-Publication Data

Names: Mukherjea, Sougata, 1965- editor.
Title: Mobile application development, usability, and security / Sougata Mukherjea, editor.
Description: Hershey PA : Information Science Reference, [2017] | Series: Advances in multimedia and interactive technologies | Includes bibliographical references and index.
Identifiers: LCCN 2016033134| ISBN 9781522509455 (h/c) | ISBN 9781522509462 (eISBN)
Subjects: LCSH: Mobile apps. | Application software--Development--Management. | Application program interfaces (Computer software)
Classification: LCC QA76.76.A65 M596 2017 | DDC 005.35--dc23 LC record available at https://lccn.loc.gov/2016033134

This book is published in the IGI Global book series Advances in Multimedia and Interactive Technologies (AMIT) (ISSN: 2327-929X; eISSN: 2327-9303)

British Cataloguing in Publication Data
A Cataloguing in Publication record for this book is available from the British Library.

For electronic access to this publication, please contact: eresources@igi-global.com.

Advances in Multimedia and Interactive Technologies (AMIT) Book Series

Joel J.P.C. Rodrigues
Instituto de Telecomunicações, University of Beira Interior, Portugal

ISSN: 2327-929X
EISSN: 2327-9303

Mission

Traditional forms of media communications are continuously being challenged. The emergence of user-friendly web-based applications such as social media and Web 2.0 has expanded into everyday society, providing an interactive structure to media content such as images, audio, video, and text.

The **Advances in Multimedia and Interactive Technologies (AMIT) Book Series** investigates the relationship between multimedia technology and the usability of web applications. This series aims to highlight evolving research on interactive communication systems, tools, applications, and techniques to provide researchers, practitioners, and students of information technology, communication science, media studies, and many more with a comprehensive examination of these multimedia technology trends.

Coverage

- Multimedia Services
- Multimedia technology
- Digital Watermarking
- Digital Technology
- Audio Signals
- Digital Images
- Mobile Learning
- Digital Games
- Gaming Media
- Digital Communications

IGI Global is currently accepting manuscripts for publication within this series. To submit a proposal for a volume in this series, please contact our Acquisition Editors at Acquisitions@igi-global.com or visit: http://www.igi-global.com/publish/.

The Advances in Multimedia and Interactive Technologies (AMIT) Book Series (ISSN 2327-929X) is published by IGI Global, 701 E. Chocolate Avenue, Hershey, PA 17033-1240, USA, www.igi-global.com. This series is composed of titles available for purchase individually; each title is edited to be contextually exclusive from any other title within the series. For pricing and ordering information please visit http://www.igi-global.com/book-series/advances-multimedia-interactive-technologies/73683. Postmaster: Send all address changes to above address.

Titles in this Series

For a list of additional titles in this series, please visit: www.igi-global.com

Intelligent Analysis of Multimedia Information
Siddhartha Bhattacharyya (RCC Institute of Information Technology, India) Hrishikesh Bhaumik (RCC Institute of Information Technology, India) Sourav De (The University of Burdwan, India) and Goran Klepac (University College for Applied Computer Engineering Algebra, Croatia & Raiffeisenbank Austria, Croatia)
Information Science Reference • copyright 2017 • 520pp • H/C (ISBN: 9781522504986) • US $220.00 (our price)

Emerging Technologies and Applications for Cloud-Based Gaming
P. Venkata Krishna (VIT University, India)
Information Science Reference • copyright 2017 • 314pp • H/C (ISBN: 9781522505464) • US $195.00 (our price)

Digital Tools for Computer Music Production and Distribution
Dionysios Politis (Aristotle University of Thessaloniki, Greece) Miltiadis Tsalighopoulos (Aristotle University of Thessaloniki, Greece) and Ioannis Iglezakis (Aristotle University of Thessaloniki, Greece)
Information Science Reference • copyright 2016 • 291pp • H/C (ISBN: 9781522502647) • US $180.00 (our price)

Contemporary Research on Intertextuality in Video Games
Christophe Duret (Université de Sherbrooke, Canada) and Christian-Marie Pons (Université de Sherbrooke, Canada)
Information Science Reference • copyright 2016 • 363pp • H/C (ISBN: 9781522504771) • US $185.00 (our price)

Trends in Music Information Seeking, Behavior, and Retrieval for Creativity
Petros Kostagiolas (Ionian University, Greece) Konstantina Martzoukou (Robert Gordon University, UK) and Charilaos Lavranos (Ionian University, Greece)
Information Science Reference • copyright 2016 • 388pp • H/C (ISBN: 9781522502708) • US $195.00 (our price)

Emerging Perspectives on the Mobile Content Evolution
Juan Miguel Aguado (University of Murcia, Spain) Claudio Feijóo (Technical University of Madrid, Spain & Tongji University, China) and Inmaculada J. Martínez (University of Murcia, Spain)
Information Science Reference • copyright 2016 • 438pp • H/C (ISBN: 9781466688384) • US $210.00 (our price)

Emerging Research on Networked Multimedia Communication Systems
Dimitris Kanellopoulos (University of Patras, Greece)
Information Science Reference • copyright 2016 • 448pp • H/C (ISBN: 9781466688506) • US $200.00 (our price)

Emerging Research and Trends in Gamification
Harsha Gangadharbatla (University of Colorado Boulder, USA) and Donna Z. Davis (University of Oregon, USA)
Information Science Reference • copyright 2016 • 455pp • H/C (ISBN: 9781466686519) • US $215.00 (our price)

Table of Contents

Detailed Table of Contents

As devices have become smaller and more pervasive, usage scenarios that have historically been common for people with disabilities are finding more general application for all users. Overall, the consideration of accessibility improves the usability of applications for all users. This chapter will discuss standards for accessibility, inclusive design, and topics related to the development of accessible mobile content and applications. The discussion will apply to mobile content, such as EPUB documents, and topics related to Web, native, and hybrid applications.

Mobile devices have revolutionized many aspects of our lives. Without realizing it, we often run on them programs that access and transmit private information over the network. Integrity concerns arise when mobile applications use untrusted data as input to security-sensitive computations. Program-analysis tools for integrity and confidentiality enforcement have become a necessity. Static-analysis tools are particularly attractive because they do not require installing and executing the program, and have the potential of never missing any vulnerability. Nevertheless, such tools often have high false-positive rates. In order to reduce the number of false positives, static analysis has to be very precise, but this is in conflict with the analysis' performance and scalability, requiring a more refined model of the application. This chapter proposes Phoenix, a novel solution that combines static analysis with machine learning to identify programs exhibiting suspicious operations. This approach has been widely applied to mobile applications obtaining impressive results.

In order to secure mobile devices, there has been movement to trust negotiation where two entities are able to establish a measure of mutual trust, even if no prior contact between either entity has existed in the past. This chapter explores adaptive trust negotiation in a mobile environment as a means to dynamically adjust security parameters based on the level of trust established during the negotiation process thereby enhancing mobile security. To accomplish this, the chapter proposes a trust profile that contains a proof of history of successful access to sensitive data to facilitate identification and authentication for adaptive trust negotiation. The trust profile consists of a set of X.509 identity and attribute certificates, where

Mobile analytics is the systematic study of mobile device and application usage, and application performance, for the purpose of improving service quality. This chapter motivates the need for mobile analytics as an essential cog in the emerging economy built around devices, applications, and communication. A taxonomy of mobile analytics problems is presented, and technical details of a typical mobile analytics solution are discussed. Scale, heterogeneity, dynamically changing environments, and diverse privacy requirements pose challenges to collecting and processing data for such analysis. This chapter examines how analytics solutions handle these challenges. The core of the chapter consists of a technical section describing the general architecture of a mobile analytics solution, procedures to collect and process data, event monitoring infrastructure, system administration processes, and privacy management policies. Case studies of a number of analytics solutions available as commercial products or prototypes are presented.

A mobile phones provides portability and personalized computing with ubiquitous connectivity. This combination makes them an ideal choice to use for various applications of personal use. The portability of mobile devices is the most important and useful feature of mobile devices. However, portability is achieved at the high cost of limited power and computation ability of the mobile device. Cloud computing fulfills the need of providing more computation power to complete the tasks that cannot be done on a mobile platform. The cloud provides an always available platform and do not have typical limitations, e.g. limited battery and computation power, of mobile platforms. Therefore combining cloud computing with mobile provides us best of both worlds i.e. we have a computing platform available for us all the time which we move, and yet we can access services and perform tasks that require high-power computation.

Preface

INTRODUCTION

It's no secret that the smart phone has become one of the most important devices in our lives. We use it for interacting with each other, entertainment as well as for performing various day to day activities. There are now more mobile devices than people in the world. Mobility is also impacting business significantly and most enterprises are providing services to facilitate their mobile workforce. More of us are using mobile technologies at work. In fact, use of mobile in the enterprise is growing at about 25% per year (Markets and Markets, 2014).

THE DIFFERENT ASPECTS OF MOBILITY

Most mobile devices use one of the two dominant operating systems: Google-developed Android and the Apple-developed iOS. Market shares of other mobile platforms like Windows and Blackberry are diminishing. Users interact with the smart phones using various mobile applications. As expected in recent days mobile applications have become extremely popular - among consumers as well as in the enterprise. There are three types of mobile application:

- Native apps live on the device and are accessed through icons on the device home screen. They are developed specifically for one Operating System and can take full advantage of all the device features. Native apps are installed through an application store (such as Google Play or Apple's App Store).
- Web apps are not real applications; they are really websites that, in many ways, look and feel like native applications, but are not implemented as such. They are run by a browser and typically written in HTML5.
- Hybrid apps are part native apps, part web apps. Like native apps, they live in an app store and can take advantage of thc many device features available. Like web apps, they rely on HTML being rendered in a browser, with the caveat that the browser is embedded within the app.

The Mobile Application Lifecycle contains different stages from development of the application, to its testing and deployment. Creating mobile applications has some unique challenges. The developer first needs to decide whether to create native, web or hybrid app (Badiu, 2013). Research shows that out of the two dozen apps each of us have on our phones, we spend 80% of our time on just five of them

(Husson, 2015). Therefore the application has to provide a superior customer experience to be attractive to the end users. Some of the requirements are being user firendly, having superior performance and not draining the battery. Testing all aspects of the app before it is released is also essential. Moreover accessibility of the application - ensuring that every person is able to access information and perform tasks regardless of that person's physical or cognitive capabilities - is another important requirement.

In these days we utilize the mobile phone to handle sensitive data; for example financial data from a banking app, corporate data from an enterprise app as well as medical data from a healthcare app. Moreover for most of us, our mobile device is within arm's reach 90% of the time; therefore we sometimes use the applications even using an insecure connection. This makes the sensitive data on our phone extremely vulnerable. Thus mobile security is an extremely important topic. In the race for mobile mind share, it is tempting to prioritize speed over security. In fact, about 65% of companies admit that the security of mobile applications is sometimes put at risk to meet customer demand. But at any given moment, malicious code is infecting 11.6 million of our mobile devices (Ponemon Institute, 2015).

One of the advantages of the smartphone is that it enables the determination of the context of the user using sensors on the smart phones. For example the GPS can give us the location of the user while the accelerometer can give the state of the user (whether she is static or in motion). This enables the creation of various very innovative context-aware apps (Chen & Kotz, 2000). However consumer concerns about the privacy of their personal sensitive information are at an all-time high. Therefore ensuring the privacy of the user's data is absolutely essential. In recent years regulators have increasingly turned a watchful eye to ensure the privacy of the mobile user (Bohorquez & Felz, 2016).

Once the application has been developed one needs to analyze the applications to determine how it is being utilized by the end user. Mobile Analytics can help determine the impact a mobile application has on a company's business. Such analysis can discover any deficiencies in the application so that the problems can be quickly rectified. The analysis can also give insights on how to make the application more useful for the consumers. Predictive analytics can discover problems even before they occur. On the other hand Prescriptive analytics provide actionable insights that will move businesses forward by not only indicating exactly what the issues are, but also suggesting which measures need to be taken to correct them. A recent report by the Aberdeen Group found companies using mobile analytics saw an 11.6 percent increase in brand awareness while those without a mobile-specific analytics strategy had a 12.9 percent decrease (Minkara, 2014).

Most mobile applications need to interact with various backend services. For example Push notifications let an application notify a user of new messages or events even when the user is not actively using the application. In many cases these backend services are hosted in the cloud. Cloud services are a good match for supporting mobile devices. Mobile applications tend to have time variable usage patterns that are well handled by the scalability and elasticity of cloud computing - increasing and decreasing the backend resources to match the level of requests from the mobile devices. It is also characteristic of mobile applications to make use of server-side data. Thus the interaction of mobility and cloud is an important topic. The Cloud Standards Customer Council (CSCC) has developed Mobile Cloud Architecture (Cloud Standard Customer Council, 2015) to showcase the interaction of mobile devices and cloud services.

OBJECTIVE OF THE BOOK

This book covers important topics from the major areas of Mobility:

- **Mobile Application Development:** We focus on the challenges of mobile application development, Mobile Testing and Accessibility.
- **Mobile Security:** Since Mobile Security is an extremely important topic several chapters are dedicated to the different aspects of Mobile Security. We look at how security can be ensured by program analysis as well as various ways to control access to sensitive data.
- **Context-Aware Applications:** We explain different techniques to track the location of the user. Some challenges of context-aware applications, specially ensuring privacy of the data, are also discussed. We also introduce an innovative participatory sensing application which makes use of mobile sensors.
- **Mobile Analytics:** A comprehensive survey of Mobile Application and User Analytics is presented.
- **Mobile and Cloud:** We discuss the synergy between Mobile and Cloud and explain the challenges and opportunities of Mobile Cloud Computing.

Some chapters survey important areas of Mobility while other chapters describe some key research challenges and present their solutions. This comprehensive publication aims to be an essential reference source and builds on the available literature in this field. Academicians, researchers, advanced-level students and mobile application developers will find this text extremely useful in furthering their research exposure to pertinent topics in this area that is becoming more and more popular and important.

ORGANIZATION OF THE BOOK

The book is organized into twelve chapters. A brief description of each of the chapters follows:

Chapter 1 discusses three key aspects Mobile application development:

1. Smart and incremental computation to improve battery consumption,
2. The trade-offs that can be made between accuracy and simplicity of the data-processing algorithms, and
3. Data storage and privacy aspects, i.e., which information should effectively leave the user's smartphone.

The authors showcase the development of an efficient life-logging Android[1] application, MEmoIt based on these three aspects.

Chapter 2 surveys a key aspect of Mobile Application development – the testing of the applications. It discusses three key forms of mobile testing. Functional testing is performed to ensure functionality, i.e. to test whether the application is performing the functions that it was designed for. Performance testing is conducted to determine how optimal the application is, in terms of its compute resource usage, battery usage and latency related issues. And finally, Usability and Accessibility testing aims to capture how easy it is for users to be able to work with the applications to execute the functions for which the application is designed.

Chapter 3 digs deeper into the important area of accessibility of the mobile applications. The mobile evolution has brought two areas of Accessibility closer together. People who rely on assistive technologies are able to participate in ways that have been historically unavailable to them. On the other hand, as people are performing more tasks on devices that are smaller and using them in situations where they may not be able to look at or hear the device, these individuals require assistive technologies that have traditionally focused on people with severe disabilities.

Chapter 4 introduces Phoenix, a novel solution that combines static program analysis with machine learning to ensure that no private information is exposed to unauthorized observers in a mobile application. Phoenix uses relatively scalable static analysis to approximate possible program behaviors, and then applies machine learning in order to identify programs exhibiting suspicious sequences of operations. This solution has been widely applied to mobile applications obtaining impressive results, with low false-positive and false-negative rates.

To secure mobile devices, there has been increasing focus on Trust negotiation, a procedure whereby two entities are able to establish a measure of mutual trust, even if no prior contact between either entity has existed in the past. Adaptive trust negotiation refers to the ability to dynamically adjust security parameters based on the level of trust established during the negotiation process. Chapter 5 explores the feasibility and utility of adaptive trust negotiation and its suitability for a mobile healthcare application.

Chapter 6 also looks into the security aspects of mobile healthcare application. It presents the benefit of utilizing role-based access control (RBAC) which allows the information owner to specify how much access other users have on the information based on their roles. For example this enables patients to grant access of their electronic health and fitness information to different individuals (e.g., primary physicians, spouse, family, emergency medical providers, etc.) at varying levels of granularity. RBAC is used by the mobile application to determine the ability of a user to view or modify medical information.

Chapter 7 proposes and discusses the Spatio-Situation-Based Access Control (SSBAC) model that combines features from existing access control models (like Role-Based Access Control, Spatio-temporal Access Control, Situation-Based Access Control, Workflow-Based Access Control, etc.) with new capabilities for the dynamic enforcement of security as a user moves among various locations with his/her mobile device and associated applications over time and distance. This concept is applied to a mHealth application, to constrain access to different health IT systems as a medical provider moves in both space/time and by the situation so that the application can dynamically adapt to the environments and allow or deny the access to specific data.

Chapter 8 focuses on the work done in the Platys project for the privacy and security aspects of context-aware mobile applications. It presents a fine-grained context driven access control mechanism for the apps. Context in Platys is generated by leveraging capabilities of smartphones. This allows an app on the phone to capture key elements of context: like the user's location and, through localization, characteristics of the user's environment, etc. The context is represented using Semantic Web technologies which allow handling of various data flow scenarios from and through users' mobile devices. Access control policies are then defined to reduce security and privacy risks.

Location-based services have become very popular both in the general public and enterprises, with the advent of smartphones and other sophisticated mobile devices. While many people are accustomed to using such services, knowledge of how they work, and which underlying technologies are most appropriate for different types of use cases, remains limited even to experienced developers. Chapter 9 describes the different types of location tracking technologies, their advantages and disadvantages as they relate to different use cases, as well as the challenges associated with the technologies and location tracking in general.

Participatory sensing empowers citizens with sensor-embedded hand-held devices to contribute to micro and macro-scale urban sensing applications. Chapter 10 focus on a novel Urban Sensing Platform

(USP) that aggregates data from an eco-system of data sources (e.g., mobile sensing data, social media, web-based public forums, as well as the civic agencies' internal data) to derive valuable insights. Challenges in terms of architecting the platform to gather and aggregate data in a scalable manner are discussed. The aggregated data is then categorized to create meaningful summaries of reports gathered by the platform. This chapter also describes a data veracity framework that determines data veracity in participatory sensing systems. Finally, a case study of a participatory sensing deployment for developing regions is presented.

Chapter 11 is a comprehensive survey of Mobile Analytics. The chapter begins with a taxonomy of mobile analytics problems and then discusses the technical details of a typical mobile analytics solution. Scale, heterogeneity, dynamically changing environments, and diverse privacy requirements, pose challenges to collecting and processing data for such analysis. This chapter examines how analytics solutions handle these challenges. The core of the chapter consists of a technical section describing the general architecture of a mobile analytics solution, procedures to collect and process data, event monitoring infrastructure, system administration processes, and privacy management policies. Case studies of a number of analytics solutions available as commercial products or prototypes are introduced.

Chapter 12 focuses on Mobile Cloud Computing which can be defined as a combination of ubiquitous connectivity of mobile device and elastic resources of the cloud to enable a computing and storage platform for providing unrestricted mobility, personalization, storage, and computing on the go. The chapter discusses the core techniques and novel applications of mobile cloud computing as well as the challenges faced by mobile cloud systems. The chapter concludes by highlighting opportunities and future research areas in the field of mobile cloud computing.

Sougata Mukherjea
IBM, India

REFERENCES

Badiu, R. (2013). *Mobile: Native apps, web apps, and hybrid apps*. Retrieved from https://www.nngroup.com/articles/mobile-native-apps/

Bohorquez, F. A., Jr., & Felz, J. N. (2016). *2016 mobile data privacy and security update and 2015 review*. Retrieved from http://www.dataprivacymonitor.com/mobile-privacy/2016-mobile-data-privacy-and-security-update-and-2015-review/

Chen, G., & Kotz, D. (2000). *A survey of context-aware mobile computing research*. Technical Report TR2000-381. Dept. of Computer Science, Dartmouth College.

Cloud Standard Customer Council. (2015). *Customer cloud architecture for mobile*. Retrieved from http://www.cloud-council.org/CSCC-Webinar-Customer-Cloud-Architecture-for-Mobile-6-16-15.pdf

Husson, T. (2015). *Five myths about mobile apps*. Retrieved from http://blogs.forrester.com/thomas_husson/15-01-30-five_myths_about_mobile_apps

Markets and Markets. (2014). *Bring your own device (BYOD) & enterprise mobility market global advancements, market forecast and analysis (2014 – 2019)*. Retrieved from http://www.marketsandmarkets.com/PressReleases/byod.asp

Minkara, O. (2014). *Mobile analytics: Precision marketing across mobile touch-points*. Retrieved from http://aberdeen.com/research/9364/RR-mobileanalytics.aspx/content.aspx

Ponemon Institute. (2015). *The state of mobile application insecurity*. Retrieved from http://www.workplaceprivacyreport.com/wp-content/uploads/sites/162/2015/03/WGL03074USEN.pdf[1]

Chapter 1
MEmoIt:
From Lifelogging Application to Research Platform

Jean-Eudes Ranvier
Ecole Polytechnique Fédérale de Lausanne, Switzerland

Michele Catasta
Ecole Polytechnique Fédérale de Lausanne, Switzerland

Ivan Gavrilovic
Ecole Polytechnique Fédérale de Lausanne, Switzerland

George Christodoulou
Cisco Systems, Inc., Switzerland

Horia Radu
Ecole Polytechnique Fédérale de Lausanne, Switzerland

Tiziano Signo'
Ecole Polytechnique Fédérale de Lausanne, Switzerland

Karl Aberer
Ecole Polytechnique Fédérale de Lausanne, Switzerland

ABSTRACT

In the recent years, smartphones became part of everyday life for most people. Their computational power and their sensing capabilities unlocked a new universe of possibilities for mobile developers. However, mobile development is still a young field and various pitfalls need to be avoided. In this chapter, the authors present several aspects of mobile development that need to be considered carefully. More specifically, this chapter covers topics like energy efficient sensing, smart computing, trade-off between accuracy and simplicity, data storage and cloud integration. These aspects are illustrated based on the authors' experience building a lifelogging application for the past two years.

INTRODUCTION

With the advent of smartphones and the evolution of mobile OSs offering ever more capabilities, the development of mobile applications became a leading activity in information technology. From mobile games to productivity-oriented applications, the market of mobile applications is blossoming. However, as a relatively novel field, it took inspiration from its computer counterpart, which led to many challenges,

DOI: 10.4018/978-1-5225-0945-5.ch001

as the computer specifications do not match those of a mobile device. In recent years for example, due to the increase of mobile web browsing, the web industry has had to depart from the classic interfaces based on wide screens and mouse interactions, in order to support vertical screens and touch-screen capabilities as well; This leads to the creation of new frameworks and new methodology.

This chapter will present three aspects of the development of mobile applications; These aspects need to be readapted in order to comply with the limitations of mobile platforms and to be compatible with fast-paced development cycles. These aspects are specifically:

1. Smart and incremental computation to improve battery consumption,
2. The trade-offs that can be made between accuracy and simplicity of the data-processing algorithms, and
3. Data storage and privacy aspects, i.e., which information should effectively leave the user's smartphone.

The mobile environment is very restricted: Although its memory is reaching that of low-end laptops, its processing power is still relatively low, and although the battery duration of a smartphone is longer than the ones of laptops, the latter are not meant to last a full day on battery. In order to improve the overall usability of the phone, mobile applications should be extremely battery efficient and smart in their computation. Furthermore, the pace of the evolution of the SDKs and mobile frameworks forces mobile development to become faster and incremental. Algorithm complexity is often the synonym of computational cost -- for this reason, it is important for application components to be simple. Finally, privacy is a rising concern. In order not to disclose the overwhelming amount of user related data that can be acquired by mobile devices, it is necessary to select which piece of data should effectively leave the user's device.

These optimizations are illustrated based on the life-logging Android application, MEmoIt, that the authors have developed in the last two years. It shows the challenges met, and explains the choices made in this project.

BACKGROUND

Local Processing and Resource-Hungry Operations

The computing capabilities of smartphones remains very limited compared to single servers or cloud infrastructures. Furthermore, digital economy analysts agree that data is the new oil. These two facts combined, led most companies to rely on the cloud to store data and/or perform heavy computation. However, recently, privacy issues have been brought to light (Chen & Zhao, 2012; Xiao & Xiao, 2013). Hence, to address users concerns about what usage is made of their data, some companies departed from this model to return to on-device computation. For instance, the health-related data collected by Apple Healthkit (Apple, n.d.) are stored and encrypted locally, and a strict access control regulates the applications that want to read it. Another advantage of local processing is the ability to provide a service that degrades gracefully when an Internet connection cannot be established. Google translate (n.d.) and Google navigation systems (Google Play, n.d.) are examples of such services, which can now be queried offline.

Location-Based Services

Smartphones gave to GPS, and localization mechanisms in general, a new purpose. However, when used, GPS sensors drain the battery significantly. Despite technological improvements at the sensor level, such as the Assisted "A-GPS", continuously acquiring the precise localization information, is generally costly. Several approaches have been taken to solve this problem, such as adapting the rate at which the GPS sensor is activated (Paek, Kim & Govindan, 2010). The Android ecosystem abstracts this problem by proposing a geofencing API that handles the calls to the GPS sensor.

In addition to being able to know her exact location, the user also gains access to additional location-based services. One of them is the detection of places of interests (PoI) as described by Montoliu (Montoliu, & Gatica-Perez, 2010). Places of interests are defined as geographical regions having a semantic meaning (e.g. a cinema, a restaurant, a landmark), where the user stayed for a certain amount of time. This definition is instrumental in the comprehension of users' behaviors at a semantic level. An application of this abstraction is the recommendation of PoI (Baltrunas, Ludwig, Peer, & Ricci, 2011) or the well-known check-in service of Foursquare (n.d.).

Local-Data Management

Data persistence is a key component of mobile applications, and choosing the right data management system is often not straightforward. Mobile operating systems propose out-of-the-box multiple persistence options, each serving a different purpose. Android, for instance, proposes a key-value system, a public and private storage on disk and an SQLite database (Developers, n.d.). In the recent years, a number of databases emerged offering alternatives to the original android implementation (Ostrovsky, & Rodenski, 2014). Couchbase lite (Couchbase, n.d.), for example, proposes an embeddable document-based database. Realm (n.d.) is an Android library proposing a relational model accessible through object-relational mapping (ORM) and is a strong contender against the SQLite system of Android.

Cloud Interaction

Despite the risks that represent sending data to the cloud, it is sometimes unavoidable: for example, to synchronize an account with a desktop application (i.e., for synchronization purposes) or to aggregate the data coming from multiple users. To facilitate the exchange of data between the cloud and the mobile device, various alternatives exist, Parse Server (n.d.) proposes to seamlessly store data objects on the cloud and access them from the mobile application, as if the storage was local. Another approach is taken by Amazon Web Service[14] (AWS) and Google Cloud plateform[15] (GC), which provide libraries to exchange data with the cloud. In addition to the storage service, they also provide powerful data processing tools at low cost.

MEmoIt PERSPECTIVES

The goal of MEmoIt is to reconstruct the daily life of the user into a digital diary. In order to detect significant activities, and to abstract these activities into routines and memorable events, the Android application uses both hardware and software sensors data. MEmoit captures the phone activities that are related to the soft sensor of the phone, specifically calendar events, pictures taken, and applications used. Also, the physical world activities are detected through the use of GPS readings in order to infer the user activity. In a second phase, MEmoit is also meant to be a research platform where researcher can conduct user experiments. For this purpose, MEmoit has the capability of running the code of the user studies directly on the mobile, and sending to a research server the results of these experiments. For privacy reasons, these results are anonymized before being presented to the researchers.

Local Processing and Resource-Hungry Operations

MEmoIt deals with very sensitive data about the user, and it would be critical to send these data to a remote server in a raw format. Therefore, a fundamental requirement of the application is to process user-data directly on device. However, in order to keep the battery consumption of MEmoIt within an acceptable range, data processing should be done in a smart way. Two aspects will be described in this chapter. The first one addresses the use of incremental and smart algorithms to only have marginal updates on the models of the user, instead of rebuilding it from zero. Although, incremental algorithms are a neat solution for reducing processing time, some operations will always be computationally expensive. The second aspect will therefore cover recommendations as to when to perform computation. In MEmoIt, the detection of memorable events is also done through software sensors (exceptional phone calls / SMSs that the user receives, a day with more pictures taken than usual, etc.) The algorithms developed to detect these events are relatively expensive and do not need to be run in real time. For these reasons, the authors exploit the Android API to detect when the phone is charging in order to run heavy tasks at that time. This implementation has the triple advantage of saving battery energy during the day while keeping the phone responsive when the user needs it; and finally, due to the short battery duration of smartphone, ensuring that the algorithms are run on average once per day.

Location-Based Services

Another aspect of the MEmoIt diary is the logging of a user's activity in the physical world. This is achieved by detecting the PoI of the user and, if no prior knowledge about this place is in the system, asking the user to describe her activity. However, existing algorithms for the detection of PoI are essentially non-incremental and require the full GPS trace to be available.

In order to keep the application battery-friendly and to be in line with the policy of incremental algorithms of MEmoIt, the authors initially studied the incremental clustering algorithm, ESOINN (Furao, Ogura, & Hasegawa, 2007). This algorithm was chosen for its capacity to cluster spatial points in an incremental fashion and without the need to know a-priori the number of clusters. However, the processing required by the algorithm and the need for continuous sensing made this approach still battery intensive. Furthermore the complexity of the algorithm made it impractical to maintain. To avoid continuous GPS sensing and to focus on transition between points of interest exclusively, the authors decided to leverage the geofencing API (Developers, n.d.) exposed by Android. Combined with a simpler algorithm based on finite state machine, this approach enables MEmoIt to achieve similar performances to ESOINN in the detection of points of interest for the user and to reduce battery consumption. Once the places of interests are detected, a final abstraction enables MEmoIt to detect a user's patterns of activities and to group them into routines, according to the algorithms presented by Ranvier in (Ranvier, Catasta, Vasirani, & Aberer, 2015).

Local-Data Management

As stated previously, MEmoIt collects very sensitive information about the user's life and in order to protect the user's privacy, the data remains on the phone and is processed locally. The initial implementation of MEmoIt used the built-in SQLite database provided by Android. It proposes a relational model suitable for storing the structured data gather by the application. However, the Android standard implementation involves many boilerplates and utility classes that decrease the readability and flexibility of the code, while proposing mediocre performances. MEmoIt recently moved from this original Android system to the Realm library. This library proposes an object-relational mapping that is based on a specific serialization. The migration greatly reduced the size of the codebase, and improved the overall performance of the app.

Cloud Interaction

MEmoIt also has a vocation to be a research platform. For this purpose, data needs to be exchanged between the research server and MEmoIt. Following the approach proposed by de Montjoye (de Montjoye et al., 2014), the authors propose an architecture in which the code that runs a user-study is sent to the device and executed; only the results, in the shape of aggregates, are sent back to the server, therefore contributing to the anonymization of the users. Due to its quality and its coherent integration with the Android ecosystem, MEmoit uses the Google Cloud platform to store and process the data sent by the users.

This shift towards the cloud, however, raises concerns about the security and the privacy of the data (Chen, & Zhao, 2012; Xiao, & Xiao, 2013; Lu et al., 2015). Without proper encryption, an attacker who gains access to the back-end could obtain a vast amount of raw data about the users. It is very well known in the literature that, given enough data points per user, datasets can be de-anonymized with a high success rate. Therefore, the authors are currently exploring solutions such as homomorphic encryption and differential privacy, which would enable the encrypted data to be stored, while being able to process the aggregates in order to extract statistics, identify correlations, etc.

Furthermore, a notion of trust needs to be established between three entities: The user, the authors and the researcher running the user studies. Indeed, users entrust the authors with sensitive information, and the authors are liable for the information they provide to the researcher and therefore should be cautious of malicious researchers. For this purpose, several tools can be applied. First, to ensure the trust of the user, the authors should provide a valid privacy-policy accessible from the application. Second, in view of transparency, the MEmoit source code should be released in open-source, in order for the user to understand what is exactly is being done with her data. By corollary, user studies designed by researchers should also be open-sourced. Furthermore, a manual approval of the studies will prevent abuse of the system. These privacy elements represent the last missing building-blocks for the promotion of MEmoIt as a privacy-aware research platform.

LOCAL PROCESSING AND RESOURCES-HUNGRY OPERATIONS

In order to be performed locally, data processing must be carefully implemented. Running an algorithm with a complexity of O(n^3) on an increasing dataset would quickly reach unacceptable runtime. This section presents a selection of solutions for keeping the computation time within an acceptable range. It focuses primarily on reducing the dataset used by the algorithm. Splitting the dataset or having a sliding window can reduce the computation cost drastically. Another solution consists in using incremental algorithms in order to compute only delta quantities that essentially take into account the most recent data to update a model. The last option discussed in this chapter concerns the time of the day at which the processing should be performed. Indeed, a heavy load on the CPU could result in slowing down applications that the users are currently using in the foreground. For this reason, and to save energy, it is worth considering running heavy computation during the night or while the phone is charging and not in use. These options are illustrated, in the case of MEmoit, in the remainder of this section.

As a digital diary, MEmoIt logs memorable events. Along with activities that the user performs, phone calls at unusual hours or with unusual acquaintances, pictures that are taken in an unusual amount qualify as being memorable. For this reason, a set of software sensors was added to the hardware sensors.

These software sensors are:

- Call logs,
- SMS logs,
- Browser logs,
- Picture gallery.

As opposed to hardware sensors, which require real-time processing, these sensors can be processed offline. For this reason, the amount of data that needs to be processed can be quite large and requires more time, both due to the quantity and the complexity of the algorithms used.

Algorithms

The authors define a software event as memorable if it does not follow the day-to-day patterns of the user. Therefore an anomaly detection approach was taken to identify these memorable events. However, the activity pattern of a user can change from one period to another. A period of calm and a normal lifestyle often follow a period of stress and intense activity. MEmoIt should be sensitive to this kind of behavior. In other words, the norm changes and MEmoIt's interpretation of the norm must change along with it.

Figure 1 presents an overview of the detection algorithm. The software sensors' logs are parsed and each piece of information is sent to the appropriate detection component. The outputs of these components are the detected memorable events, that are then stored in the database in order to present them to the user at a later time.

Each component behaves similarly. An initial pass over the data builds a distribution of the events at hand and detects a pattern in the user's behavior. Subsequently, in order to detect unusual events, the components compare the new data to the previously computed norm. Finally, the norm is updated to take into account the new data.

The four components, however, differ in granularity. Phone calls and SMS are usually received throughout the day, but are less frequent during nighttime. A call at 2am would be unusual. Conversely pictures are taken on a regular basis, but an unusual event such as a visit to the zoo can increase the number of pictures taken during this time. This number will most likely be significantly higher than for the rest of the day. The same reasoning can be applied to web browsing. This is the main difference between the two flavors of the detection algorithm: One keeps a per-day norm, and the other keeps a

Figure 1. Memorable events detection

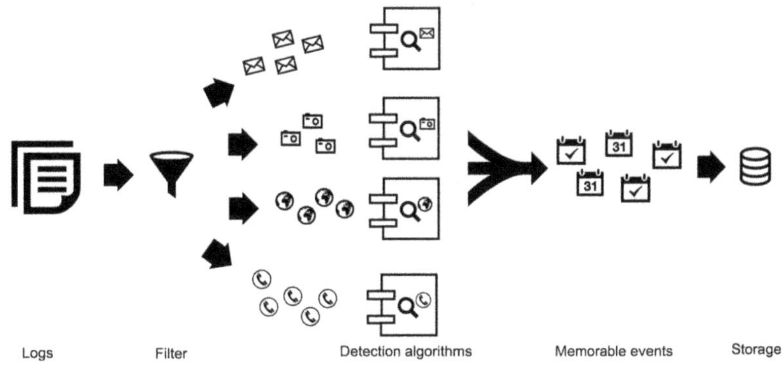

per-hour one. The similarity of the approach, however, led to a generic solution that provides flexibility and factorization.

Without loss of generality, the detection algorithm for memorable calls is analyzed, but similar reasoning can be applied to the three other dimensions.

The goal of this algorithm is to detect if the number of calls received in a predefined time interval, on a given day, is out of the ordinary. For simplicity, the time dimension is discretized by splitting days into 24 bins of one hour each. The adaptability of the algorithm, defined by how fast the norm can change, is empirically set to one month. In other words, it is sufficient to keep track of the last 30 days in order to detect abnormal behavior. The following features provide the information required to define the norm and perform the detection:

- **Number of Hours:** Intervals for which the algorithm currently has data.
- Calls per time interval.
- **The Last Checked Date:** The time at which the algorithm last ran the algorithm. Only the data that was generated since the last time the algorithm was run is necessary.
- The number of calls on the last checked time interval.

Upon retrieving the call logs, the algorithm averages the amount of calls, per hour and per day (in order to distinguish between the two events: four calls from 10:00 to 10:59 on Monday and two calls in the same time interval on Tuesday, for example).

An event occurring in a certain hour bucket h where the user has received n calls is defined as important if: $n >$ average(h)$\pm\delta$ where δ corresponds to the standard deviation of the distribution.

After going through all the logs and detecting all the memorable events, the algorithm makes them available to the application by storing them in the database. Finally, the algorithm updates the status information to keep track of the last log entry that has been processed.

Moment of Analysis

For each software sensor, the corresponding detection component must retrieve a potentially large number of records and analyze them. This operation can be time consuming and computationally intensive. Moreover, mobile devices have considerably less resources than a computer, although, the user expects to do, more or less, the same things and to do them at the same quality of service: browsing the internet, using applications, taking pictures, making phone calls; all these operations must be very fast and responsive for a good user experience. The Android OS has this as one of its primary goal and employs a lot of safety mechanisms to protect the user's experience. For example, if a process blocks the user interface for longer than a couple of seconds, it is automatically killed. Furthermore, the number of threads that can be launched at the same time is limited and trying to exceed this limit will result in tasks being rejected from execution.

Based on this observation, it is clear that the computation should be scheduled, as much as possible, when the phone is idle. Furthermore, as opposed to a computer, the phone runs most of the time on battery and battery lifetime is very important for the user. If an application is seen as a high-energy consumer, then the risk of the user abandoning it is greater. Thus, launching these algorithms while the phone is on battery would be a bad decision. The battery might run out while the algorithms are running and this would increase the risk of corrupt data. Hence, extra effort should be invested in a more robust, fault tolerant algorithm. For this reason, the detection algorithms run only when the phone is plugged and charging. This is not a disadvantage because these sensors do not need to be processed in real time. Also, the user charges his phone on a regular basis, usually once a day, and this will guarantee the detection of the memorable events at regular intervals and can present to the user, each day, the memorable events which she has performed on that particular day. It could even become a daily routine: in the evening, they plug their phones and will be notified with the memorable calls or text messages of that day. They are offered an overview, in terms of important events, of the past day.

The technical solution presented in Figure 2 fulfills these requirements. It is a task queue implementation. This way, the tasks can be submitted to the queue, and later on, a service pulls the tasks and accomplishes them step-by-step. The tasks can also be serialized in order for them to be stored on a disk, and later retrieved, when the service is launched again.

LOCATION BASED SERVICES

Location-based services are at the core of most mobile applications on the market. It can be to simply understand the user's location, for example with the Facebook check-in option, or to infer more com-

Figure 2. Overview of the task submission and execution mechanism

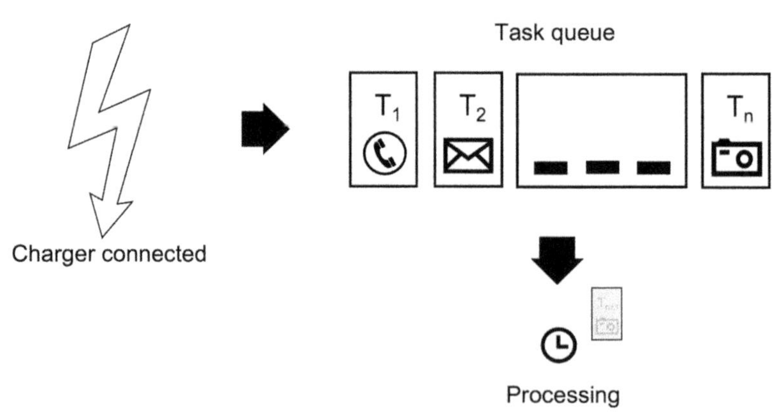

plex concepts about the user, for example with the Google now (n.d.) personal assistant that infers the information to be displayed to the user, based on her habits, time and location. The activities performed by a user are tightly bound to the places where she performs them. The intensity of this connection varies, but it is always there. For example, a person can eat at a wide variety of places, ranging from her own dining room, to the most fancy restaurant in town. However, she can only attend a football game at football stadium.

Taking advantage of the fact that most people have their mobile phones with them at all times, it is possible to define an initial building block: *a tracking system*. This system is responsible for detecting whenever the user is in a certain location, because this usually means that she is performing some activity there. Disregarding her movements, the tracking system focuses essentially on her idling periods.

About Location

Today's smartphones have several built-in features that allow applications to access the current location of the phone:

1. Simple location detection via network triangulation.
2. 3G based location detection.
3. GPS (global positioning system).

Triangulation-based detection is very inexpensive in terms of network traffic and battery usage, but it is the least accurate of the three. Also, it is always available for a mobile phone that is connected to a cell-phone network. Using the 3G data network service is better in terms of accuracy, but more expensive in terms of battery usage. Also, not all the users have this feature enabled at all times. The last location provider available, GPS, is the most accurate of them all, but the most expensive in terms of battery energy.

The recent developments in modern mobile OSs (Android, IOS) make this decision seem much easier than it actually is. An application can specify what provider it wants, but it will receive location information from the best currently enabled one. Because of this, the most accurate option, GPS, was selected. If the GPS sensor is not enabled, the application receives information via network triangulation.

Nevertheless, the GPS coordinates provided by the location service (regardless of the option chosen) are not enough to determine the places of interest (PoI), also known as staypoints, the user visited. The next section describes how to abstract the raw GPS coordinates into PoI.

From Coordinates to Places of Interest

Once the GPS coordinates are acquired, they need to be clustered in order to define the user's PoI. Two incremental approaches are compared: A first one, based on the ESOINN algorithm (Furao, Ogura, & Hasegawa, 2007; Aberer, Catasta, Radu, Ranvier, Vasirani, & Yan, 2014), provides an incremental clustering algorithm to detect points of interests from GPS coordinates. The second approach uses the Android location API's geofences, as well as a finite state machine (FSM), to detect places where the user spends a certain amount of time to perform an activity.

ESOINN

ESOINN is a clustering algorithm with particularly interesting features for the purpose of mobile applications. It is *incremental*, which means that adding new data to the model will require only a marginal update. It is *online* and therefore does not need the entire GPS trace of the user (i.e., the succession of GPS coordinates she visited) to start detecting clusters. Furthermore, it is *unsupervised* and does not need labeled data. Finally, it does not require knowing *a priori* the number of clusters, which is particularly suitable as an application should be able to cluster GPS signals from new places as the user visits them. The general idea behind the algorithm is to build a network of nodes corresponding to GPS coordinates and to aggregate multiple signals to the same node if these signals are close to each other. The partitioning of the network defines the different clusters.

MEmoIt uses ESOINN in order to build, from the GPS trace of the user a list of clusters corresponding to the PoI, whereas transition signals between PoI are considered as noise and discarded. To be more suitable for mobile computation, the original version of the algorithm was slightly modified to simplify the generated network.

The evaluation of this approach was based on the Nokia dataset (Laurila, Gatica-Perez, Aad, Bornet, Do, Dousse, Miettinen, 2012; Kiukkonen, Blom, Dousse, Gatica-Perez, Laurila, 2010) containing GPS traces of more than 100 users and collected over several months. ESOINN was compared to the non-incremental clustering algorithm DBSCAN (Ester, Kriegel, Sander, & Xu, 1996). Although the precision and recall of both algorithms are quite similar (precision: 0.92, recall: 0.87), the ESOINN has the advantage of being incremental. After gathering data for 10 months, retraining DBSCAN with an extra day would take on average 36 minutes, whereas ESOINN would update its model within an average of 2.6 seconds. However, the ESOINN approach requires accurate GPS data in order to perform correctly, leading to high battery consumption.

Finite State Machine

The alternative approach for the detection of PoI is based on finite-state machine (FSM), the geofencing API and different granularity of location tracking services. The gist of this approach is that coarse-grained and energy-efficient location tracking should be used on a regular basis to detect if the user is still within the boundaries of a geofence, and fine-grained and more expensive location tracking should be only activated when necessary.

The decision about the granularity level to be used is managed by a state machine containing the following states:

- **IDLE:** The user is still in the radius of the last know PoI location
- **PRE_MOVE:** The coarse-grained location-tracking service has triggered an updated because the new location is displaced outside of the geofence that was centered on the last known PoI location. At this point, the fine-grained location service is started in order to provide a more precise location. This way, it is possible to confirm that the user actually moved (change state to MOVE), or that it was just a location update noise (return to IDLE). After the fine-grained location service returns a result, it stops, in order not to use too much battery energy.

- **MOVE:** In this state, if the coarse-grained location service sends an update, it means the user is moving. Once these updates stop for a period of IDLE_TIME, the fine-grained location service is triggered for the update, and the machine current-state advances to the PRE_IDLE state.
- **PRE_IDLE:** After receiving the update from the location service, that location is compared to the last known location received while in MOVE state. If the distance between the two is less than the displacement threshold (i.e., the geofence radius), it is safe to assume the user stopped. A new PoI is detected, and the machine state returns to IDLE. Otherwise, the machine's state reverses to MOVE.

The difference in sensor accuracy for some devices can lead the GPS coordinates to change, although the device has not changed its location. This defect, which might occur for various reasons, was the cause of false positive PoI generation. These corner cases are handled by keeping track of the time the last PoI was entered. In a realistic situation, user will most likely not visit two different PoI in a very short period of time, and this is something that can be exploited in order to solve this problem. The minimum time difference between the two events is approximated as being twice the IDLE_TIME.

This solution was evaluated against the ESOINN approach and performed with similar accuracy. As it is much simpler and more battery efficient, the finite-state machine approach was retained for the implementation of MEmoIt.

Semantic Enhancement

Once the place of interest detected, a semantic tag can be associated with it to provide an extra abstraction level. For this, an external service being able to reconcile GPS coordinates with their respective semantic location is required. Two main service providers exist: Google Places and Foursquare. Both providers supply similar services through their own API. Although MEmoIt originally used the Foursquare API because of its richer dataset, the application recently switched to Google Places. The reason for this change of provider is the improvement of the Google Places dataset, the seamless integration of its visual components of and its ease of deployment on Android. Several components of the application that were dedicated to the integration of the Foursquare API were discarded, which simplified the location-related functionality significantly.

Routine Detection

In the context of user modeling, another important feature is the detection of daily routines in user's habits. A routine is defined as the repetition of a pattern in the list of activities performed by the user. The pattern can be sequential, unordered, or a mixture of the two.

Several approaches exist, as for example the work of Ye (Ye, Zheng, Chen, Feng, & Xie, 2009). However, due to the mobile computing constraints, the algorithm used to detect these routines needs to be incremental. Two approaches are presented to solve this problem (Ranvier et al. 2015).

Finite-State Machine

The first approach is based on finite-state machines (FSM). It defines different states between the different activities performed by the user, and the transitions between these states represent the sequential order in which these activities are executed. Each routine is modeled by a different FSM and branching in the models account for small variations of the same routine. (e.g., Once in a while a user goes to the supermarket before going home in the evening.)

In order to have a uniform representation, each day of the user, represented as a succession of activities, is discretized in time. Then, the discretized day is tested against the existing FSMs in the system as follows:

1. Each FSM is updated temporarily with the sequence of activities of the day.
2. The system compute a degree of match (DoM) between the day and the updated FSMs. DoM is defined as the product of the transitions probabilities that build the path required to reconstruct the day. (e.g., in the FSM presented in Figure 3: the day Home -> Office -> Office -> Shop -> Home has a DoM = 1 * 1 * 1 * 2/5 * 2/5 = 0.16).
3. Existing FSMs are ordered based on their DoM. If the highest FSM has a DoM below a certain threshold θ, it means that none of the existing FSMs represent correctly the day at hand; therefore another FSM is generated based on the day.
4. If the highest FSM has a DoM above θ, it is updated with the sequence of activities of the day, changing the transition probabilities between the states and potentially adding more states.

The threshold θ is defined as (1/NFSM)^λ where NFSM is the number of days that have been matched with a given FSM, and λ is an indicator of how much diversity is tolerated in the definition of routines and how much difference is tolerated between the day at hand and the given FSM.

This approach has the advantage of being computationally inexpensive and proposes a clean visual representation of the routines. However, the time-discretization of the days, necessary for the definition of the FSM, induces some loss of information and adds noise. Another problem is that the FSM approach does not handle unordered activities easily (e.g., going to the gym and then to the supermarket can be part of the same routine as going to the supermarket and then to the gym). These flows triggered a second approach based on frequent patterns mining.

Figure 3. Example of finite state machine

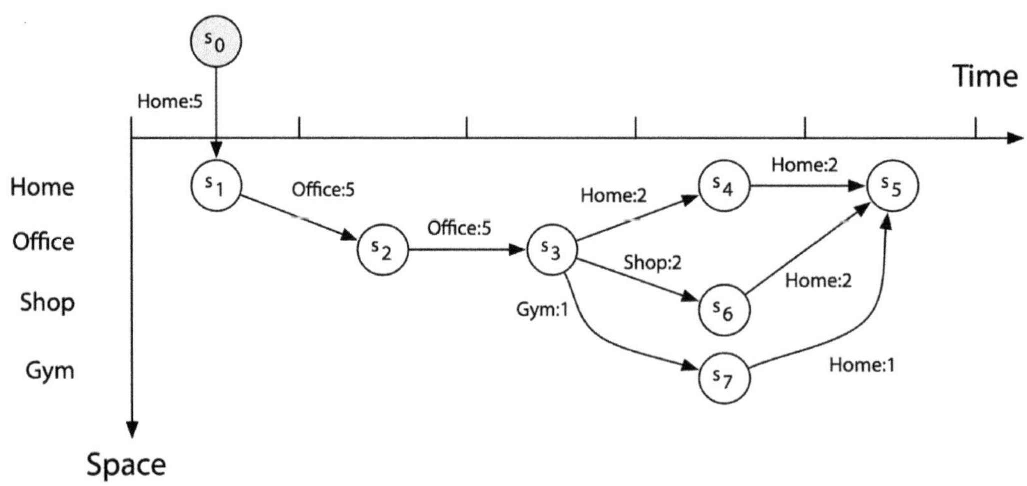

Frequent Pattern Mining

This approach uses data mining techniques used for frequent-pattern mining on the list of activities of the user. The term frequent pattern is used here to encompass

- Unordered sets of activities that are present in more than *Si* transactions. *Si* being the minimum support required to be defined as frequent.
- Sequential lists of activities, that are present in more than *Ss* transaction. *Ss* being the minimum support for a sequence to be defined as frequent.

In this way, the system accounts for the ordered nature of the routines, and enables activities to be swapped but still remain coherent in the definition of a routine.

The main challenge of this method is that standard algorithms for computing frequent itemsets, such as the Apriori algorithm, are usually non-incremental and not efficient. This challenge is addressed by using the incremental algorithm FUP (Tsai, Lee, & Chen, 1999). Although more recent and efficient algorithms have been devised, FUP has the advantage of being based on the Apriori algorithm, making it useful for both sequences and itemset mining.

Even if FUP addresses directly the problem of frequent itemset mining, it does not support ordered sequences mining. Frequent sequences can be mined in a non-incremental fashion by using the GSP algorithm (Srikant, & Agrawal, 1996). However, as the GSP algorithm is itself based on the original Apriori algorithm, the GSP algorithm can use the FUP algorithm to mine frequent sequences in an incremental fashion. This enables the system to perform the update of frequent sets and sequences in one pass, further improving the time complexity of the algorithm.

Once the frequent patterns of activities are detected, each day in the user's history is assigned a boolean feature vector for which the element at position i is 1 if the day matches pattern i, and 0 otherwise. This vector provides a representation of the day in the frequent patterns space. This enables the clustering of days, based on their features vector, therefore generating routines. Indeed, days having a high number of patterns in common should belong to the same routine.

Evaluation

These two approaches were evaluated both in terms of accuracy, by comparing them to a state-of-the-art algorithm, T-patterns (Giannotti, Nanni, Pinelli, Pedreschi, 2007), by using the Nokia dataset (Laurila et al., 2012; Kiukkonen et al., 2010) defined in the previous section. They were also evaluated in term of performance on Android by comparing them with a non-incremental frequent pattern-mining algorithm.

Due to the lack of ground truth for the Nokia dataset routines, the routines mined by the three algorithms (the two aforementioned approaches and T-patterns), were provided to a crowd-sourcing platform for an evaluation in which the crowd was asked, to detect among four days belonging to the same users, the outlier that did not belong to the same routine as the three others. A six-class answer was devised: The user could select one of the 4 days as the outlier, answer that all the 4 days belong to the same routine or that none of the days belong to the same routine.

The experiment resulted in 3800 tasks performed by 279 different workers. The performances of the three approaches are presented in Table 1. These results present FPM as the winner in terms of accuracy

Table 1. Crowdsourced evaluation of T-Patterns, FSM and FPM

	Accuracy	Sensitivity	Specificity
T-patterns	21.6%	40.6%	8.8%
FSM	44.0%	59.1%	49.7%
FPM	56.3%	74.3%	51.7%

sensitivity and specificity. The reader can note that a randomized baseline among the six available options offered to the crowd would yield 16.7% accuracy.

The performance of the algorithms was tested on a Nexus 4 smartphone (quadcore CPU at 1.5GHz, 2GB of RAM, Android 4.4), by adding data to the history of the user, one week at a time. The results are displayed in Figure 4 and clearly state the importance of having incremental algorithm to sustainably perform regular tasks on increasing datasets on a smartphone.

LOCAL DATA MANAGEMENT

One of the main problems of mobile application is the performance. A badly performing application with much waiting time and freezing interactions can lead to a user's irritation and result in the application being uninstalled. Mobile systems have multiple hardware and software limitations that other systems do not present, such as CPU clock, available memory and power consumption. The database system is often one of the heaviest components of mobile applications, and good performances are necessary to conserve a good user experience.

Figure 4. Time required to process a new week of data for both SEM generation methods, compared to a non-incremental algorithm based on the original Apriori and GSP

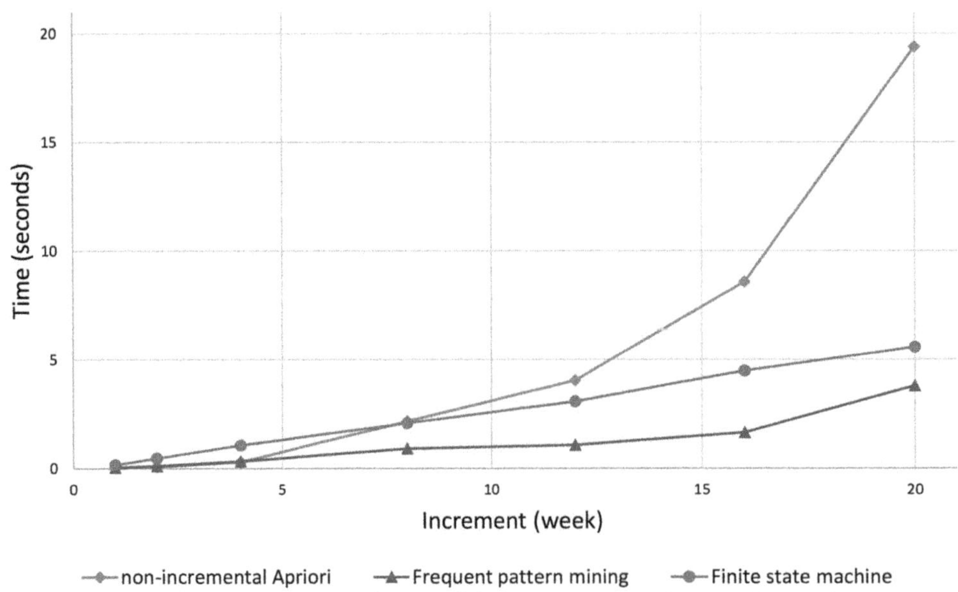

Realm

The performance of the default database system (SQLite) in heavy operations (e.g., selecting all the records) is critical. Some improvements could be made with different techniques such as prefetching and caching. However, a downside is that the margin of improvement is limited and it comes at the expense of other resources (memory and battery consumption). Another downside is the complexity of the code used to support operations on the database. In the specific case of MEmoIt, the frequent turnover of developers combined with a high code complexity leads to a very steep learning curve, a high number of bugs in the code and a broad set of coding styles that periodically have to be fixed and refactored.

For all these problems, it is important to find a solution that could solve or improve the application on many levels. Realm offers such a possibility.

Realm is a mobile database-engine that can be used as an alternative to SQLite. The strengths of this system, presented by its developers, are the following:

- **Ease of Use:** As Realm does not work on top of SQLite, it provides a complete alternative to the engine provided in Android. It uses Java objects as a model and a Fluent Interface for the queries, as described in Table 2.
- **Fast and Optimized for Mobile Platforms:** Due to its zero-copy design, Realm is much faster than an ORM, and often faster than raw SQLite.
- **Cross-Platform:** By the means of wrappers, Realm is available for Android and IOS ecosystems.

Performance Analysis

The main use-cases where an application interacts with the database, and the types of operation associated with it, are listed in Table 3. In the context of MEmoIt, write operations are very rare and essentially affect one or only a few records, whereas the application heavily uses read operations, especially the selection of all the events performed every time the timeline interface (one of the main one) is shown.

*Table 2. Example of a Realm query and an Android SQLite query, equivalent to the SQL select *from Event where id = eventId*

```
//Realm query
int recordId = generateRandomId();
RealmResults<Record> records = mRealm.where(Record.
class).equalsTo("id", recordId).findAll();
//Android SQLite query
ContentResolver contentResolver = getContext().
getContentResolver();
String[] proj = {RecordExtendedView.COLUMN_ID};
String sel = RecordtExtendedView.COLUMN_ID + " = ?";
String[] selArgs = {String.valueOf(recordId)};
Cursor cursor = contentResolver.query(RecordEventExtended
ContentProvider.CONTENT_URI, proj, se, selArgs, null);
```

Table 3. Use cases per database operations

Use Case	Database Operation
Add a new record	Single insert
Update an existing record	Single update
Load records from file	Bulk insert
Show single record details	Single read
Show all records	Bulk read with join
Upload records to the cloud	Bulk read
Show aggregated overview	Aggregate (sum/count)

To be complete, tests were run on different devices, in order to study how different hardware capabilities affect the performances of the two systems. The tables were populated with a different number of records, ranging from 1000 to 10000.

The metric used to define performances was the execution time of a single operation. Multiple iterations of the tests were run and a confidence interval over the median was used to evaluate the average performances.

Results

The main operations tested were

- **Write Records (from 1 Record to 100 Records at a Time):** Write operations show a difference in the performances with different hardware, consistently showing that the mobile device with better hardware performed slightly faster. No difference could however be identified between the two databases that performed similarly (on the same hardware).
- **Read a Single Record:** The results in this area were too small to identify any correlation with the hardware configurations, the number of records in the tables or even with the database engine. In both cases the average execution times were shorter than 5ms.
- **Read All the Events (the Main Operation that Needs to Be Measured):** This operation yielded more interesting results.

Two different types of tests were run, the first one performing only the query, the second one performing the query and reading all the return records afterwards.

The results showed that:

- Better hardware consistently leads to better performances, with a meaningful difference for SQLite and a much smaller difference for Realm.
- In the case of Realm, due to the lazy evaluation this engine implements, the first type of tests averaged execution times at 1ms, independently of the number of returned records.
- The performances of SQLite for the first type of test were linear to the number of records and very slow when this number was too high (e.g., more than 1s for 10000 records).
- The performances for the second type of tests proved that both SQLite and Realm execution times were linear to the number of records retrieved; but with very different growth, much steeper for SQLite than Realm.

The overall results, as depicted in Figure 5 and Figure 6 prove that Realm performs much better than SQLite (more than 10 times faster). Based on this, MEmoIt was reimplemented in Realm, simultaneously applying some improvements and refactoring. This implementation had the side effect of reducing the codebase of the database component from 28000 lines of code for the SQLite implementation to 13000 lines for the Realm implementation, reducing the complexity of the code by more than a half, and therefore illustrating the importance of the choice of a database system for a mobile application. These values were measured using Sloccount (Wheeler, 2014).

Figure 5. Boxplot of the median showing execution times for Realm on a Google Nexus 5 for both types of operations (type 1 left, type 2 right)

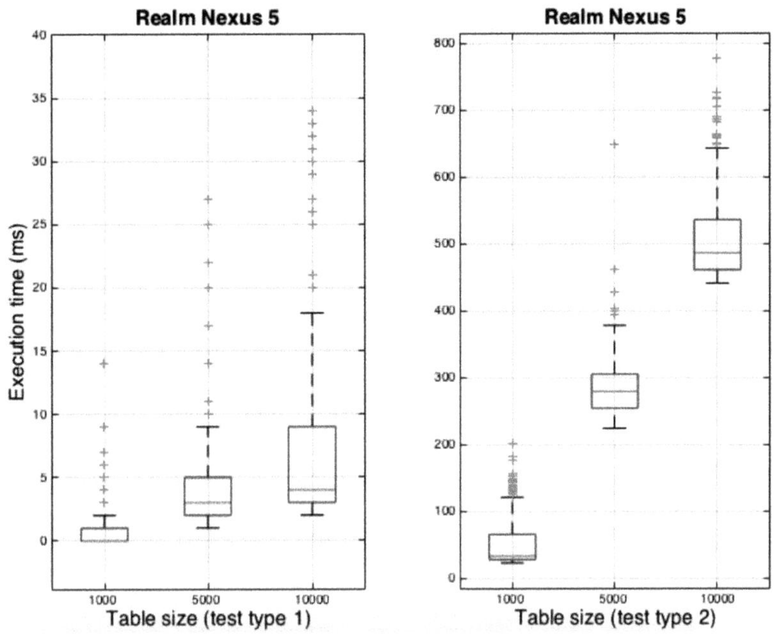

Figure 6. Boxplot of the median showing execution times for SQLite on a Google Nexus 5 for both types of operations (type 1 left, type 2 right)

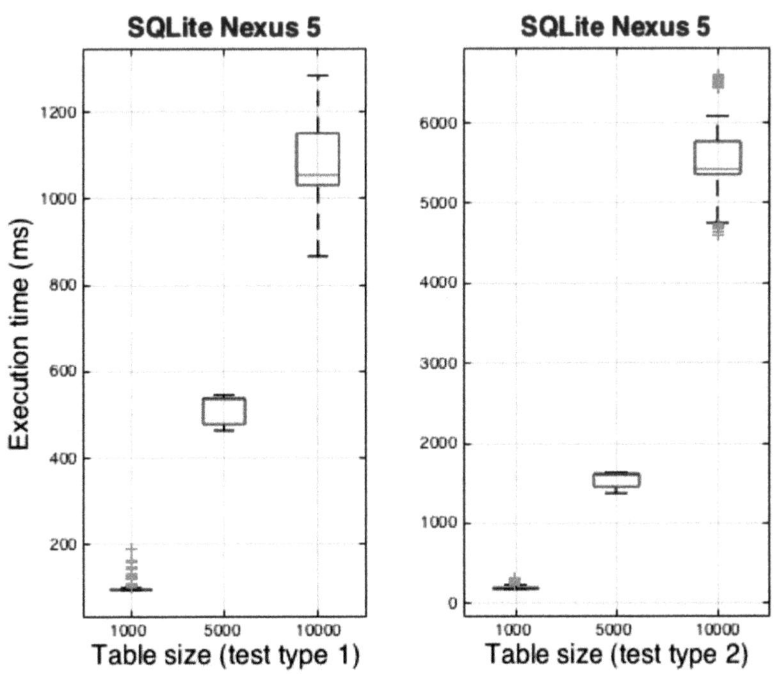

CLOUD INTERACTION

An alternative to keeping computation and data local is to outsource these services to the cloud. This has the advantage of lifting the restrictions on memory and computational power, but it presents a threat for the privacy of the data. Indeed, data will be stored and processed remotely by a service provider, that is not controlled by the user. However, interactions with the cloud are sometimes unavoidable. MEmoIt has a scientific vocation, and in order to be able to conduct experiments, some data still need to be sent to the servers. For privacy reasons, it is important to make sure that data is properly handled.

The guidelines presented in this section correspond to a use case where a back-end system collects and stores sensitive data about users and makes it available to third parties in a privacy friendly fashion. Designing such a system raises two concerns: First the scalability, considering the important amount of data that users produce; and second the privacy concerns of the users who will not be easily convinced to share their data.

A Reference System

The authors take inspiration from OpenPDS (de Montjoye et al., 2014), a system for collecting metadata from users; IT enables easy access by different entities in a safe manner. Each user has her own private data repository under her full control, and different applications store the user's metadata on his repository. The user also controls to which part of the metadata each application has access. When an application wants to extract useful information, it can run a query on part of the metadata that the user authorized it to access; no application has direct access to the metadata. OpenPDS has a distributed nature; each user's storage space is completely independent of other users' spaces. The user creates the storage herself, for example, by installing OpenPDS on the cloud, or by using a service that provides such storage spaces; She is the sole owner of her data. When the user uses applications that want to use her private data, she explicitly authorizes them to have controlled access to part of her metadata. These applications can freely add data to the user's storage but cannot read them. They can only query the data through the SafeAnswers module, which guaranties that the queries do not violate the user's privacy.

Choice of Platform

There exist, at the time of writing, two main contenders for integrating cloud computing in mobile applications. The first one, Amazon Web Services displays an important catalog of services ranging from a notification system, to data storage and processing. The second, Google Cloud Platform, is a complete set of services for developing applications to the cloud. Its capabilities include high-replication cloud storage, an auto scaling platform, App Engine, for deploying web applications, and big data services that include Google Could Storage, BigQuery, Dataflow. The main advantage of Google Cloud Platform is the seamless integration of its different components. Moving data between different services is free and requires much less time. Its "pay for what you use" pricing scheme is especially useful for emerging projects. Another reason for choosing Google Cloud Platform for the back end is the compatibility between it and Android applications, which allows secure and reliable communication with little effort. In the case of MEmoIt, the back end uses Google Cloud Storage to store user data, Dataflow to transform the data and BigQuery to access and query the transformed data.

Google Cloud Storage

Google Cloud storage (GCS) is a high-availability storage service for application developers; regular users should not use it. GCS uses the concept of buckets and objects to store data. Buckets are at the root of a project and enable the storage of objects within them. Objects within a bucket are immutable and can be only overwritten and not appended. The data in a GCS object are opaque to the cloud storage and are encrypted by default. The objects consist of the actual data and a set of metadata. The object metadata component is a collection of name-value pairs that describe various object qualities.

All operations on the GCS objects are atomic, thus any object uploaded is immediately available for reading and any object deleted will immediately become unavailable. For GCS buckets only eventual consistency is guaranteed; for example, if a request for a list of all objects in a bucket is made after a deletion operation a deleted object might be returned but it will still be inaccessible.

Google Dataflow

Dataflow[18] is a unified programming model and a managed service (in the Google Cloud infrastructure) for developing and executing a wide range of data processing patterns, including ETL (extract, transform and load), batch computation, and continuous computation. Cloud Dataflow frees the end user from operational tasks such as resource management and performance optimization.

Dataflow implements a superset of features of the widely-known MapReduce framework (introduced by Google in 2004). With Dataflow, Google provides its users with a processing layer that can seamlessly transfer (and transform) the data, among all the components of the Google Cloud infrastructure.

It has been recently released as an open-source framework, effectively reducing the lock-in factor of adopting a full-stack solution based on Google components. Dataflow comes with a Java API and an experimental Python API.

Google BigQuery

Google BigQuery is a Google service enabling SQL queries against append-only tables, and the processing is done using Google's infrastructure.

BigQuery is similar to a regular SQL database in terms of capabilities but with the ability to handle more complex data and in larger quantities (at the scale of terabytes). A table in BigQuery is a regular two-dimensional array where each row represents one entry and each column represents the data. A column can contain fields of a standard type such as string, integer and float. But can also contain nested/repeated fields, for example a column that contains telephone numbers can contain multiple telephone numbers in each row and/or can represent a telephone number as a set of values such as a country/area prefix and the actual telephone number. These entries are flattened when a query, that uses the nested fields, is executed. To perform any type of action on BigQuery, a job needs to be created. Jobs handle running queries, and importing and exporting data. A job can be synchronous or asynchronous and can be queried to test its completion. BigQuery offers interoperability with other Google services. It is very easy to import and export data into a table from the Google Cloud Storage simply by providing the fully qualified URI (i.e., the bucket and object from which to load or export the data).

MEmoIt Implementation

Compared to OpenPDS (de Montjoye, et al. 2014), MEmoIt has a simpler architecture and a centralized nature. The authors propose an architecture in which the code that runs a user study is sent to the device, executed, and only the results, in the shape of aggregates, are sent back to the server and are stored on Google Cloud Storage, thus contributing to the anonymization of the users. Periodically, a Dataflow job is run, which aggregates the data. Finally, a BigQuery job is run to present the data to the researchers.

Data Storage

The REST service, used by the application to send data to the server, exposes two API calls: Synchronize User and Upload Data. Each communication with the REST API is authenticated with the user's Google username and also contains the random salt or user defined passphrase that is used to hash the username to secure the user's anonymity.

In the first transmission, the application performs a Synchronize call to notify the service that there is a new user and gives the salt for the hashing so that the storage bucket is created. The service hashes the username with the random salt and obtains a unique hash that will become the name of the bucket that represents the specific user. In the case where the salt/passphrase is changed for a user, the service that creates the new bucket with the hashed username copies the contents of the previous bucket and then deletes it. In the case the user opts out of the data collection, she can choose to have her data retained on the cloud or deleted -- if she wants the data to be deleted the Synchronize call notifies the service to delete the data.

To upload data, the application uses the Upload Data API sending the data to the back-end service. It is worth noting that data transferred between the mobile application and the rest service is structured. The Endpoints API enables annotations for the API calls and data packet classes and automatically creates the corresponding Android Java objects.

The REST service is responsible for storing the received data in the Google Cloud Storage. When data arrives for a user, a new object is created in the user's bucket with the data for that upload session. The filename of each object is the timestamp of the upload. This makes it easier to search and delete parts of the data. When the REST service is notified to delete the data, it only deletes the bucket for the corresponding user and its contents, and it does not modify the BigQuery table as it is append-only.

Data Processing

On a regular basis, a DataFlow job is responsible for aggregating the data and loading it into BigQuery tables. This is required because of limits in the Google BigQuery API calls. There is a limit on the maximum files per load job of 10,000 files and a limit of 1,000 load jobs per table. With these limits, 1 million users would exceed the limit in 10 weeks. Also, the user id needs to be added to each row of data, and additional transformations can be applied (e.g., data obfuscation for privacy-preservation purposes).

FUTURE WORK

This chapter developed several aspects of mobile-application development that need to be re-adapted and proposed guidelines as to how to implement them. Multiple problems, however, are still unanswered. The location-based problem presented in this paper is not restricted to MEmoIt and an API should be provided by mobile platforms to address this problem without customized algorithms. Another aspect that the authors remarked during the development of MEmoIt, was the lack of lightweight machine-learning libraries for smartphones. With the quantity of data at hand and the increase in computational power, such libraries would open the way to most interesting on-device data-analytics. Finally, the ongoing effort to implement privacy mechanisms for the cloud, based on strong theoretic foundations such as differential privacy or k-anonymity (Sweeney, 2002), should be further pursued, as it represents one of the most promising solutions in this field.

CONCLUSION

Using their experience acquired while developing the life-logging application, MEmoIt, the authors have described in this chapter different aspects of mobile development that need to be carefully considered when building an application. Mobile device resources such as memory, computing or battery power are scarce and should be optimized in order to provide a good user-experience. Simplicity of the different components of an application is also essential. As seen for the places-of-interest detection algorithm, it can be interesting to build simple solution using built-in functionalities, rather than designing relatively complex algorithms that would perform equally well but increase the complexity of the codebase. Similarly, the authors show that a key enabler to implementing advanced features on a smartphone is to design algorithms that process data incrementally. In this way, the application can overcome excessive running times, even if it collected data for months in a row. Finally, cloud integration becomes increasingly appealing as it provides services that simply cannot be performed on a mobile device. Extra care, however, should be taken regarding the data that are transmitted to remote servers.

ACKNOWLEDGMENT

The authors would like to thank all the EPFL students who participated in the creation of the MEmoIt application. The MemorySense Project was sponsored by Samsung Research America and by Haslerstiftung in the context of the Smart World 11005 (MEM0R1ES) Project. Portions of the research in this paper used the MDC Database (owned by Nokia and made available by the Idiap Research Institute, Switzerland).

REFERENCES

Aberer, K., Catasta, M., Radu, H., Ranvier, J. E., Vasirani, M., & Yan, Z. (2014, March). Memorysense: Reconstructing and ranking user memories on mobile devices. In *Pervasive Computing and Communications Workshops (PERCOM Workshops), 2014 IEEE International Conference on* (pp. 195-198). IEEE.

Apple. (n.d.). *Healthkit*. Retrieved from: http://developer.apple.com/healthkit

Baltrunas, L., Ludwig, B., Peer, S., & Ricci, F. (2011). Context-aware places of interest recommendations for mobile users. In Design, User Experience, and Usability. Theory, Methods, Tools and Practice (pp. 531-540). Springer Berlin Heidelberg. doi:10.1007/978-3-642-21675-6_61

Chen, D., & Zhao, H. (2012, March). Data security and privacy protection issues in cloud computing. In *Computer Science and Electronics Engineering (ICCSEE), 2012 International Conference on* (Vol. 1, pp. 647-651). IEEE. doi:10.1109/ICCSEE.2012.193

Couchbase. (n.d.). *Couchbase Mobile*. Retrieved from; http://www.couchbase.com/nosql-databases/couchbase-mobile

de Montjoye, Y. A., Shmueli, E., Wang, S. S., & Pentland, A. S. (2014). openpds: Protecting the privacy of metadata through safeanswers. *PLoS ONE*, *9*(7), e98790. doi:10.1371/journal.pone.0098790 PMID:25007320

Developers. (n.d.a). *Saving Data*. Retrieved from: http://developer.android.com/training/basics/data-storage

Developers. (n.d.b). *Creating and Monitoring Geofences*. Retrieved from: http://developer.android.com/training/location/geofencing.html

Dwork, C., & Aaron R. (2014) The algorithmic foundations of differential privacy. *Foundations and Trends in Theoretical Computer Science, 9*(3-4).

Ester, M., Kriegel, H. P., Sander, J., & Xu, X. (1996). A density-based algorithm for discovering clusters in large spatial databases with noise. In KDD (Vol. 96, No. 34, pp. 226-231).

Foursquare. (n.d.). Retrieved from: http://foursquare.com

Furao, S., Ogura, T., & Hasegawa, O. (2007). An enhanced self-organizing incremental neural network for online unsupervised learning. *Neural Networks*, *20*(8), 893–903. doi:10.1016/j.neunet.2007.07.008 PMID:17826947

Giannotti, F., Nanni, M., Pinelli, F., & Pedreschi, D. (2007). Trajectory pattern mining. In *Proceedings of the 13th ACM SIGKDD international conference on Knowledge discovery and data mining* (pp. 330-339). ACM. doi:10.1145/1281192.1281230

Google Now. (n.d.). *Landing Now*. Retrieved from: https://www.google.com/landing/now

Google Play. (n.d.). *Maps*. Retrieved from: http://play.google.com/store/apps/details?id=com.google.android.apps.maps

Google Translate. (n.d.). Retrieved from: http://play.google.com/store/apps/details?id=com.google.android.apps.translate

Kiukkonen, N., Blom, J., Dousse, O., Gatica-Perez, D., & Laurila, J. (2010). Towards rich mobile phone datasets: Lausanne data collection campaign.*Proc. ICPS*.

Laurila, J. K., Gatica-Perez, D., Aad, I., Bornet, O., Do, T. M. T., Dousse, O., . . . Miettinen, M. (2012). The mobile data challenge: Big data for mobile computing research. In Pervasive Computing (No. EPFL-CONF-192489).

Lu, K. et al.. (2015). *Checking More and Alerting Less: Detecting Privacy Leakages via Enhanced Data-flow Analysis and Peer Voting*. NDSS; doi:10.1145/1899475.1899487

Montoliu, R., & Gatica-Perez, D. (2010). Discovering human places of interest from multimodal mobile phone data. In *Proceedings of the 9th International Conference on Mobile and Ubiquitous Multimedia* (p. 12). ACM.

Ostrovsky, D., & Rodenski, Y. (2014). Couchbase Lite on Android. In *Pro Couchbase Server* (pp. 283–292). Apress.

Paek, J., Kim, J., & Govindan, R. (2010). Energy-efficient rate-adaptive GPS-based positioning for smartphones. In *Proceedings of the 8th international conference on Mobile systems, applications, and services* (pp. 299-314). ACM.

Parse. (n.d.). *Parse Server*. Retrieved from: http://parse.com

Ranvier, J. E., Catasta, M., Vasirani, M., & Aberer, K. (2015). RoutineSense: A Mobile Sensing Framework for the Reconstruction of User Routines. In *2th International Conference on Mobile and Ubiquitous Systems: Computing, Networking and Services* (No. EPFL-CONF-208793). doi:10.4108/eai.22-7-2015.2260055

Realm. (n.d.). Retrieved from: http://realm.io

Srikant, R., & Agrawal, R. (1996). *Mining sequential patterns: Generalizations and performance improvements*. Springer Berlin Heidelberg.

Sweeney, L. (2002). k-anonymity: A model for protecting privacy. *International Journal of Uncertainty, Fuzziness and Knowledge-based Systems*, *10*(05), 557–570. doi:10.1142/S0218488502001648

Tsai, P. S., Lee, C. C., & Chen, A. L. (1999). An efficient approach for incremental association rule mining. In *Methodologies for Knowledge Discovery and Data Mining* (pp. 74–83). Springer Berlin Heidelberg. doi:10.1007/3-540-48912-6_10

Wheeler, D. A. (2004). *SLOC count user's guide*. Retrieved from http://www.dwheeler.com/sloccount/sloccount.html

Xiao, Z., & Xiao, Y. (2013). Security and privacy in cloud computing. *IEEE Communications Surveys and Tutorials*, *15*(2), 843–859. doi:10.1109/SURV.2012.060912.00182

Ye, Y., Zheng, Y., Chen, Y., Feng, J., & Xie, X. (2009, May). Mining individual life pattern based on location history. In *Mobile Data Management: Systems, Services and Middleware, 2009. MDM'09. Tenth International Conference on* (pp. 1-10). IEEE. doi:10.1109/MDM.2009.11

KEY TERMS AND DEFINITIONS

Cloud: The physical infrastructure composed by many computers housed in massive warehouses all over the world.

Data Storage: The software and hardware layers to retain digital data.

Finite State Machine: A mathematical model of computation used to design both computer programs and sequential logic circuits.

Incremental Algorithm: A software algorithm which, whenever a piece of input data changes, attempts to save time by only re-computing those outputs which depend on the changed data.

Location Services: The software and hardware layers used by a smartphone to retrieve the geographical coordinates of a user.

Privacy: The aspect of Information Technology that deals with the ability an organization or individual has to determine what data in a computer system can be shared with third parties.

Routine: A sequence of actions regularly followed.

Chapter 2
Mobile Application Testing

Vijay Ekambaram
IBM Research, India

Vivek Sharma
IBM Research, India

Nitendra Rajput
IBM Research, India

ABSTRACT

Statistics hold that 80% of the mobile applications are deleted after just one-time use. A significant reason for this can be attributed to the quality of the mobile application, thus impressing on the need for testing a mobile application before it is made available on the app stores. At the same time, the mobile application lifecycle time is shrinking. So while operating systems used to get release about once in a couple of years, mobile operating systems get updated within months. And talking of apps, new apps are expected to be built and released in a matter of weeks. This impresses the need for automated mechanisms to do mobile testing. The space of mobile application testing is challenging owing to the variety of phone devices, the operating systems and the conditions under which an app can be used by the user in the wild. This chapter is focused on tools and techniques that are used for automated testing of mobile applications.

INTRODUCTION

Testing mobile applications is an emerging research area that faces a variety of challenges due to increasing number of applications getting developed and a plethora of new devices being released into the market. These new devices have varied form factors, screen size, resolution, OS, hardware specification etc. which increases the difficulty to effectively test an application. Also, in comparison to conventional Desktop and Web applications, mobile applications have shorter release cycles (lesser time-to-market) and the update frequency is high, making it necessary for the tester to perform additional testing quite often. Due to these factors, testing a mobile application becomes a very expensive, laborious and time consuming process. This chapter primarily focusses on explaining the inherent challenges and solutions associated with different types of mobile testing.

DOI: 10.4018/978-1-5225-0945-5.ch002

The goal of any mobile testing solution is two folds. The first aim is to ensure that all possible challenges with respect to the application are detected. Detection of such challenges, which could be functional issues in the application, or usability issues that make the application difficult to use, or performance issues that make the application frustrating to use due to resource constraints. The second, optional but preferred, aim is to determine the cause of such a challenge.

As has been discussed in earlier chapters, there are at least three different ways of building a mobile application: native, hybrid, and web. Each of these mechanisms differ in the way where and how most of the processing of the application happens. To complicate matters further, a mobile application can access backend data and services over the network. In such a complex scenario, it is important to determine the challenges in a mobile application, not only from within the mobile device, but across entities that are eventually enabling the application to execute. Such issues make mobile application testing an interesting and a challenging software engineering problem to look at.

ORGANIZATION OF THIS CHAPTER

Testing a mobile application has some obvious key goals. The main goal being that the application should work well. However, from a software engineering perspective, we need to ensure that the application does the task for which it is designed, and does it easily enough so that users can use it easily and does it with optimal compute resources. These main characteristics of a desirable good mobile application then raise three key forms of mobile testing. Functional testing is performed to ensure functionality, i.e. to test whether the application is performing the functions that it was designed for. Performance testing is conducted to determine how optimal the application is, in terms of its compute resource usage, battery usage and latency related issues. And finally, Usability and Accessibility testing aims to capture how easy is for users to be able to work with the applications to execute the functions for which the application is designed. All of these three different forms of testing have an implication to each other and so are not strictly orthogonal. However, a division of purpose enables to determine the source of the problem in a mobile application, thus making the possible solutions and the resources required to fix it. As an example, if there is an issue detected in usability of the application, then perhaps designers are best at fixing them. On the other hand, if there is an issue with respect to the performance of the application, such as the response is too slow (which itself can lead to usability issues), then a software engineer having knowledge of the mobile operating system should get involved. For such reasons, in this chapter, mobile testing will be studied under these three umbrella topics: functional, performance, and usability & accessibility. At the end of this chapter, we discuss detailed literature survey on these three umbrella topics with various discussion on latest trends followed in mobile testing. But before studying each one of these in detail, let us discuss the basics of mobile testing in the next section.

KEY COMPONENTS OF MOBILE TESTING

Mobile Applications are tested in 3 main ways: unit, manual and automated testing. This chapter is mostly focused on automated testing and its associated challenges and standards, but we will also discuss below about the unit and manual testing process at high level, from a completeness perspective.

Unit Testing Process

In unit testing process, developers test individual software artifacts (i.e. code module, libraries, UI, etc.) to determine their readiness to integrate with other code components. This testing phase happens in parallel during development and mostly conducted by developers. The goal of unit testing is to isolate and test each part of program to test its correctness. To achieve this, developers write and execute various unit test-cases to test various individual components.

Manual Testing Process

In manual testing process, a test engineer manually executes the complete Application Under Test (AUT) in various scenarios and record the observed output. Then, Test Engineer compares the observed output with the expected output to discover defects/issues in the AUT.

In mobile testing, it is highly recommended to test the AUT in various devices, contexts and configurations. So, a Test Engineer has to repeat the same testing process multiple times in different devices/configurations which is time-consuming and tiring process. Automation of the testing process is extremely helpful for Test Engineers to speed-up this otherwise repetitive testing process.

Automated Testing Process

In this section, we explain the basic components involved in a Mobile Test Automation Framework (in Figure 1). These are generic components that are used across the different type of automated testing (functional, performance and usability) and so we will discuss these at a high level first:

Figure 1. Components of mobile test automation framework

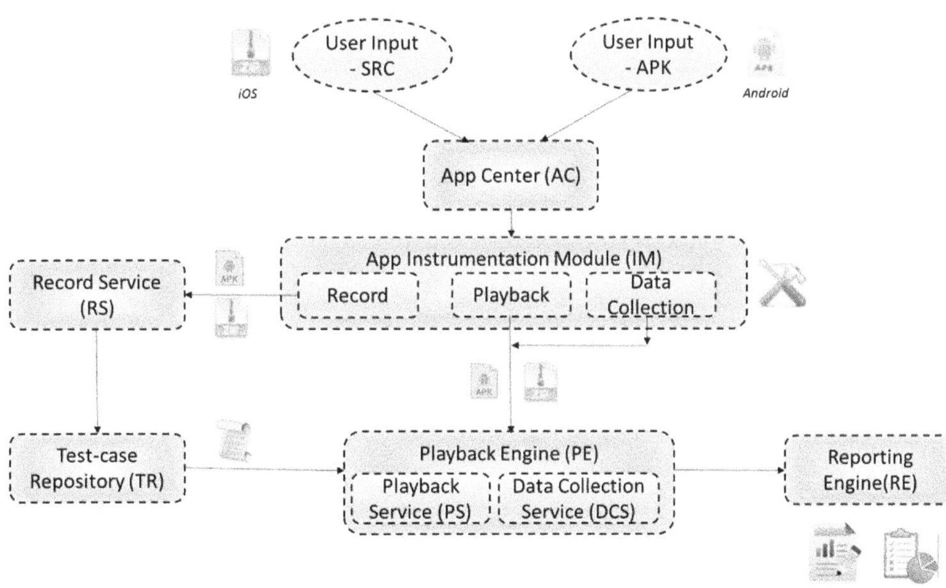

1. **App Center (AC):** Contains a collection of AUTs for instrumentation and testing. For Android, AUT is same as an Application Package Kit (APK) file and for iOS, AUT same as the Source-code Zip file.
2. **App Instrumentation Module (IM):** The IM transfers the input AUT to an instrumented AUT by injecting specific code snippets for various purposes. The process of instrumentation is to ensure that the tester platform has certain hooks through which it can capture the parameters that can help in identifying whether the test was successful or not. Following are the different types of instrumentation techniques that are commonly used.
    a. **Record Instrumentation:** In Record instrumentation, Code-snippets are injected into an AUT, with the goal that it could record all events and interactions with the AUT. This instrumentation process enables construction of test-cases.
    b. **Playback Instrumentation***:* In the testing phase, it is also important to enable playback of a test case. Code-snippets are injected into AUT which could apply specific events on the AUT. This instrumentation process enables automatic execution of events on AUT without human involvement.
    c. **Data Collection Instrumentation***:* This instrumentation is focused on code-snippets that are injected into AUT to collect various logging events with respect to functional, performance, usability and accessibility testing. This instrumentation process collects data for defect analysis.
    d. More details on the instrumentation process is explained in the next section.
3. **Record Service (RS):** The Record Service provides a framework to enable Test Engineer to manually execute and record sequence of actions on the AUT. When user interacts with the AUT, RS captures all the user/system events with respect to the AUT. These captured user/system events are then converted to a test-case. Test-case contains all the user/system events with additional meta-data (like timestamp). Example of a Test-case is depicted in Figure 2.
4. **Test-Case Repository (TR):** This is a repository that maintains a collection of test-cases for each AUT. These test-cases are constructed using the above record service.
5. **Playback Engine (PE):** Given an instrumented AUT and a test-case, the Playback Engine enables execution of the test-case on the AUT without human involvement and also collects necessary logging details for defect analysis. The Playback Engine depends on the following two services to enable the automatic playback of test cases:
    a. **Playback Service:** Enables execution of recorded events (as depicted in test-case) on the AUT.
    b. **Data Collection Service (DCS):** Collects various logging details for defect analysis based on the type of testing (i.e. functional, performance, usability, etc.)
6. **Reporting Engine (RE):** The Reporting Engine is the key component that eventually generates test reports for the test case executions on the AUT by the Playback Engine. An example of test-report is depicted in Figure 3. Based on the type of testing, various interactive visualizations and reports could be generated. Ideal test-report should not only pin-point defects in the application, but also suggest root-cause for the defects.

Figure 2. Sample test-case

- Launch application: IBM NUC Sensing Ver 1 4
  - Click on **Button 'Start Reporting'**
  - Click on **Button 'Neighbourhood Issues'**
  - Enter text **'testing first time'** in **Edit text** below Text view **'Within City Limits'**
  - Enter text **'no details available'** in **Edit text** below Text view **'In the suburbs'**

Figure 3. Sample test report

Test:	sample_test [executed on 7/22/13 12:53 PM]
Categories to test:	Functional, Performance, Accessibility
Application:	IBM NUC Sensing App (version 4; created on 7/22/13 12:50 PM; executed on 7/22/13 12:53 PM)
Execution Status:	**Functional** – All steps passed **Performance** – Passed **Accessibility** - Failed. 1 or more elements failed the test
Device:	samsung GT-P3100 (Android; API level 15)

FUNCTIONAL TESTING

Proper functioning is a basic need of an app to stay in App world. If an app does not function well, it gets bad reviews and less user attention. That is why functional testing of a mobile application is very important. In this phase, the app goes under a process to check app's functionality thoroughly in terms of every screen or widget is responding as expected, transition between screens is also as expected in all kind of end user possible scenarios. Functional Testing can be done in both fashions - manual or automated. Let's talk about automation of functional testing in detail below.

Automating the process of testing an app in terms of its overall functionality is Functional Test Automation. It is very helpful to reduce time and effort required to test an app, especially in Regression Testing model where apps have to repeatedly tested. There are different phases of Functional Test Automation – Instrumentation Phase, Record Phase, Playback Phase and Reporting Phase (Functional Test Reports).

Instrumentation Phase

Since the instrumentation of an application depends heavily on the operating system for which the application is authored, we will cover the instrumentation for android and iOS in separate sections.

Instrumentation in Android

In the case of Android, the AUT is decompiled, reverse engineered and instrumented. The detailed instrumentation process in Android is described in Figure 4. The following are the steps involved in the Android instrumentation process.

Figure 4. Android instrumentation

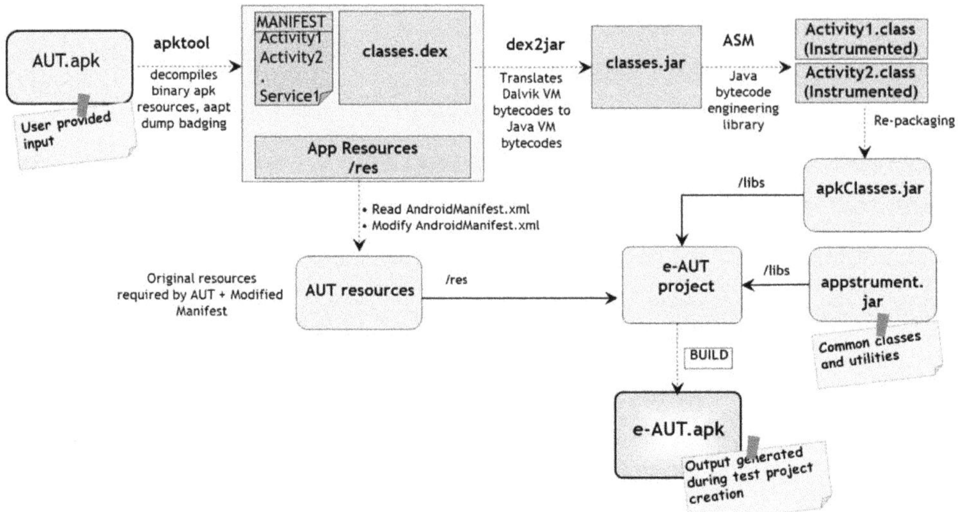

The Android application package referred to as AUT.apk file is given as input by the tester. Using the Dump Badging command of the Android Asset Packaging Tool (AAPT) (Ubuntu manuals, n.d.), high level information about the .apk file like package name, versionCode, versionName, permissions, features, minSdkVersion, targetSdkVersion, app label, app icon file path, main launchable activity name etc. are obtained.

The AUT .apk file is then reverse engineered using the Android apktool (Apktool, n.d.). This tool unzips the .apk file generating all the resources (images, layout XMLs, string resource XMLs, manifest file, etc.) and the compiled classes (classes.dex) of the AUT. Classes.dex is nothing but byte-code of all AUT classes compiled for the Dalvik virtual machine.

In the next step, dex2jar (Bitbucket, n.d.). tool is used to decompile classes.dex to decode and understand the underlying Java classes. This tool translates the Dalvik VM byte-codes to the Java VM byte-codes. The resultant file is classes.jar which contains corresponding Java classes from classes.dex. The individual files from classes.jar is extracted by using the Java jar tool (Jar-The Java Archive Tool, n.d.). From the Manifest file, the declared Activity information is obtained and using the ASM byte-code engineering library (ASM-Guide, n.d.), the Java byte-code is loaded into the memory and altered in different ways.

Once the bytecode is loaded, for each Activity class declared in the manifest, modifications are performed on the Android activity life cycle methods viz. onCreate(), onStart(), onResume(), onPause() and onDestroy(). These methods are modified to include an additional line towards the end of each method, which enables a call-out to Appstrument Instrumenter class, thereby establishing control over the complete application during runtime. Appstrument Instrumenter class contains code to log or inject the required activities/events. This method of instrumentation is a standard process followed across android testing frameworks where we do not have access to the source code of the android apps.

Instrumentation in iOS

In the case of an iOS application, due to complexity of code signing and other security reasons, reverse engineering is not very successful and also considered as illegal to use for any product. That is why with most of the test automation tools available in market, there is a requirement of AUT's source code to carry out the instrumentation process for iOS.

The Appstrument library is an instrumentation library that gets linked with the AUT's source code and the AUT is built to get the instrumented version of it. However, in this process there is no modification in source code of the original AUT. Objective-C Method Swizzling concept (Method Swizzling, 2014) is used to force a method in AUT to call an alternate implementation which is different from the original implementation.

In the below example (as in Figure 5), since directly overriding the sendEvent method (called by UIApplication to dispatch events to views inside the window) would break the responder chain sequence, the Objective-C Method Swizzling concept is used to hook into our Appstrument adapter code on the iOS event bus using the original sendEvent method.

Once the sendEvent method is swizzled, Appstrument gains access into the event bus at runtime and can push a new event during playback of the AUT. Figure 6 explains the overall component model required for iOS instrumentation.

Figure 5. Method swizzling example

```
// called by UIApplication to dispatch events to views inside the window

-    (void)sendEvent:(UIEvent *)event;
```

Figure 6. iOS instrumentation

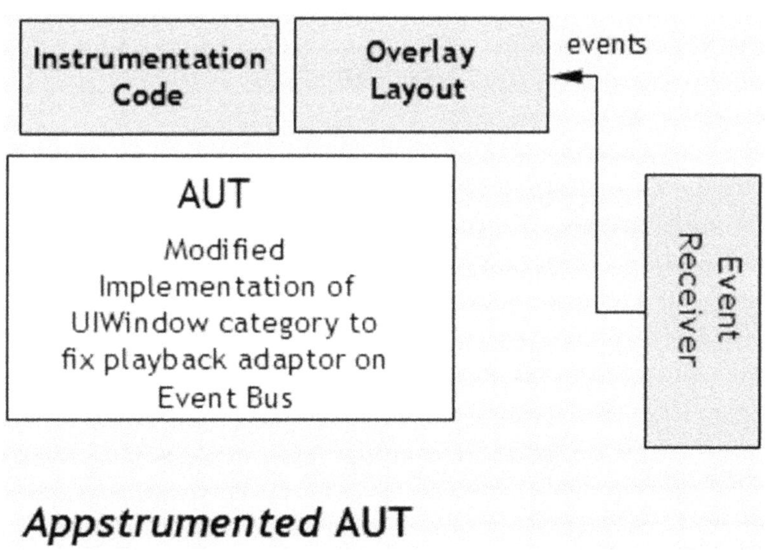

Record Phase

In this phase a tester performs actions on instrumented version of the AUT and this instrumented version of AUT records all actions as Test Steps i.e. Launch Application, Click on button "Sign In" etc. These recorded Test Steps are combined and get stored into a persistence memory and these sets are called as Testcases. It's a primary input for a playback Phase.

Figure 7 shows a small use case of an AUT where the app just has a three screens, one login screen, home screen on successful login which shows account specific discount coupons along with options to change username or password, these options takes user on third screen to edit username or password. The application flow is also shown in Figure 7.

Once this app is instrumented then a tester may perform one or a set of actions from below possible actions for this application.

1. Launch the application.
2. Fill username and password.
3. Click login (Successful login takes the tester to the home screen)
4. On home screen tester sees different discount coupons and taping any of them copies the coupon code.
5. Tester clicks "Change Username" button which takes him to settings screen.
    a. (It has a text field to edit username with two buttons, "Update" button and "Cancel" button. There is another text field with same set of buttons to edit password)
6. Tester edits the username text field and clicks on "Update" button.
7. Tester clicks logout and closes the app.

Figure 7. Example use case

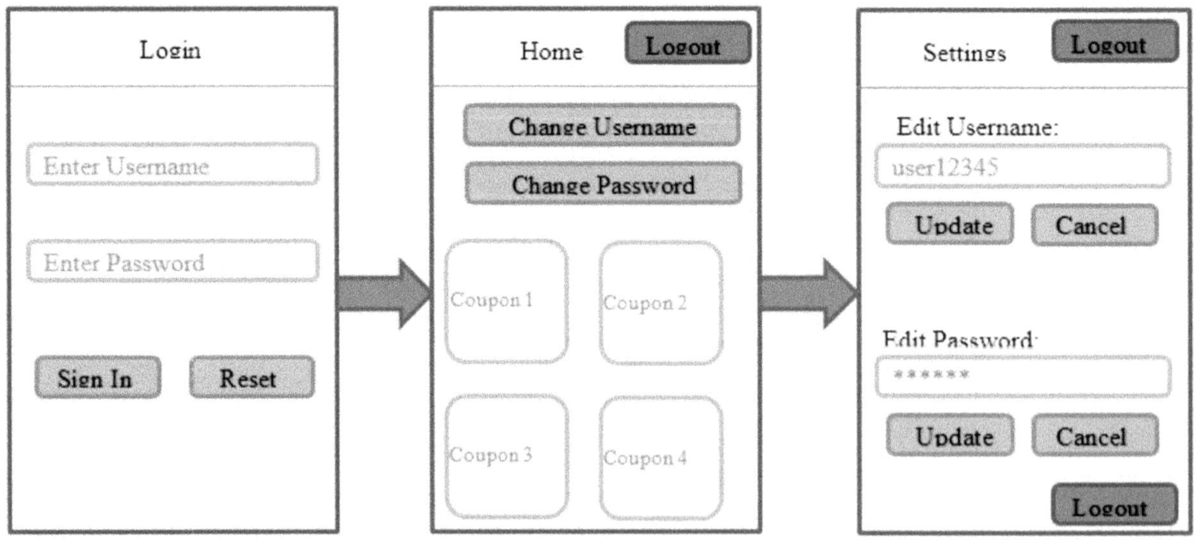

Now when a tester is performing these actions in recording phase on an instrumented AUT, the recording engine interprets these actions in small machine level steps as below.

1. Launch application: Sample App V1.
2. Enter text "user12345" in "Enter Username" edit text.
3. Enter text "passw0rd" in "Enter Password" edit text.
4. Click on button "Sign In".
5. Click on button "Coupon 1".
6. Click on button "Change Username".
7. Enter text "user12789" in "user12345" edit text.
8. Click on button "Update".
9. Click on button "Logout".
10. Close application.

In above testcase, all test steps are written in easy human readable English language which is referred as clearscript. There are clearscript parsers available which can convert these steps back into a machine understandable format. There are few components of a clearscript test step as described in Table 1.

Looking at test step 7 and 8, which are highlighted in red, in step 7 the widget identifier is captured as the text written in the edit text that is "user12345". Similarly, in step 2 the widget identifier is captures as the "Enter Username" which is place holder text for that edit text. Point to be noted here is that the "Enter Username" place holder is a static string and cannot be modified by tester, so every time when tester reaches to that screen, "Enter Username" would be there. That makes it easy and clear for playback engine to identify the widget to enter parameter "user12345" in "Enter Username" edit text while executing step 2 in playback phase but in step 7 the identifier captured is "user12345" which is dynamic and shown according to the logged user. Assume, in recording phase username - user12345 was used but in playback phase some other username was used, like "user56765". In that case while executing step 7 in playback, the playback engine will never find a text edit with identifier "user12345". So we can consider that the identifier captured in step 7 is insufficient for a guaranteed playback.

Similarly, in step 8 there are two "Update" buttons on the screens and it can be difficult for playback engine to identify that on which widget this step 8 needs to executed. "Update" button click of "Edit Username" or "Update" button click of "Edit Password". This problem is called as Dynamic Find Problem.

To solve this problem and to capture a most appropriate and unique identifier for a widget of a test step, "Dynamic Find Algorithm" is used.

Table 1. Clearscript test step

Step #	Action	Parameter	Widget Identifier	Widget Type
3	Enter	"passw0rd"	Enter Password	Edit text
4	Click		Sign In	Button

Dynamic Find Algorithm

Dynamic Find is not a problem restricted to 'dynamically finding objects' at playback time, it has implications at record time also. However, only if we generate the best possible clearscript (CS) tuples at record time, will we be able to properly find objects at playback time.

Crux of dynamic find problem is to resolve disambiguation for CS tuples. There are several ways to disambiguate. Couple of important ones are:

1. Disambiguate by global proximity,
2. Disambiguate by hierarchy,
3. Disambiguate by Global Proximity is about identifying neighboring objects that uniquely describe the chosen object, and then moving on to describe the neighboring object in turn (and so on, until we have completely resolved the object under consideration). Figure 8 shows an example for resolving disambiguation by global proximity.

This method is more human-friendly and less machine-friendly. It is difficult and sometimes impossible to encode unambiguously the position/identity of an object using global proximity.

Disambiguate by Hierarchy is about using the view hierarchy of the layout. This method is more machine-friendly and less human-friendly. Now, tester needs to know names of layout elements (like ListView, Gallery, etc.). However, disambiguation is more accurate and replay can be easy and effective. Figure 9 shows an example of disambiguate by hierarchy where imageView (target object) is referred with respect to its view hierarchy (ex. listview and contained textview).

A combination of disambiguation by global proximity and hierarchy is recommended. The following algorithm is then used to perform the disambiguation.

Figure 8. Disambiguate by global proximity

Figure 9. Disambiguate by Hierarchy

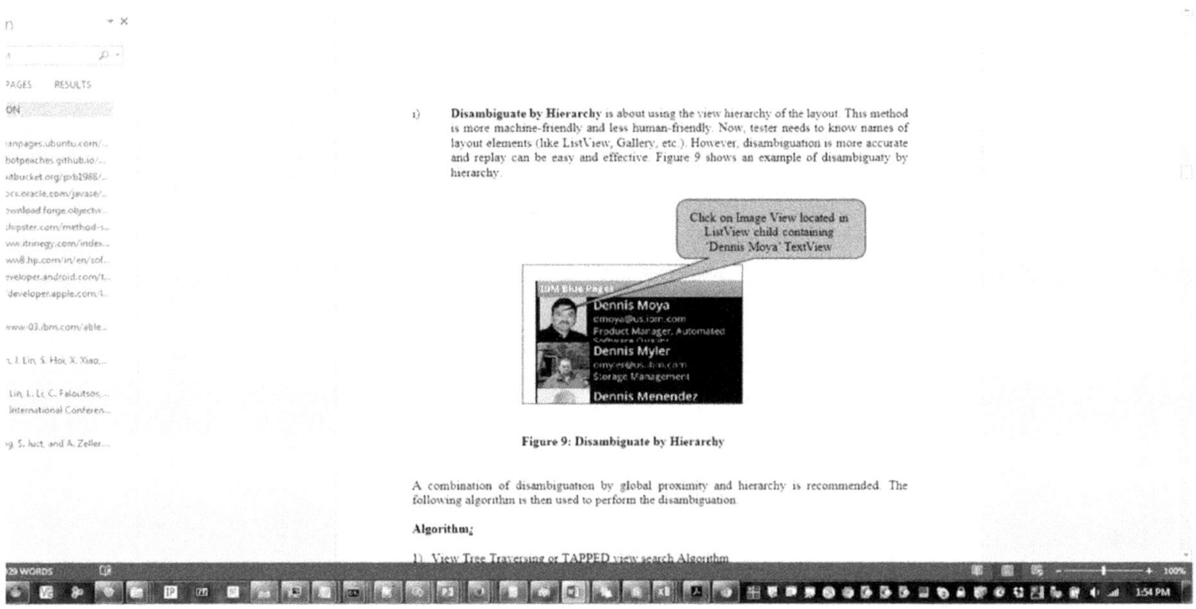

Algorithm

1. View Tree Traversing or TAPPED view search Algorithm:
    a. Enqueue the root object.
    b. Dequeue an object and examine it whether it contains the TAP coordinates (x, y).
        i. If the TAP coordinates is found in this object, check whether this object contains any child object or not. If It contains child object, then set the above object as root and repeat from step a. If not then above object is the TAPPED object, now move to execute 2nd level examination on this object (Search for object title or nearby label).
        ii. Otherwise enqueue any successor (the direct child object) that have not yet been discovered.
    c. If the queue is empty, every object on the view tree has been examined – quit the search, return "not found" and ignore that TAP event.
    d. If the queue is not empty, repeat from step b.
2. Associated title/label search algorithm (2nd Level examination)
    a. Collect and enqueue all the labels in the same parent view.
    b. Dequeue a label and examine its direction in respect of the TAPPED object.
        i. If the label exists on CENTER / LEFT / ABOVE / RIGHT / BOTTOM side of the TAPPED object, then validate it and again queue the label separately by tagging it as CENTER / LEFT / ABOVE / RIGHT / BOTTOM with 1 to 5 priority respectively.
        ii. Otherwise invalidate the label, discard it and repeat step b until any label is in queue.
    c. If the queue is not empty, repeat from step b.
    d. Examine the count of CENTER queue.
        i. If queue is empty, then move to the step e.

  ii. Otherwise If count is 1 then dequeue the label from CENTER queue and set it as required associated label and move to the step f.
e. Examine the count of next priority queue. (i.e. LEFT -> ABOVE -> RIGHT -> BOTTOM)
  i. If queue is empty, then repeat step e with next priority queue.
  ii. Otherwise If count is 1 then dequeue the label from queue and set it as required associated label and move to the step f.
  iii. Otherwise is count > 1 then sort the queue in respect to the distance in between the label and the TAPPED object then dequeue the 1st label of sorted queue and set it as required associated label and move to the step f.
f. Examine if it found any nearest associated label
  i. if YES then generate a clearscript statement using found associated label for the view object.
  ii. if NO then repeat step a with any successor parent (branches) that have not yet been discovered.

So after applying dynamic find algorithm step 7 and 8 should be modified as:

1. Enter text "user12789" in text edit below label "Edit Username".
2. Click on button "Update" below label "Edit Username".

Figure 10. Associated title/label search algorithm (2nd level examination)

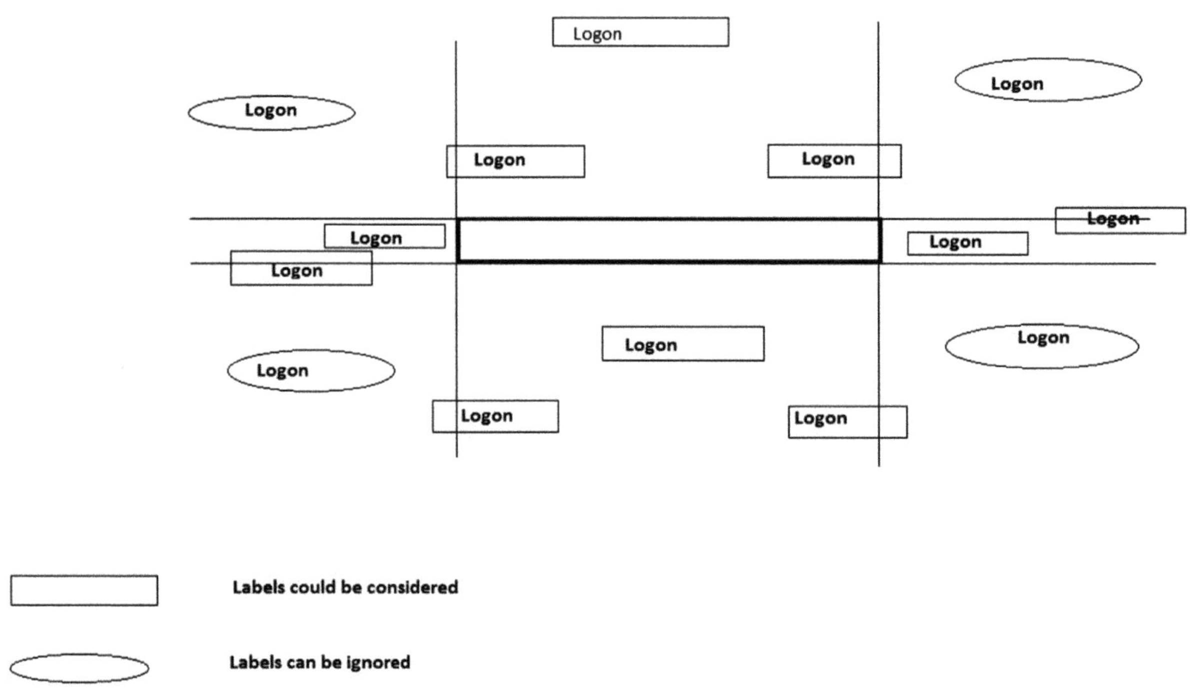

Playback Phase

In this phase playback engine takes a testcase from testcase repository and executes it step by step. Let's take an example of step 7 in above test case.

Enter text "user12789" in text edit below label "Edit Username".

Playback engine parses above test step using CS parser and provides the widget identifier to dynamic find engine, then dynamic find engine looks for the targeted widget in view hierarchy. Here "Edit Username" is the identifier of label which is above to the targeted widget. So dynamic find searches for a label with title "Edit Username" in the view hierarchy and then looks for the targeted edit text field below "Edit Username" label. Once it's found playback engine reads what action needs to be performed on this widget from the test step, which is "Enter text" here. Now playback engine creates an event to feed the targeted edit text with text parameter "user12789" and will push this event into application event bus. Then OS will take care of this injected event and will execute it.

Similarly, playback engine moves to the next step and executes it, until either a test step fails in execution or all test steps are executed successfully. Playback engine plays an important role in automation process, and collects set of data for defect analysis on every test step execution like time taken in execution of an individual step, test step result i.e. failed or successful etc. Playback engine passes this collected data to reporting engine, which get utilized in Reporting phase by putting it in formatted test report.

Functional Test Reports

Post playback phase, all collected data goes to Reporting Engine (RE) where it prepares a formatted report for this test case execution. In Figure 11 we can see a sample function test report prepared by Reporting Engine.

Figure 11. Sample test report

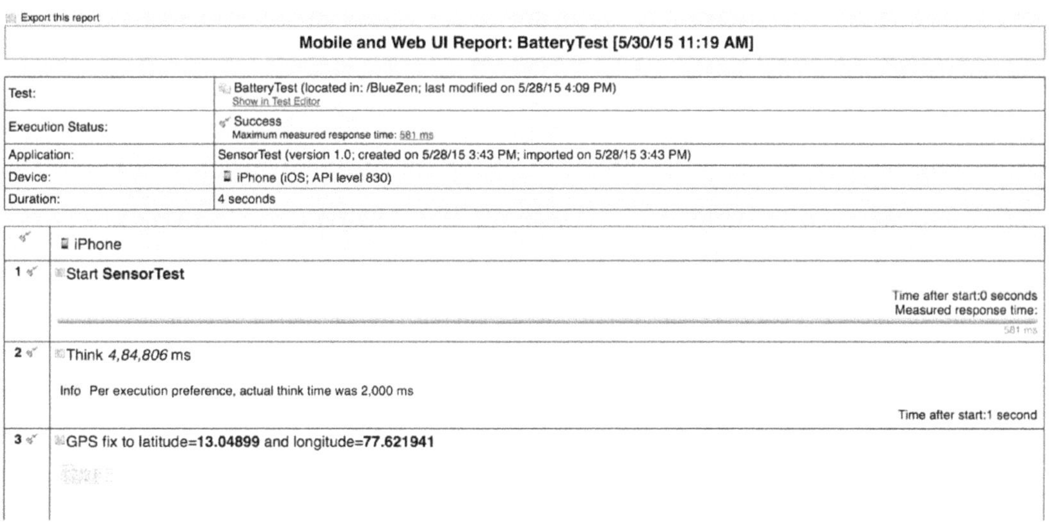
Export this report

Mobile and Web UI Report: BatteryTest [5/30/15 11:19 AM]

Test:	BatteryTest (located in: /BlueZen; last modified on 5/28/15 4:09 PM) Show in Test Editor
Execution Status:	Success Maximum measured response time: 581 ms
Application:	SensorTest (version 1.0; created on 5/28/15 3:43 PM; imported on 5/28/15 3:43 PM)
Device:	iPhone (iOS; API level 830)
Duration:	4 seconds

	iPhone
1	Start **SensorTest** Time after start:0 seconds Measured response time: 581 ms
2	Think *4,84,806* ms Info Per execution preference, actual think time was 2,000 ms Time after start:1 second
3	GPS fix to latitude=**13.04899** and longitude=**77.621941**

A test report should have information of testcase(s) executed in playback. As shown in above sample report "Battery Test" is the Test Case name. It can have additional information of a test case like location of test case and last modified timestamp. A test report should show execution status, which can be success or failure. It also can display maximum/minimum/average response time of an individual test step. Response time is the time taken in playback of a test step from start to end. A test report should have application under test details, like name of the application along with application version information and creation or import timestamps. A test report should have information about the device used for testing as in above sample report, an iPhone with iOS 8.3 was used. Also, it should have information of overall duration of playback. Later part of the report should have details of individual test steps, like test step text along with respective response time which helps to analysis an individual test step.

Change Impact Analysis

One of the classical problem with respect to the test automation is that, when a Test Engineer receives a new build to test, user has to repeat all the test-cases which were tested and verified with respect to the earlier build of the application to ensure test completeness. Considering the scenario of mobile development life cycle where new versions get released very often, it is not practically feasible to retest all the test-cases for every version of the build. Even automated testing is time consuming as the scale of test-cases to validate for every release will be very high high for industrial mobile apps. The practical solution followed in the industry is to randomly choose the test-cases to test the new build. In-order to provide a more consistent solution, user could filter test-cases which were affected due to the version code changes and test only the affected test-cases. By comparing the source code of the old and the new build, user could obtain the methods which were present in the old version but were modified in the new version. Then, by analyzing the instrumentation runtime logs of the test-cases which were executed on the old build, user could filter out the affected test-cases which executed these identified affected methods (obtained from code difference between old and new build). Testing the new version of the build with the affected test-cases ensures test completeness.

PERFORMANCE TESTING

Performance of Mobile application highly determines its success. Performance issues like app freezing, unresponsive application, app lags, poor responsiveness, etc. are more frequent in a mobile app and have a high impact on its users. So, Mobile Performance Testing is extremely important in the mobile app software life-cycle. This section highlights various challenges involved in the mobile performance testing and also discusses various solution approaches for performing effective mobile performance testing.

Challenges in Mobile Performance Testing

Most of the native/hybrid mobile applications in market (like Ticket Booking Apps, Banking Apps) are online applications. Performance analysis of these applications is very challenging and complicated because it depends on the inter-relationship between the client, network and server performance. The client side performance refers to that of the application installed in the mobile device, independent of the network or the server with which it communicates. Client-side content rendering, thread execution, native/kernel call execution, mobile device resource availability are some of the factors which decides the client side performance. Network performance refers to the quality of the network to which the mobile device is connected. Network bandwidth, signal strength, interference are some of the factors which depend on the quality of the connected network. Server side performance refers to the performance of the servers/cloud with which the AUT communicates. User Load (i.e. number of requests served by the server per unit time) and server resource utilization are some of the factors which determines the server performance. Bottleneck in any of these components (i.e. client, network and server) can lead to poor application performance. Identifying these performance issues in the mobile applications and mapping each of these with the corresponding bottleneck component across client, network and server with root-cause is very challenging; however, it is necessary for effective performance testing.

Further, performance issues of mobile applications are highly dependent on the test environment/ context in which the application is tested. Test contexts can be further classified as:

1. Device Context,
2. Network Context, and
3. Server Context.

Device Contexts such as CPU Load, Memory Load, Network States, etc. strongly determine the performance of mobile applications. Generally, mobile applications tend to perform with lags under high CPU/Memory load. However, it is essential for the application developers to optimize the code such that, applications withstand these unfavorable scenarios and they do not crash. Simulating various real time device contexts accurately in an automated fashion is a challenging task. Network Context refers to the quality of the network to which the device is connected. Emulating various properties of networks like bandwidth, signal strength, data rate and then stress testing the application help testers to identify rare performance issues. Server Context refers to the user load on the server. Server load testing refers to the capturing of various real-time traffic requests (of all protocol types) of the AUT and simulating the server load as realistic as possible. Simulation of these real-time end-to-end context (as in Figure 12) is of high interest to the mobile software testing community which provides realistic on-field testing environment for effective testing.

Figure 12. End-to-end context simulation

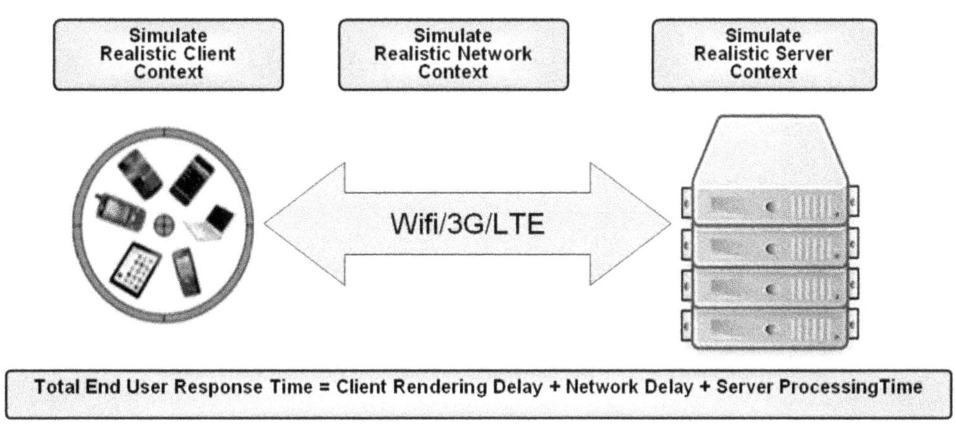

It is very difficult to manually configure these realistic contexts and analyze end-to-end performance monitoring. An alternate and easy approach is to use automation tools which can programmatically emulate contexts and collect required logs for analysis. In the next section, we discuss various steps involved in the mobile performance test automation for effective identification of performance issues.

Steps Involved in Mobile Performance Test Automation

Mobile Performance Test Automation has 3 phases:

1. **Record Phase:** Test Engineer records several test-cases (GUI test-cases and HTTP test-cases) and maintains the test-case repository.
2. **Context Simulation Phase:** Test Engineer simulates various realistic client, network and server context for testing in real-scenarios. After setting up the required test environment/context, Test Engineer can playback test-cases. Since test-cases are executed in real-life context, there is chance of more issues to get discovered.
3. **Playback Phase:** Test Engineer can choose a test-case and play it back on the AUT without any human involvement. During playback phase, various data logging happens which is analyzed for effective performance testing.

Figure 13 explains the various components involved in a generic mobile performance test automation framework.

Figure 13. Mobile performance test automation framework

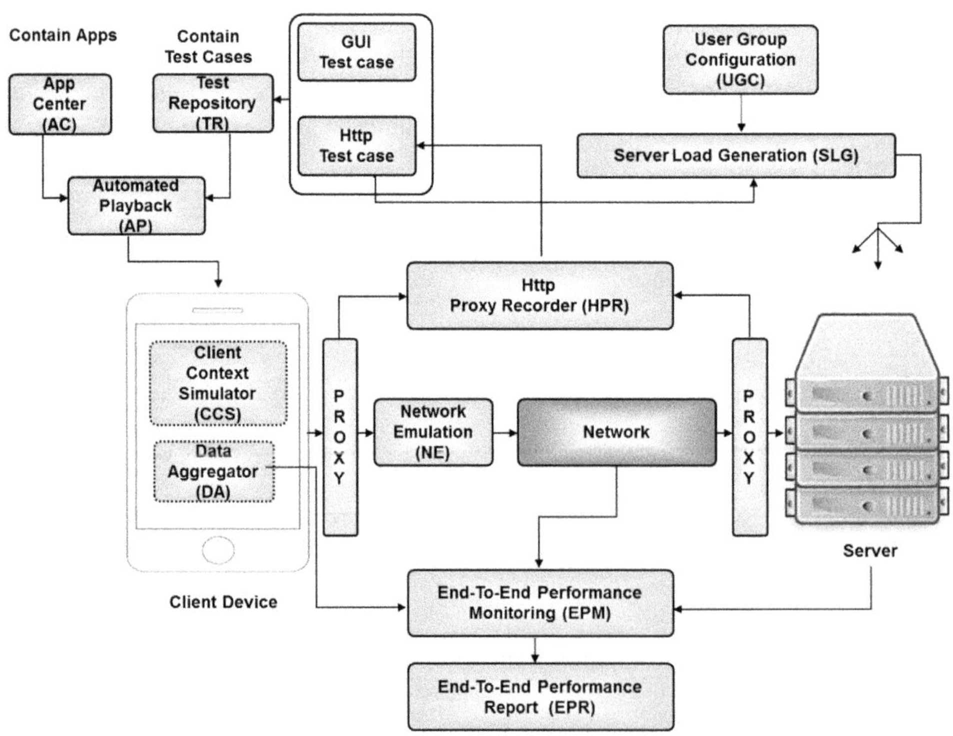

Record Phase

During record phase, Test Engineer manually executes sequence of actions to test the AUT. In the background, Record Service (RS) component (in Figure 1) captures all the user/system events with respect to the AUT during the test-case execution. These captured user/system events are then converted to a GUI test-case. The GUI Test-case contains all the user/system events with additional meta-data (like timestamp). Example of a GUI Test-case is depicted in Figure 2. In addition to the GUI test-case, RS component also captures HTTP requests/responses associated with the AUT along with its triggered time stamps to generate the HTTP test-case. Example of a HTTP Test-case is shown in Figure 14. The HTTP test-case is used to create different kinds of realistic server loads. RS component could also support other web protocols including Citrix, SAP, TCP Socket, Web Services (SOA) and Siebel.

Realistic Context Simulation Phase

In this section, we explain the techniques involved in simulating realistic client, server and network context for the effective performance testing.

- **Realistic Client Context Simulation:** Client Context refers to the actual state of mobile (with respect to CPU, memory, Network, Sensors, etc.) during a test-case execution. Performance issues does not occur all the time. It occurs only in certain contexts/scenarios. For example, if an application heavily uses internet, then it is necessary to test the application under various network conditions to effectively discover the performance issues. Since mobiles have high resource constraints, mobile client contexts highly determine the performance of the AUT. So, it is essential and necessary for Test Engineers to test the AUT under various mobile contexts to identify performance issues. Moreover, it is really a tricky, tiring and challenging task for Test Engineers to identify contexts and to simulate them accordingly on a real device. Client Context Simulator (CCS) Component as shown in Figure 13 is responsible for realistic client context simulation. Now, we explain how CCS component simulates various realistic contexts on the real device for a realistic production-like testing environment.

Figure 14. Http test-case

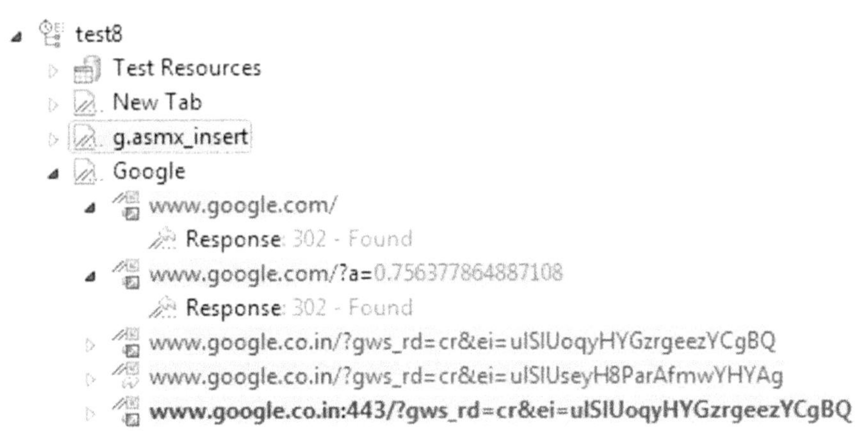

Various Mobile Services/factors that affect the mobile contexts are describes as follows:

- **CPU Load:** CPU Load refers to the percentage of CPU used in the mobile. Applications with high computational tasks (like gaming apps) are required to be tested under various CPU loads. It is necessary to observe the behavior of these apps in various CPU loads (from light load to heavy load) to identify performance issues. CCS component simulates the necessary CPU load by forking dummy processes. These processes are dynamically configurable which hogs the CPU accurately as required by the user. This context is a good example to illustrate how one app can affect the performance of another app.
- **RAM Availability:** RAM Availability refers to the free memory available in the RAM for the AUT to use. Memory consumption is an important aspect to be considered while designing mobile apps as the resource availability is relatively very limited as compared to desktop apps. So, developers have to be very careful in avoiding memory leaks. Applications which does not follow proper memory management often get crashed when it is tested under poor RAM availability context. Static code analyzer, for ex. LINT (Lint (software), n.d.) spots memory leaks associated with the program but not all memory problems. A classic example to illustrate the above problem is an application which downloads image for every two seconds without following any replacement policies. This example programmatically does not have any memory leaks, but logically has memory problems. This application crashes at some point of time when there is no space for the downloaded image in the RAM. This logical error can be easily detected if the application is tested with lesser RAM availability. In order to simulate realistic RAM availability, CCS Component could categorize RAM Availability in to 5 levels (where level 5 represents very less RAM availability). User can choose the required RAM level which can be simulated in the real device.
- **GPS:** An application that makes use of GPS has to be tested under various GPS affecting contexts. GPS does not work indoors and its location accuracy also varies based on the mobile location. So it is necessary to understand the behavior of the AUT (which uses GPS) in various contexts like: Context when GPS service is not available, GPS results are not precise, user moves very fast leading to fast co-ordinate changes, etc. CCS Component simulates these scenarios by morphing the GPS coordinate values with the required values at the application level.

The other services/factors that affect the mobile contexts are Telephony, Camera, Bluetooth, Near Field Communication (NFC), WiFi-Direct, SQLite, Vibrator Service, Notification service, Alarm Service, Accelerometer and other sensor systems. CCS Component could be programmed accordingly to simulate several real-life context scenarios to test the AUT to identify performance issues.

- **Realistic Server Context Simulation:** It is essential to test the behavior of the AUT (having connecting servers) under various server load scenarios to identify more performance issues. An application which works pretty fast in normal scenarios can hang or crash, if the connecting server is responding slowly. So, it is essential to stress the connecting servers of the AUT with various traffic loads and then test the behavior of the AUT. So, in this section we explain Server Load Generation (SLG) (as in Figure 13) which is responsible for stress testing the server with different server loads.
    - User has to provide the URL of the server and number of user instances as input to SLG. Based on the number of user instances, SLG creates various virtual agents which can play-

back recorded HTTP test-cases on the mentioned server URL. Recorded HTTP test-cases (as depicted in Figure 14) captures all the HTTP traffic with associated meta-data with respect to the AUT for a particular GUI test-case. Since HTTP-test-cases trigger sequence of HTTP requests in an automated fashion when played back, SLG could load the server in a controlled manner. For Ex. User can choose 10 users in SLG and associate each user with a particular HTTP test-case. When SLG starts playing back the test-cases, server gets loaded with real http traffic as if, there were 10 real users interacting with the AUT and generating HTTP traffic to the server.

  - By this mechanism, we can test the behavior of the AUT when the server is loaded by specified number of users executing various http test-scripts which simulates realistic server contexts.
- **Realistic Network Context Simulation:** Network load implies the bandwidth availability for the AUT to use. An application which makes use of the internet has to be rigorously tested in various network load conditions. The following is a classic example to illustrate the necessity of this context testing. In Android, executing time-consuming actions in the main UI thread leads to application crash. If developer accidentally has used some network operation in the main UI thread, then the application crashes if Internet is slow in the mobile, but not in all the cases. So, it is essential to test the AUT under various networking conditions. There are various network emulator tools like Network Emulator by iTrinegy (Network Emulators from iTrinegy, n.d.), HP Network Virtualization (Network Virtualization, n.d.) which could create various network conditions based on user's input. In order to test the AUT in various network conditions, Test Engineer has to do the following:
  - Test Engineer has to host a hotspot for a Network Emulator (NE).
  - Test Engineer connects the testing mobile device to this hotspot.
  - Test Engineer then configures various network conditions (with respect to bandwidth, response time, packet loss, geo-locations, speed, etc.) in the network emulator and then test the AUT in the device connected to the NE's hotspot.

By this technique, Test Engineer can create various network loads to stress test the mobile applications.

Playback Phase

Once Test Engineer has configured the required client, network and server context, Test Engineer can playback test-cases on the AUT using Playback Engine (as depicted in Figure 1) to detect performance defects. Since, AUT is tested in unfavorable external contexts, there is more possibilities of defects getting revealed.

During Playback, Data Aggregator (DA) (as in Figure 13) collect all the resource metrics (such as CPU, memory, battery, network, etc.) at application as well as system level. Mobile Operating System provide various performance profiling tools like TraceView (Profiling with Traceview and dmtracedump, n.d.), Instruments (Instruments User Guide: About Instruments, n.d.) which could collect resource consumption of the application at various levels (method, class, thread, application, system). By analyzing the resource usage logs, Test Engineer can determine performance defects with the AUT. In the next section, we explain End-to-End Performance Monitoring (EPM) which helps in measuring and analyzing page load delays.

Higher page load delay is one of the most common performance issues in mobile applications. EPM Component (as depicted in Figure 13) helps tester to measure page load delays and also helps them to understand the complex interactions between client, network and server during an application page load. EPM provides page load delay breakdown which analyzes the performance issues involved during this process. During automated playback process, the time between execution of the last user-action on the current screen and complete rendering of the next page/screen is referred as page load time (refer Figure 15). High page load in network-dependent mobile applications can arise because of many reasons such as poor client content rendering, poor network, or high server processing delays. EPM breaks-down each page load delay in to 2 components (refer Figure 15) namely:

1. Effective content transfer delay (time during which the AUT waits for any HTTP response from server to load the next page).
2. Effective client rendering time (time between the reception of last HTTP response and the completion of the page load).

Higher effective content transfer delay indicates that the bottleneck might be with the network or the server. Higher effective content rendering delay indicates that the bottleneck likes in the client end. Also, EPM breaks down each HTTP transaction into 3 components:

Figure 15. Page load delay breakdown

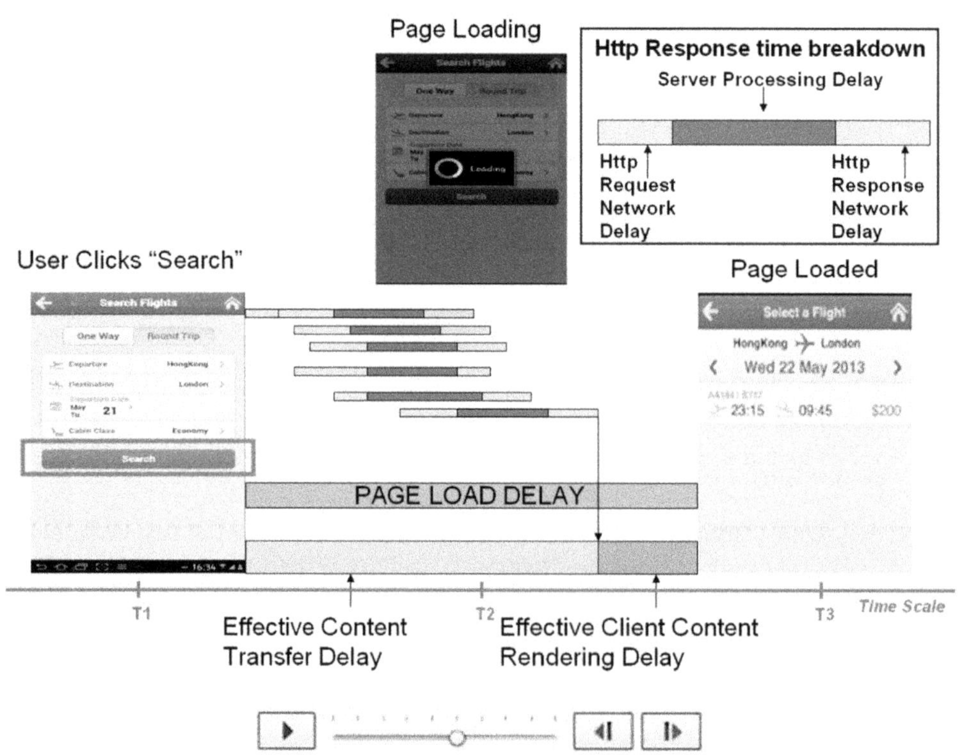

1. HTTP request network delay (network delay to transfer the HTTP request to the server).
2. Server processing delay.
3. HTTP response network delay (network delay to transfer the corresponding HTTP response back to the mobile).

Since, EPM breaks down the page load time in to its various components as depicted in Figure 15, it is very easy for testers to identify the bottleneck component (either client, network or server) for further root-cause analysis.

We now explain one possible method using which EPM could capture fine-grained page load delay breakdown. EPM can configure a client proxy server and server proxy server (as depicted in Figure 13). Proxy server logs all the HTTP requests and responses flowing through them. EPM configures AUT and communicating servers to direct its traffic through the proxies such that, the complete HTTP trace with time logs gets captured. By correlating the HTTP traces collected at the client and server end (PTME synchronizes the clock of both client and server), PTME can breakdown each of the HTTP transaction delay in client delay, server processing delay and network delay. This proxy recording technique can also be used for protocols other than HTTP.

At the end of playback, automation tool will provide a consolidated report which contains information about the resources consumed by the AUT (at method, class, thread, etc.), page load delays and breakdown during the test-case execution. By analyzing the consolidated report, Test Engineer can easily detect various performance defects with the AUT.

USABILITY AND ACCESSIBILTY TESTING

Accessibility (Developer guidelines, n.d.) in mobile applications is about removing barriers that inhibit the access of certain groups, including people with disabilities, mature users, and non-native language learners. To make an app accessible we need to design or modify the app to allow access by the greatest number of people. It is no longer just compliance, it is a major component for a successful app. Therefore, most of the apps in app store today are accessibility enabled.

Technologies used to increase, maintain, or assist the functional capabilities of people with disabilities are called assistive technologies. However, it can also benefit other user groups, such as mature people with age-related disabilities and non-native language learners. In short, assistive technology can be any device or technique that assists people in removing or reducing barriers and enhancing their everyday life activities. Here are examples of assistive technology include:

- Screen readers that use text-to-speech technology to read software to people who are blind.
- Screen magnifiers to enlarge information on the screen for people with low vision.
- Closed captioning displays for people who are deaf or hard of hearing.
- Special keyboards and input devices for people with limited hand use or mobility impairments.

It can only be effective if the software and hardware it interfaces with are accessible. A screen reader cannot read web pages unless the developer has followed common accessibility standards. For example, a screen reader cannot read informational graphic images on the Web unless the developer has provided alternative (Alt) text for those images. If the alternative (Alt) text is missing, the screen reader cannot provide the information and the page is inaccessible.

There are different accessibility properties for every widget on screen like Accessibility Label, Accessibility Hint & Accessibility Traits. If a widget does not have any of these properties settled or filled then it should be considered as in-accessible widget, and if a widget has few properties settled or filled but not all then it should be considered as partially accessible.

Various accessibility standards have been defined for mobile and web apps. WAI-ARIA (WAI-ARIA Overview, n.d.). defines various standards and guidelines for improving accessibility of mobile and web apps. These guidelines have to be programmatically or manually checked to make apps more accessible to end-users with disabilities.

Similarly, Usability of an app is all about how much of its every component is usable in terms of ease of using them and to understand the app flow with them.

There are different user interface guidelines provided by mobile operating systems which helps to make an app more usable. For example, in iOS any widget should have at least 44x44 px size to provide a better touch response, distance between two widgets should be at least 44 px. Any text on screen should not be leaser then 8 px, etc.

An app under test goes through below steps for accessibility and usability testing as shown in Figure 16.

- An app need to be instrumented to support Accessibility & Usability testing Solution.
- A local agent sitting in mobile, collects the view hierarchy as combination of accessibility & usability properties for every widget, then this local agent sends it to the data Collector Agent at

Figure 16. Sample architecture of a usability and accessibility tester

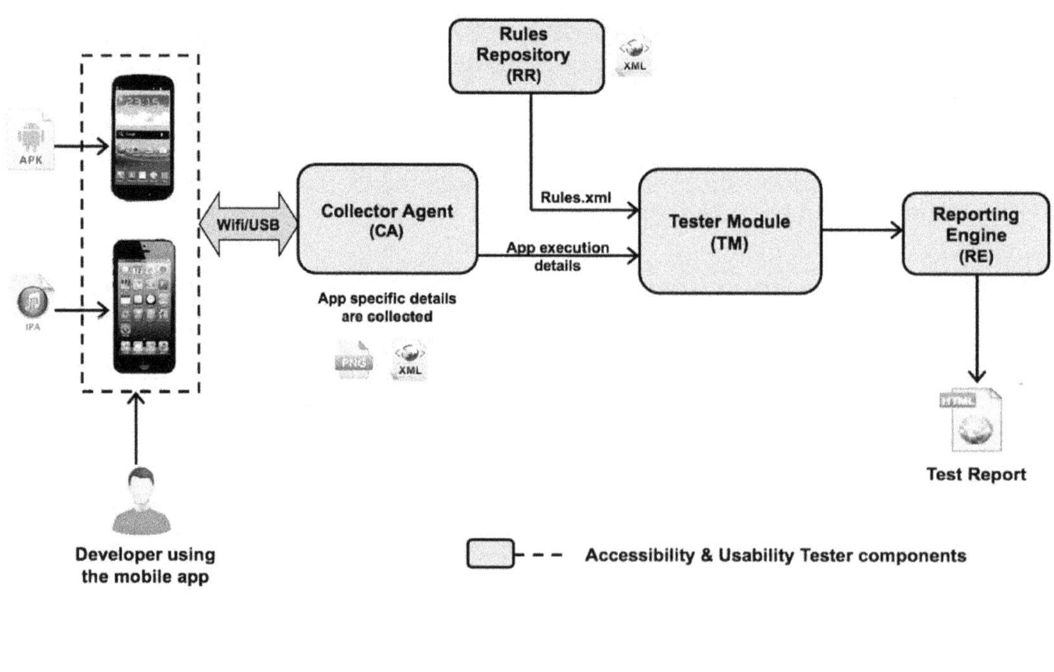

server side. Later Tester Module receives the collected set of information from CA along with rules or guidelines written for Accessibility and Usability, from Rules Repository.

- TM module analysis the collected data and tags the widgets which are not falling under accessibility & usability guidelines and sends that tagged data to Reporting Engine. So that Reporting Engine prepares reports for accessibility & usability on behalf of collected information. Figure 17 and 18 shows sample reports for accessibility & usability testing respectively.

Figure 17. Sample accessibility test report

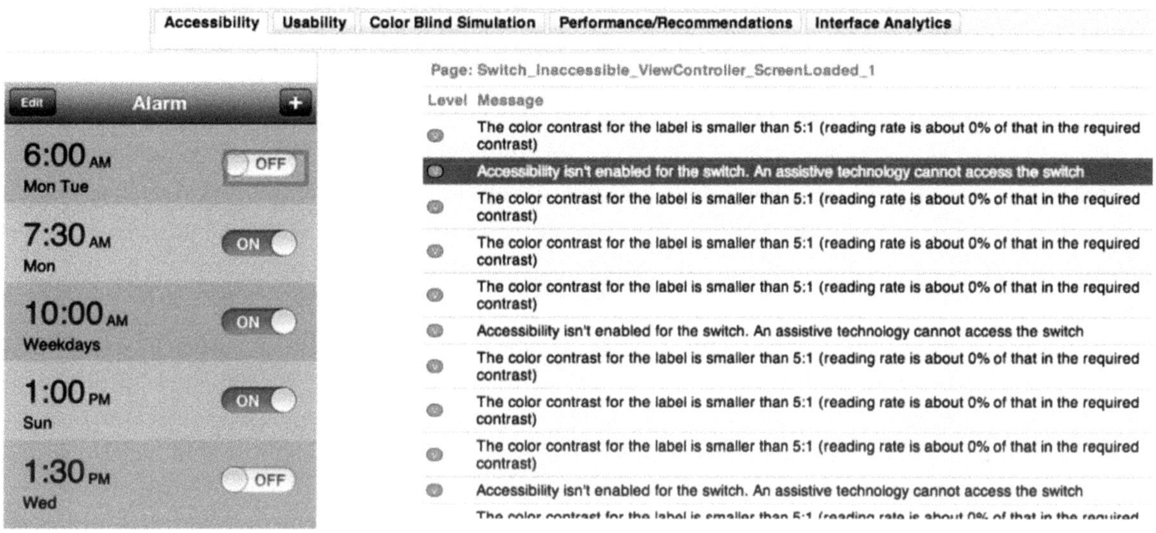

Figure 18. Sample usability test report

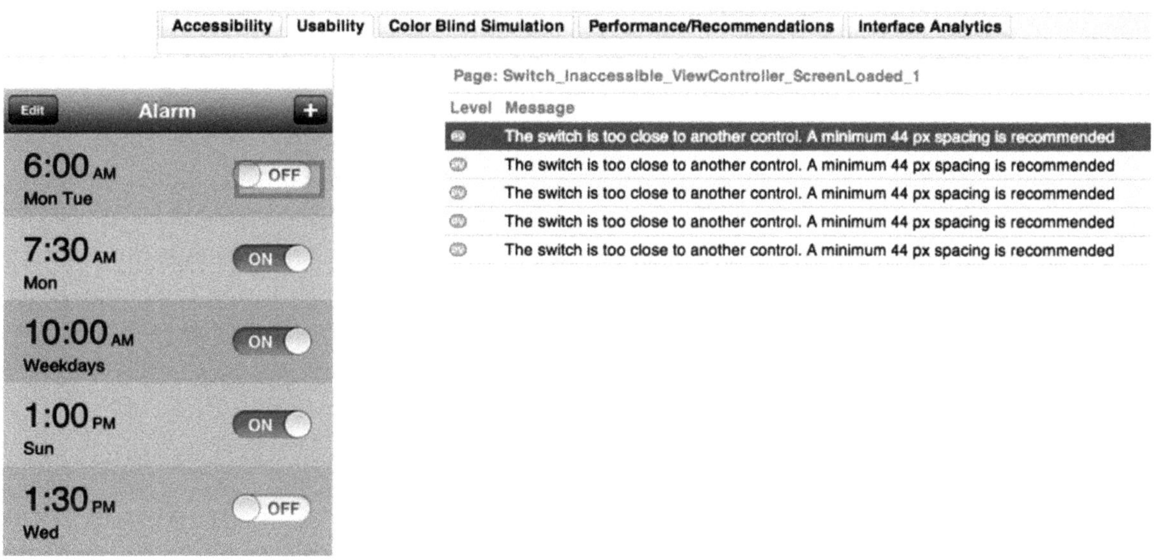

LITERATURE REVIEW

In this section, we briefly review various research efforts and practices in the space of mobile testing. For better understanding, we categorize the review of relevant art into three categories: Functional Testing, Performance Testing, and Usability Testing. In what follows, for each category, we present a comprehensive view of the relevant work.

Functional Testing

Automating mobile apps based on GUI objects and its associated meta-data is the most common approach followed in mobile test automation tools for test-case record and playback capability. In this context, authors of Amalfitano et.al. (2011, 2012), Hu and Neamtiu (2011) present various challenges involved in automating mobile apps at GUI layer and also propose techniques and solutions for efficient record and playback of test-cases based on GUI objects. It is very challenging to effectively record and playback time and touch sensitive interaction and gestures in the mobile apps. To address this specific problem, Gomez et.al. (2013) proposed RERAN, a sophisticated tool capable of capturing and reproducing complex gestures and mobile interactions for effective mobile test automation. Also, instrumentation is a standard step involved in the most of the automation process. Nandakumar et.al. (2013) propose various instrumentation techniques for automation considering the performance overload factors for Android and iOS OSes.

Multi-device testing is very important factor in the context of mobile testing. Due to huge number of devices with various characteristics (Signal, 2013), it is difficult to guarantee that an application will work flawlessly in all devices. So, it is necessary to test the mobile application in various devices with different characteristics to ensure test-completeness. In this regard, Vilkomir et.al. (2015) propose techniques to choose minimum subset of devices having maximum device characteristic coverage for effective heterogeneous testing of mobile application. Since it is a very costly process for companies to purchase all devices for effective multi-device testing, authors of Baride & Dutta (2011), Starov et.al. (2013), Kaasila et.al. (2012) propose device cloud approaches where heterogeneous mobile devices are hosted in cloud and tester can either perform manual testing with the devices in cloud or perform automated testing by pushing test-cases to the cloud and retrieving test reports from the cloud services. These testing services hosted in cloud provide cost effective solutions to do multi-device testing of mobile apps to discover device specific defects for ensuring test-completeness.

Another aspect of test-automation is to automatically generate test-cases through source code or bytecode analysis. Authors of Mirzaei et.al. (2012), Azim et.al. (2013) propose techniques for test-case generation through symbolic execution. They also propose various exploration techniques at code level to discover data and control flows for generating critical test-cases. Test-cases generally have fixed and dynamic inputs. Dynamic inputs are generally provided by the user during test-case execution. In order to avoid this manual test, Machiry et.al. (2013) propose techniques to auto-generate user inputs through code analysis. These auto-generated inputs could lead to effective execution of test-cases with minimal or no human involvement.

Performance Testing

Testing mobile apps in various real-life contexts helps to discover several critical and rare defects. Authors of Liang et.al. (2013), Liang et.al. (2014) propose emulator techniques where emulators can simulate various real-life contexts. Applications are tested in the emulators after these contexts are simulated to discover more defects with the application.

Automatically discovering performance issues is another important aspect of mobile application. Towards this, Ravindranath et.al. (2014) propose VanarSena, a tool for automatically discovering defects/bugs by emulating various user, network and sensor data behavior. In similar thread, Ravindranth et.al. (2012) proposed AppSight, a sophisticated tool for detecting performance issues in the mobile application through static code analysis.

Energy consumption is an important performance metric to be tested for mobile apps. Authors of Pathak et.al. (2012), Mittal et.al. (2012) propose solutions to effectively identify energy consumption of mobile apps at finer granular level. Using the proposed approaches, tester could identify components in the mobile app which is power-hungry and needs power optimizations.

Huang et.al. (2010) present systematic study on various critical parameters to consider for benchmarking network dependent mobile apps. Various key factors such as carrier networks, device capabilities, and server configurations are considered in their study and they also explain the impact of them in affecting the performance of mobile applications.

Usability and Accessibility Testing

Accessibility and Usability guidelines vary based on device and OS characteristics. Authors of Schulz et.al. (2015), White et.al. (2015) study various accessibility guidelines for developing Android and iOS apps. They also discuss various best practices to be followed while developing good accessible mobile apps.

Coursaris et.al. (2011) provide a very detailed systematic analysis on the usability aspects to consider while developing mobile apps. This work not only talks about the traditional usability standards but also provides interesting study around human factors in HCI, various cognitive and physiological parameters to consider in developing high quality Mobile UI screens.

It is very essential to have a formal framework to evaluate the usability score of the mobile application. In this regard, Hussain and Kutar (2009) study various metrics to be following for evaluating the usability of mobile apps. This work also differentiates the challenges involved in usability testing of mobile apps as compared to desktop apps. Likewise, there are various methods and techniques proposed in literature to come up with usability score for mobile apps. Hussain and Ferneley (2008) propose a goal question metric based approach to derive usability score for mobile apps.

There are also domain specific solutions for evaluating usability issues in mobile apps. Gafni (2009) discuss various challenges and proposes solutions for evaluating usability of apps in mobile-wireless domain. Similarly, Hubert (2006) propose techniques for improving usability of mobile apps in health-care domains. Since usability is highly coupled with business aspects, it is essential to have dedicated domain specific standards and metrics for evaluating usability of mobile apps.

Usability testing for mobile apps are also exposed as cloud services to avoid local deployment. Liang et.al. (2011) propose a framework for remote usability testing of mobile apps in cloud. These services discover the usability issues in the UI of the mobile app and also provide recommendations.

In the next section, we explain the various new trends followed in mobile testing as compared to traditional testing approaches.

NEW TRENDS IN MOBILE TESTING

In this chapter, we discussed the traditional mobile test automation approaches to detect functional, performance, accessibility and usability issues. As an alternate approach, the recent trend observed in mobile industry is to quickly release the mobile application with minimal testing but perform effective analysis on the App-Store reviews to identify and resolve various defects logged by the end-users. Since App-Store reviews/ratings is one of the key indicators for the success or failure of Application, Test Engineers (TE) spend significant efforts to mine defects from the user feedback to improve the comprehensive experience of the App users. This approach could either be automated or manual. If the scale of reviews is in order of hundreds, then manual analysis of the reviews would easily reveal all the defects logged by the end-users. However, if the scale of reviews is very high, then automation tools are required to analyze the reviews to detect defects. Since app-store reviews are very noisy and end-users could represent same defect in different phrases, automation of defect mining from app-store reviews is very challenging to realize. Authors of Chen et.al. (2014), Fu et.al. (2013), Herzig et.al. (2013) address techniques for automated app-store review analysis to tackle these challenges. Going forward in the future, it is very likely that the development, testing and review analytics together form a much tighter mechanism to enable better applications rather than treating them as separate divisions in a mobile application lifecycle.

CONCLUSION

This chapter was focused on providing insights into the key elements that are tested in a mobile application, mostly looking at automated testing solutions. The key components of an automated mobile application testing framework were first described. Then the chapter discussed detailed techniques of functional, performance and usability testing. A study into these techniques should provide a reader with enough information on the importance of mobile testing, the different types of mobile testing required and the methods to perform automated mobile testing. While mobile application lifecycle has evolved with a tremendous speed, much needs to be done in the mobile testing domain, to keep up with the pace of the application development and the maturity of development tools.

REFERENCES

Amalfitano, D., Fasolino, A. R., & Tramontana, P. (2011, March). A gui crawling-based technique for android mobile application testing. In *Software Testing, Verification and Validation Workshops (ICSTW), 2011 IEEE Fourth International Conference on* (pp. 252-261). IEEE. doi:10.1109/ICSTW.2011.77

Amalfitano, D., Fasolino, A. R., Tramontana, P., De Carmine, S., & Memon, A. M. (2012, September). Using GUI ripping for automated testing of Android applications. In *Proceedings of the 27th IEEE/ACM International Conference on Automated Software Engineering* (pp. 258-261). ACM. doi:10.1145/2351676.2351717

Apktool. (n.d.). Retrieved May 31, 2016, from https://ibotpeaches.github.io/Apktool/

ASM-Guide. (n.d.). Retrieved May 31, 2016, from http://download.forge.objectweb.org/asm/asm4-guide.pdf

Azim, T., & Neamtiu, I. (2013, October). Targeted and depth-first exploration for systematic testing of android apps. In ACM SIGPLAN Notices (Vol. 48, No. 10, pp. 641-660). ACM. doi:10.1145/2509136.2509549

Baride, S., & Dutta, K. (2011). A cloud based software testing paradigm for mobile applications. *Software Engineering Notes*, *36*(3), 1–4. doi:10.1145/1968587.1968601

Bitbucket. (n.d.). Retrieved May 31, 2016, from https://bitbucket.org/pxb1988/dex2jar

Chen, N., Lin, J., Hoi, S. C., Xiao, X., & Zhang, B. (2014, May). AR-Miner: mining informative reviews for developers from mobile app marketplace. In *Proceedings of the 36th International Conference on Software Engineering*(pp. 767-778). ACM doi:10.1145/2568225.2568263

Coursaris, C. K., & Kim, D. J. (2011). A meta-analytical review of empirical mobile usability studies. *Journal of Usability Studies*, *6*(3), 117–171.

Developer Guidelines. (n.d.). Retrieved May 31, 2016, from http://www-03.ibm.com/able/guidelines/index.html

Fu, B., Lin, J., Li, L., Faloutsos, C., Hong, J., & Sadeh, N. (2013, August). Why people hate your app: Making sense of user feedback in a mobile app store. In *Proceedings of the 19th ACM SIGKDD international conference on Knowledge discovery and data mining* (pp. 1276-1284). ACM. doi:10.1145/2487575.2488202

Gafni, R. (2009). Usability issues in mobile-wireless information systems. *Issues in Informing Science and Information Technology*, *6*, 755–769.

Gomez, L., Neamtiu, I., Azim, T., & Millstein, T. (2013, May). Reran: Timing-and touch-sensitive record and replay for android. In *Software Engineering (ICSE), 2013 35th International Conference on* (pp. 72-81). IEEE.

Herzig, K., Just, S., & Zeller, A. (2013, May). It's not a bug, it's a feature: how misclassification impacts bug prediction. In *Proceedings of the 2013 International Conference on Software Engineering* (pp. 392-401). IEEE Press. doi:10.1109/ICSE.2013.6606585

Hu, C., & Neamtiu, I. (2011, May). Automating GUI testing for Android applications. In *Proceedings of the 6th International Workshop on Automation of Software Test* (pp. 77-83). ACM. doi:10.1145/1982595.1982612

Huang, J., Xu, Q., Tiwana, B., Mao, Z. M., Zhang, M., & Bahl, P. (2010, June). Anatomizing application performance differences on smartphones. In *Proceedings of the 8th international conference on Mobile systems, applications, and services* (pp. 165-178). ACM.

Hubert, R. (2006). Accessibility and usability guidelines for mobile devices in home health monitoring. *ACM Sigaccess Accessibility and Computing*, (84), 26-29.

Hussain, A., & Ferneley, E. (2008, November). Usability metric for mobile application: a goal question metric (GQM) approach. In *Proceedings of the 10th International Conference on Information Integration and Web-based Applications & Services* (pp. 567-570). ACM. doi:10.1145/1497308.1497412

Hussain, A., & Kutar, M. (2009). *Usability metric framework for mobile phone application*. PostGraduate Symposium on The Convergence of Telecommunications, Networking and Broadcasting.

Instruments User Guide. (n.d.). *About Instruments*. Retrieved May 31, 2016, from https://developer.apple.com/library/prerelease/mac/documentation/DeveloperTools/Conceptual/InstrumentsUserGuide/index.html

Jar-The Java Archive Tool. (n.d.). Retrieved May 31, 2016, from http://docs.oracle.com/javase/7/docs/technotes/tools/windows/jar.html

Kaasila, J., Ferreira, D., Kostakos, V., & Ojala, T. (2012, December). Testdroid: automated remote UI testing on Android. In *Proceedings of the 11th International Conference on Mobile and Ubiquitous Multimedia* (p. 28). ACM. doi:10.1145/2406367.2406402

Liang, C. J. M., Lane, N., Brouwers, N., Zhang, L., Karlsson, B., Chandra, R., & Zhao, F. (2013). *Contextual fuzzing: automated mobile app testing under dynamic device and environment conditions*. Microsoft.

Liang, C. J. M., Lane, N. D., Brouwers, N., Zhang, L., Karlsson, B. F., & Liu, H. et al. (2014, September). Caiipa: automated large-scale mobile app testing through contextual fuzzing. In *Proceedings of the 20th annual international conference on Mobile computing and networking* (pp. 519-530). ACM. doi:10.1145/2639108.2639131

Liang, H., Song, H., Fu, Y., Cai, X., & Zhang, Z. (2011, June). A remote usability testing platform for mobile phones. In *Computer Science and Automation Engineering (CSAE), 2011 IEEE International Conference on* (Vol. 2, pp. 312-316). IEEE.

Lint. (n.d.). Retrieved May 31, 2016, from https://en.wikipedia.org/wiki/Lint_(software)

Machiry, A., Tahiliani, R., & Naik, M. (2013, August). Dynodroid: An input generation system for Android apps. In *Proceedings of the 2013 9th Joint Meeting on Foundations of Software Engineering* (pp. 224-234). ACM. doi:10.1145/2491411.2491450

Method Swizzling. (2014). Retrieved May 31, 2016, from http://nshipster.com/method-swizzling/

Mirzaei, N., Malek, S., Păsăreanu, C. S., Esfahani, N., & Mahmood, R. (2012). Testing android apps through symbolic execution. *Software Engineering Notes*, *37*(6), 1–5. doi:10.1145/2382756.2382798

Mittal, R., Kansal, A., & Chandra, R. (2012, August). Empowering developers to estimate app energy consumption. In *Proceedings of the 18th annual international conference on Mobile computing and networking* (pp. 317-328). ACM. doi:10.1145/2348543.2348583

Nandakumar, V., Ekambaram, V., & Sharma, V. (2013). Appstrument-A Unified App Instrumentation and Automated Playback Framework for Testing Mobile Applications. In Mobile and Ubiquitous Systems: Computing, Networking, and Services (pp. 474-486). Springer International Publishing.

Network Emulators from iTrinegy. (n.d.). Retrieved May 31, 2016, from http://www.itrinegy.com/index.php/products/network-emulators

Network Virtualization. (n.d.). Retrieved May 31, 2016, from http://www8.hp.com/in/en/software-solutions/network-virtualization/

Pathak, A., Hu, Y. C., & Zhang, M. (2012, April). Where is the energy spent inside my app?: fine grained energy accounting on smartphones with eprof. In *Proceedings of the 7th ACM european conference on Computer Systems* (pp. 29-42). ACM. doi:10.1145/2168836.2168841

Profiling with Traceview and dmtracedump. (n.d.). Retrieved May 31, 2016, from http://developer.android.com/tools/debugging/debugging-tracing.html

Ravindranath, L., Nath, S., Padhye, J., & Balakrishnan, H. (2014, June). Automatic and scalable fault detection for mobile applications. In *Proceedings of the 12th annual international conference on Mobile systems, applications, and services* (pp. 190-203). ACM. doi:10.1145/2594368.2594377

Ravindranath, L., Padhye, J., Agarwal, S., Mahajan, R., Obermiller, I., & Shayandeh, S. (2012). AppInsight: mobile app performance monitoring in the wild. In *Presented as part of the 10th USENIX Symposium on Operating Systems Design and Implementation (OSDI 12)* (pp. 107-120).

Schulz, T., Gladhorn, F., & Sæther, J. A. (2015). *Best Practices for Creating Accessible Mobile Applications.* Report Norsk Regnesentral–Norwegian Computing Center.

Signal, O. (2013). *Android fragmentation visualized.* Retrieved from opensignal. com: http://opensignal.com/reports/fragmentation-2013

Starov, O., Vilkomir, S., & Kharchenko, V. (2013). Cloud Testing for Mobile Software Systems Concept and Prototyping.*8th International Conference on Software Engineering and Applications (ICSOFT-EA)*

Ubuntu Manuals. (n.d.). Retrieved May 31, 2016, from http://manpages.ubuntu.com/manpages/wily/man1/aapt.1.html

Vilkomir, S., Marszalkowski, K., Perry, C., & Mahendrakar, S. (2015). Effectiveness of multi-device testing mobile applications. In *Mobile Software Engineering and Systems (MOBILESoft), 2015 2nd ACM International Conference on* (pp. 44-47). IEEE. doi:10.1109/MobileSoft.2015.12

WAI-ARIA Overview. (n.d.). Retrieved May 31, 2016, from https://www.w3.org/WAI/intro/aria

White, K. (2015). Determining Accessibility for iOS Applications: Piloting a Checklist for Practitioners. Theses and Dissertations. University of Wisconsin-Milwaukee.

Chapter 3
Accessible and Inclusive Content and Applications

Tom Brunet
IBM, USA

P. G. Ramachandran
IBM, USA

ABSTRACT

As devices have become smaller and more pervasive, usage scenarios that have historically been common for people with disabilities are finding more general application for all users. Overall, the consideration of accessibility improves the usability of applications for all users. This chapter will discuss standards for accessibility, inclusive design, and topics related to the development of accessible mobile content and applications. The discussion will apply to mobile content, such as EPUB documents, and topics related to Web, native, and hybrid applications.

INTRODUCTION

The field of Accessibility is focused on ensuring that every person is able to access information and perform tasks regardless of that person's physical or cognitive capabilities at any given time. Practitioners in this field consider a wide range of capability types and impairment severities, including not only severe physical or cognitive disabilities, but also contextual limitations such as noisy environments.

The history of accessibility in the technology industry dates back to at least 1914, when IBM® hired its first employee with a disability. IBM® also led the evolution of assistive technologies with a Braille printer in 1975, a talking typewriter in 1980, and the IBM® Screen Reader product in the early 1980's (IBM, 2015). While this level of inclusion has been ingrained in companies like IBM®, legislation in this area is fairly recent - the Americans with Disabilities Act (ADA) was passed in 1990 and the Section 508 amendment to the Rehabilitation Act was passed in 1998.

For physical access, most local governments did not build accessible sidewalks and pathways until they were mandated by the ADA. In many cases, the need for these features was not immediately obvious since they do not impact everyone on a day-to-day basis. However, when the need arises, retrofit-

DOI: 10.4018/978-1-5225-0945-5.ch003

ting can be difficult and costly. Due to cost, historical preservation, and other issues, retrofitting is still occurring for many buildings.

Technology access has been a different environment. Due to the rapid turnover of hardware and software, and the relative ease of modification, accessibility has evolved much faster than in the physical space. Additionally, relatively recent technologies, such as the World Wide Web, have improved access to information like never before.

The power of the Web is in its universality. Access by everyone regardless of disability is an essential aspect. (Tim Berners-Lee, 1997)

Now that mobile devices are commonplace, the universality of access has been unshackled. It is no longer available only from the desktop in your home, it is also available from the phone in your pocket.

This mobile evolution has brought two areas of Accessibility closer together. People who rely on assistive technologies are able to participate in ways that have been historically unavailable to them. On the other hand, as people are performing more tasks on devices that are smaller and using them in situations where they may not be able to look at or hear the device, these individuals require assistive technologies that have traditionally focused on people with severe disabilities.

In a less-obvious manner, our perception and abilities have not evolved, but we are using devices that are smaller than ever before. Therefore, there is a strong desire from all users for day-to-day interactions with their devices to become personal, aware, and cognitive. When outdoors, our displays need to be brighter, and when in noisy environments, the sound may need to be louder or captions may need to be provided.

Technology should adapt to humans, not the other way around. Multimodal interactions provide another interaction pattern to consider. If the same operation can be performed via various modalities, users have more flexibility and choice in how they perform the operation. During multi modal interactions, the systems also need to prioritize and be selective about how to respond to and process interactions in order to avoid accidental input. Humans are easily able to switch and prioritize. Devices will have to be smarter about processing. To some extent, multimodal interaction is already available. However, the means to enable these interactions needs to be further simplified. Development tends to prefer simplicity, so the means to enable accessibility will have to be simplified to the extent that this blends in with new development paradigms.

Wearable devices present another technology that can improve access for the general population and also for people with disabilities. Early wearable devices are similar to wristwatches and add abilities to provide feedback via vibrations and provide input via tap. Some companies are already building devices that can be woven into fabric, allowing for the integration of sensors and feedback in clothing, shoes, jewelry, glasses, etc. Each of these provides new opportunities for a user to interact with devices.

While the possibilities are endless, it is up to the technical community to prioritize simplification and enablement of accessibility. Sometimes more gadgets is not the answer; simplifying the existing ones and investing the time to understand the usage patterns can yield good results.

Many of the technologies that have enabled the mobile evolution have their roots in accessibility. Those with vision impairments leverage magnifiers and zoom technologies. Text-to-speech technology has provided additional information channels for those who are blind or have visual or cognitive impairments. Speech recognition has provided an input mechanism for those who cannot type, such as people with physical impairments.

In this chapter, we will discuss the importance of accessibility considerations throughout the content and application lifecycles, and applicable standards, techniques, and tools.

MOBILE ACCESSIBILITY

Business Imperative

In the 2010 US Census, 56.7 million people (19%) reported a disability, and more than half of these reported a severe disability (US Census Bureau, 2012). 40 million individuals were over the age of 65, and that number is expected to increase to 72 million by 2030. It is estimated that 1 billion people worldwide have a disability (WHO and World Bank, 2011).

In past decades, accessibility was largely a niche market, focused on by a handful of companies and advocacy groups. As world demographics shift to older populations and as more companies are sued to improve access to retail, services, and other offerings beyond brick-and-mortar locations, accessibility has shifted into a more visible mainstream role.

Recognition of the business imperative began with companies that were doing business with government entities, starting with the passage of the Section 508 Amendment to the US Rehabilitation Act of 1973. This amendment added accessibility considerations for US federal procurement, requiring that any company selling assets to the US government report their compliance. Many states followed suit, with similar requirements for sales to government entities at many levels. Therefore, companies that sold to US government entities were the first to develop accessibility practices, whereas other industries, such as retail, lagged behind. Other governments passed related legislation, such as the Ontarians with the Disabilities Act in 2005, and more recently due to the UN Convention on the Rights of Persons with Disabilities.

For retail entities, compliance was often deprioritized, with arguments that brick-and-mortar locations offered equivalent facilitation. This argument was used to deflect application of the Americans with Disabilities Act to Websites and other digital offerings. However, that changed with class-action lawsuits, such as National Federation of the Blind (NFB) v. Target® Corporation in 2006. These lawsuits aimed to highlight that the convenience of online shopping and discounts in online-only deals should not be available only to a subset of the population – brick-and-mortar locations are not suitable alternatives.

As more companies have reviewed their design and development processes to consider accessibility issues, these companies have recognized that design for accessibility also forces their design teams to create more organized and thoughtful user experiences. This provides critical impact for people with severe disabilities, major impacts for people with minor disabilities, and improved user experiences for all users.

Standards

There are multiple guidelines and standards that apply to the interface design of a mobile application. Individual device platforms provide basic recommendations, such as the iOS Human Interface Guidelines and the Android™ Guidelines. Following these guidelines can improve the usability and accessibility of an application, but they do not cover all aspects of accessibility.

The World Wide Web Consortium (W3C®) provides the current overriding standard for accessibility. When the W3C® started on the Web Content Accessibility Guidelines (WCAG) v2.0 (W3C, 2008), they set a goal to create a set of standards that was not focused on any particular technology and therefore would not become obsolete as technology evolved. This approach has been evident in that other standards have moved to align with WCAG 2.0 after it became a W3C® Recommendation in 2008. These standards focus on four key principles:

Principle 1 – Perceivable: Information and user-interface components must be presentable to users in ways they can perceive.
Principle 2 – Operable: User-interface components and navigation must be operable.
Principle 3 – Understandable: Information and the operation of user interface must be understandable.
Principle 4 – Robust: Content must be robust enough that it can be interpreted reliably by a wide variety of user agents, including assistive technologies.

When the checkpoints and guidelines of these principles are satisfied, applications and content can be consumed and interacted with reliably by people with a broad range of abilities to perceive and operate content.

In addition to platform guidelines, specific legislation may also apply for a given application or audience. Examples include the US Americans with Disabilities Act (ADA), Section 508 of the US Rehabilitation Act (US Access Board, 2000), Accessibility for Ontarians with Disabilities Act (AODA), EU Mandate 376, UK Disability Discrimination Act of 1995, and others. Generally speaking, following the WCAG guidelines will help you meet the content and applications components of this legislation, but often the legislation also comes with a reporting component that is specific to each.

Accessibility Starts with Design

Accessibility is a consideration of every stage of the application and content lifecycle, affecting design, development, test, and deployment.

For many development teams, their first introduction to accessibility starts with an audit, lawsuit, or other notification of accessibility issues in a released application. The common reaction to these events is remediation, which triggers a QA-driven process of identifying issues and sending them back to development to be fixed. Meanwhile, this expensive process is making changes without revisiting them in the context of the original design.

It is important that these teams take a step back and look at the larger development process. Accessibility, like all user experience and user interaction specification, starts with design. Designers should consider the user experience for a broad range of personas. Designers may already have training to consider personas in different job roles and with different lifestyles, but they also have to consider personas that prefer different interaction styles. Inclusive Design refers to a design process that includes all of these personas.

When Inclusive Design is practiced, accessibility issues are considered at design time, and developers and testers include accessibility as part of the regular process of implementing the design.

There are a number of issues that designers should consider. The following list is not complete, but provides a significant review of issues that a designer might consider:

- **Alternatives for Non-Text Content:** All non-text content (for example, images) must have appropriate text alternatives. In some cases, images are used to supplement existing text, and therefore an empty alternative is appropriate to avoid redundancy. Additionally, for videos, captions must be provided.
- **Text Contrast and Color:** Ensure that text provides enough contrast to be readable per WCAG 2.0 1.4.3 and that links or hotspots are distinguishable from the surrounding text by color contrast or other formatting. Also, do not use color alone to convey information.
- **Multi-Device Access:** Different users may utilize a variety of different input devices or styles: mouse, keyboard, touch, voice, or sip/puff switch. Often, starting with basic mouse and keyboard access will assist with the other input methods, with additional fine-tuning for a specific device as desired.
- **Container Sizes:** Designs are often created for specific sizes or amounts of content. When content is altered by translation, zoom, or altered by a particular device, content may grow or shrink in unexpected ways. Designers must consider how content containers respond in these situations.
- **Semantics and Labels:** UI frameworks often provide ways to specify the semantics and labels for content. For example, in HTML, h1, h2, and other elements specify headings, ARIA landmarks mark important regions on the page, label elements help label form elements, and more. When content and applications choose elements based on semantics rather than default look and feel, the user interface and assistive technologies can use these semantics to provide a robust and consistent user experience across applications. For example, if a framework provides a checkbox widget, focus on making that checkbox adopt your look and feel rather than re-creating a partial checkbox behavior with a set of changing image widgets.
- **Form Flow and Guidance:** Ensure that there are instructions on how to fill out forms, that there are error messages when input is not properly specified, and that there are mechanisms to review or reverse submissions to avoid accidental submissions.

Mobile Application Development

An accessible mobile solution involves multiple components, consisting of at least an app, the operating system (OS) (for example, iOS, Android™), and assistive technologies (ATs) (for example, VoiceOver, Siri, TalkBack). In order to ensure that the app is accessible, the developers must ensure that the app communicates necessary information with the OS and ATs through the platform APIs. In this section, we will outline many of the significant platform APIs that affect accessibility.

iOS Native Applications

iOS provides a variety of properties that affect how VoiceOver reads content. Basic properties include:

- **isAccessibilityElement:** VoiceOver will only read elements when this property is set to true. The default value is true for standard UIKit controls and false otherwise. Usually, this value should be set to true. However, some exceptions are outlined below.
- **UIAccessibilityContainer:** If a widget is a container of accessible controls (for example, a table containing rows), it should implement UIAccessibilityContainer. In this case, isAccessibility-

Element should be set to false on the container and true for the enclosed elements. This allows VoiceOver users to visit the elements of the container.

- **accessibilityLabel:** The accessibilityLabel is spoken by VoiceOver for controls. For UIKit controls, this label is derived from the title value. In the case of a labelled input control, you may wish to set isAccessibilityElement to false on the related text element and specify the equivalent text as the accessibilityLabel of the input control.
- **accessibilityValue:** The accessibilityValue is used with elements with dynamic values, such as sliders and switches. For example, a switch that toggles visibility might use the following:

```
visibilitySwitch.accessibilityValue = bIsOn ? "Visible": "Invisible"
```

- **accessibilityTraits:** accessibilityTraits is a function that indicates the roles that a user interface elements implements, such as a button, static text, a search field, and more.
- **accessibilityHint:** accessibilityHint is used when the label is not descriptive enough and should be a brief phrase that starts with a verb. A button with the label "Add" probably does not need a hint. In a list of songs, the label might be the song name and the hint might be "Play the song".

There are also properties for how interfaces respond to user settings:

- **UIAccessibilityIsBoldTextEnabled:** Returns whether or not the user preference for bold text is set. For example it can be used as follows:

```
return UIAccessibilityIsBoldTextEnabled() ? "Avenir-Medium": "Avenir-Light"
```

- **UIAccessibilityIsReduceTransparencyEnabled:** Returns the setting for preferences regarding transparency. For example:

```
return UIAccessibilityIsReduceTransparencyEnabled() ? UIColor.whiteColor():
UIColor(white:1.0, alpha:0.8)
```

- **UIAccessibilityDarkerSystemColorsEnabled:** Determines if the user has asked for darker system colors.

```
return UIAccessibilityDarkerSystemColorsEnabled() ? UIColor.blackColor():
UIColor.grayColor()
```

- **UIAccessibilityIsReduceMotionEnabled:** Determines if the user has asked to reduce motion. If this is set, avoid using animations or other moving user interfaces that may be distracting to the user.

In addition to these basic properties, there are some more advanced characteristics for interacting with VoiceOver. Applications should avoid using these mechanisms for the purpose of providing a separate user interface for VoiceOver users, since some users may just be using VoiceOver as a cognitive assistant, or they may be seeking help from another user who is not using VoiceOver. These additional properties are:

- **UIAccessibilityIsVoiceOverRunning:** Determines if VoiceOver is turned on.
- **UIAccessibilityPostNotification(UIAccessibilityScreenChangedNotification, self.searchText):** Sends a notification for VoiceOver, which can be useful for elements that appear or disappear.
- **accessibilityElements:** Can be used to specify the order VoiceOver should read through elements if the default order is not working as desired.
- NSNotificationCenter.defaultCenter().addObserver(self, selector: "onVoiceOverStatusChanged", name: UIAccessibilityVoiceOverStatusChanged, object: nil): **This provides a mechanism to detect when VoiceOver is turned on and off.**

This listing is a sampling of the accessibility capabilities that iOS provides. For additional documentation and properties, see the UIAccessibility Protocol Reference in the UIKit Framework Reference (Apple, 2016).

Android™ Native Applications

The default screen reader for Android™ is TalkBack. Android™ supports third-party screen-reader applications, which can be found in the Play Store. Similar to iOS, Android™ provides applications with an API in order to interact with these assistive technologies. Basic properties include:

- **importantForAccessibility:** This property tells screen readers whether the given element should be read. It can have one of four values: IMPORTANT_FOR_ACCESSIBILITY_AUTO (0), IMPORTANT_FOR_ACCESSIBILITY_YES (1), IMPORTANT_FOR_ACCESSIBILITY_NO (2), or IMPORTANT_FOR_ACCESSIBILITY_NO_HIDE_DESCENDANTS (4). noHideDescendants means that both the current element and all of its descendants are not important.
- **android:focusable:** Sets whether or not the element is focusable. Focused elements have a number of related properties which are defined below.
- **nextFocusUp, nextFocusDown, nextFocusLeft, nextFocusRight:** Provides ways to indicate which element should receive focus if the user requests to move from the current element in the indicated direction. Some devices provide directional pads that can take advantage of all of these options, or they can be exercised via the keyboard.
- **requestFocus:** A view can request to gain focus. In some cases, this request can be blocked if the element is unable to gain focus.
- **android:contentDescription:** The contentDescription is spoken by TalkBack when the user interacts with controls. For labels that will not change, the label can be defined in the XML layout:

```
android:contentDescription="@string/prev"
```

Otherwise, the setContentDescription method can be used to set the label.

- **android:labelFor:** This allows TextViews to indicate that they act as a label for other controls. For example:

```
<TextView android:id="@+id/user_nicklabel" android:labelFor="@+id/user_nick"
android:text="@string/txt_nickname"></TextView >
<EditText android:id="@+id/user_nick "></EditText>
```

- **android:hint:** User hints will display and will be conveyed to TalkBack when a field is empty. This can provide additional information to the user to help them understand the purpose of the field.

Web Applications

When discussing mobile applications, many people think of native applications, but disregard the many Web applications that are designed for mobile devices. With responsive design techniques via CSS media queries, user interface designers are finding ways to ensure that their applications work well across different device formats: desktop, tablet, and phone.

This section will highlight some significant techniques that can improve accessibility, but only covers a subset. WCAG 2.0 provides hundreds of techniques that can be found at http://www.w3.org/TR/WCAG20-TECHS/. Note that some of these techniques are noted as Advisory in the checkpoints, meaning that they are not required for compliance with WCAG, but can improve accessibility nonetheless.

Additionally, it is important to follow the HTML specification where possible. While Web browsers allow developers to deviate from the specification, they also provide default behaviors when the most appropriate elements are utilized to develop Web content and applications. When the most appropriate elements are used, and used properly, developers will get a variety of behaviors 'for free.' For example, when using a select element, the widget is styled based on the OS preferences specified by the user. This widget has a variety of keyboard behaviors: it can be reached by pressing the Tab key, it changes value with the up and down arrows, etc. In addition, JavaScript can respond to onchange events. When defining a custom version of a widget, it must consider these features that may be used by some users, but may not be obvious to the developer.

The following properties and techniques, in addition to issues addressed in the design, will help you to make your mobile applications accessible:

- **img/@alt:** The alt attribute is intended to be an inline replacement of the image, meaning that if the image is unable to load, the text provides a meaningful replacement of that image. If the image is decorative or redundant, the alt text should be empty. For example:

```
<a href="..."><img alt="Site Map" src="..."></a>
<a href="..."><img alt="" src="...">Site Map</a>
```

- **label/@for:** The for attribute of the label element is an id reference, which allows you to specify that this label is associated with the indicated input element. For example:

```
<label for="email">E-Mail:</label><input id="email">
```

- **th@scope/td@headers:** The scope attribute of a table header may be set to a value of 'row' or 'col'. This indicates whether the header cell applies to elements in this row or column, respectively. In more complex tables, developers may need to use the headers attribute of a table cell, which takes a space-separated list of ids that act as headers for the specified cell. These techniques help assistive technologies understand the relationships between data table headers and data cells.
- **@role:** The role attribute is a mechanism provided by ARIA 1.0 to override the default semantics of an element. This is particularly helpful for conveying the behaviors of custom made widgets that may be created with a combination of div/span tags, CSS, and JavaScript. For example, role='checkbox' on a div would indicate that the div acts like a checkbox. The role attribute can also be used to indicate landmarks, such as main, navigation, and search. These landmarks help users of assistive technologies find key areas of a website without having to wade through large amounts of repetitive information. For a complete list of ARIA roles, see http://www.w3.org/TR/wai-aria/roles.

Hybrid Applications

Hybrid applications are native applications with embedded Web applications. For these types of applications, both the native and Web application sections above may apply.

Mobile Content Development

There are a variety of formats that content developers can use for electronic publications. Typically, for mobile content, developers require that the format support reflowable documents, so that the content will fill the screen, potentially using multiple columns, while not requiring scrolling on smaller devices. They also need the content to be viewable on a wide range of devices. The two most common formats for electronic publications that meet these needs are tagged PDF and EPUB®.

EPUB® is a free and open standard published by the International Digital Publishing Forum (IDPF). It has a number of characteristics that make it attractive for accessible mobile content development, many of which it inherits because it is based on HTML5:

- **Reflowable:** Aside from HTML being reflowable by default, EPUB® documents can leverage CSS media queries to precisely control the layout of content based on the width of the document. For example, decorative images might be hidden for smaller formats. This behavior not only assists with readers on mobile devices, but can also provide accessibility for users who need to zoom content.
- **Variety of Readers and Editors:** EPUB® documents utilize standard formats, so they are essentially a zip file of HTML documents with an XML description. Content developers can utilize any editor they prefer to modify the content. Additionally, there are various readers for EPUB® documents, so users can use the reader that works best for them.
- **Accessibility Standards and Documentation:** Since EPUB® is based on HTML, it is straightforward to apply WCAG 2.0 techniques to EPUB® documents.

There is also a W3C® Working Draft, which defines roles, states, and properties specific to the digital publishing industry. The latest version can be found at http://www.w3.org/TR/dpub-aria-1.0/. For example, this chapter of the book in an EPUB® representation might be wrapped with:

```
<body role="doc-chapter">Accessible and Inclusive Content and Applications</
body>
```

Tools

There are a variety of tools to aid in the detection of accessibility issues. The capabilities and the level of automation provided depend on the platform of the application that is being tested. For some accessibility issues, tools are unable to automate detection because they are unable to infer intent. For example:

If an input error is automatically detected and suggestions for correction are known, then the suggestions are provided to the user, unless it would jeopardize the security or purpose of the content. (WCAG 2.0 3.3.3, 2008)

In this situation, the tools may be able to prompt a tester to assess whether a suggestion for correction is known or whether the correction would jeopardize the security or purpose of the content, but it is unlikely to be able to determine this automatically. Due to this range of complexity, some testing tools are just assistive technologies, or provide more assistance for performing manual analysis, while others provide full automation for some subset of the issues that need to be tested.

When assessing a tool, it is important to consider the role of the tool user, and the usage of the tool in the overall development lifecycle. Many tools are built by auditing teams who are performing post-deployment audits. These tools may provide wider coverage, but they also tend to have a higher false-positive rate because the results are reviewed by an expert who can further filter the list. Other tools are intended for developers and focus on high-impact issues with high reliability. These tools may not flag everything since quality assurance teams are expected to do a more thorough deep dive. These tools ensure that developers are considering accessibility issues and providing a solid foundation, and reduce the noise that may need to be sifted through as compared to other tools. In general, it is difficult to determine a "percent coverage" that can be reliably automated by a given tool since the set of possible user interfaces and situations that can cause accessibility issues is unbounded.

In this section, we will highlight some of the tools the authors have found helpful while developing mobile applications, with a short description of the ideal situation in which the tool provides value.

Screen Readers

The catchall test for any application is to perform a complete test of the application as an end user would experience it, whether without any additional assistive technologies or with assistive technologies such as screen readers.

For iOS, the built-in screen reader is called VoiceOver. If you are not primarily a VoiceOver user, it is useful to change in the Settings under the General, Accessibility section "Triple-click Home" to VoiceOver. With this setting in place, you can press the home button three times to toggle VoiceOver.

Without this setting enabled, it can be difficult for novice VoiceOver users to turn VoiceOver off again. For more information about VoiceOver, refer to VoiceOver for iOS (Apple, 2016).

For Android™, the built-in screen reader is called TalkBack. The settings for enabling and disabling Talkback are in the Accessibility, Vision section. The Accessibility Shortcut section explains methods for toggling TalkBack so that it can be quickly enabled when needed. For more information on TalkBack, refer to Google TalkBack (Google, 2016)

For mobile Web applications, desktop screen readers, such as JAWS or NVDA, can also provide some assistance, but may not accurately reflect the mobile screen-reader environment.

Accessibility Inspector (iOS)

The Accessibility Inspector is available in the iOS Simulator (Apple, 2013). When the Accessibility Inspector is turned on, the user can click on elements of the interface and see a pop-up with a listing of significant accessibility information. This will show event notifications and accessibility-related attributes such as label, traits, frame, etc.

Since VoiceOver is not available in the iOS Simulator, the Accessibility Inspector can help developers who are working in the iOS Simulator determine how VoiceOver might behave, or determine why VoiceOver is behaving a certain way on an actual device.

Lint (Android™)

The Android™ SDK includes a static code analysis tool called lint (Google, 2016). When you compile your program, these checks run, but you can also run them manually via Analyze, Inspect Code. The lint tool, in addition to checking items for correctness, will check for some issues related to security, performance, usability, accessibility, and internationalization.

Only three rules are categorized as accessibility: ClickableViewAccessibility, ContentDescription, and LabelFor. However, there are various other rules that improve the accessibility of applications, such as those categorized as Usability. These rules may not provide significant coverage of accessibility standards, but provide quick feedback to Android™ developers so that accessibility issues are detected early.

Accessibility Scanner (Android™)

The Accessibility Scanner is an app provided by Google® that is available in the Google Play™ store (Google, 2016). This app provides a button that allows the user to scan apps on their mobile device. It will report common accessibility issues such as color contrast, small buttons, and missing labels.

The Accessibility Scanner is useful for checking individual states of an app, such as when testing a specific page or UI element. This tool requires the user to push the button to obtain a report, so it is not ideal for large scale automated testing.

AMP for Mobile

SSB Bart's Accessibility Management Platform (AMP) for Mobile uses a component that is built into an iOS or Android™ native or hybrid application (SSB Bart Group, 2015). The component just needs to be linked with your app and does not require any code changes. With the component linked with an app, the app can be used and tested normally. Each time the user interface changes, the user interface is evaluated in the background and the results are sent back to the AMP server.

AMP for Mobile is useful for developers and testers since it can be run in both a simulated environment or on an actual device.

The AMP for Mobile solution relies on technology from the IBM® Mobile Accessibility Checker, which provides the instrumentation, rules, and scanning of the mobile applications. AMP then provides the reporting and business process related to that data.

IBM® Automated Accessibility Tester

The IBM® Automated Accessibility Tester is a service available from IBM® Bluemix™(IBM, 2015). The Automated Accessibility Tester wraps the Selenium WebDriver, which is a browser-automation framework. Using Selenium, you can programmatically drive your mobile Web applications in standard Web browsers, such as Firefox® and Chrome™. During the automated testing process, accessibility checks can be run and the results are returned to the Automated Accessibility Tester.

This service can be used for testing at various stages in the development process, such as for unit testing or system testing. It is primarily focused on usage in a continuous integration process. In these continuous integration processes, after a developer checks in a change, the application is automatically rebuilt and deployed to a test or stage environment. Once deployed, tests, which can include the Automated Accessibility Tester, can be automatically run. In these scenarios, the functionality and accessibility can be tested together with every change that a developer commits.

This service provides good coverage of mobile Web applications, but is unable to test mobile native and hybrid applications.

IBM® Digital Content Checker

The IBM® Digital Content Checker is also a service available from Bluemix™, and is focused on accessibility checking of digital publications (IBM, 2015). It supports HTML and EPUB® formats. The service can be used by submitting an individual file to the service, or by utilizing the available API.

Using the API, accessibility testing of EPUB® documents can be included in a variety of publishing processes. For example, an extension can be built for rich-text editors such as CKEditor™, allowing content authors to check their content as they develop it. In an automated workflow, an accessibility check might be performed as an automated review process. In other cases, teams may have a build process that pulls in content from a repository and builds an EPUB® document. In this case, the build process can submit the document to the API and include the results of the check in the build logs.

FUTURE OF ACCESSIBILITY TESTING

Accessibility incorporated into design of software is the best approach that is going to yield exceptional results. Retrofitting software after testing is not only time consuming but also prone to delayed release

Table 1. Summary of available tools

Tool	Target	Test Usage
VoiceOver	iOS Device	• Learn VoiceOver specific user interactions • Learn expected behavior of VoiceOver • Manually interact with applications and identify unexpected behaviors
Accessibility Inspector	iOS Simulator	• Learn expected API values • Manually interact with applications and identify unexpected API values
TalkBack	Android	• Learn TalkBack specific user interactions • Learn expected behavior of TalkBack • Manually interact with applications and identify unexpected behaviors
Lint	Android, Compile Time	• Review results of static analysis • Note: Coverage is relatively small
Accessibility Scanner	Android	Push button and review report
AMP for Mobile	iOS or Android	• Interact with application normally – reports gathered automatically • Review report after interacting with the application
Automated Accessibility Tester	Mobile Web Applications	• Create a test script to automatically interact with application • Run automated script • Review report after automated script runs
Digital Content Checker	EPUB and static HTML content	• Submit content manually or via API • Review report of results

schedules. As it evolves some of the tools will be embedded into design and how each user with a unique perspective can have a delightful experience. Software projects are becoming more and more about composing and reusing rather than building from scratch, the need for prolonged regression testing is less. Pretty soon the big difference between mediocre software and exceptional software will be based on how much accessibility and usability is built in during design rather than the length and extensiveness of test cycles.

REFERENCES

Apple Inc. (2013, April 23). *Debug Accessibility in iOS Simulator with the Accessibility Inspector.* Retrieved from https://developer.apple.com/library/ios/technotes/TestingAccessibilityOfiOSApps/TestAccessibilityiniOSSimulatorwithAccessibilityInspector/TestAccessibilityiniOSSimulatorwithAccessibilityInspector.html

Apple Inc. (2016a). *VoiceOver for iOS.* Retrieved from http://www.apple.com/accessibility/ios/voiceover/

Apple Inc. (2016b). *UIKit Function Reference.* Retrieved from https://developer.apple.com/library/ios/documentation/UIKit/Reference/UIKitFunctionReference/#//apple_ref/doc/uid/TP40006894-CH3-SW39

Apple Inc. (2016c) *UIAccessibility Protocol Reference.* Retrieved from https://developer.apple.com/library/ios/documentation/UIKit/Reference/UIAccessibility_Protocol

Google Inc. (2016). *Lint.* Retrieved from http://developer.android.com/tools/help/lint.html

Google Inc. (2016a). *Google TalkBack*. Retrieved from https://play.google.com/store/apps/details?id=com.google.android.marvin.talkback&hl=en

Google Inc. (2016b). *Accessibility Scanner*. Retrieved from https://play.google.com/store/apps/details?id=com.google.android.apps.accessibility.auditor&hl=en

IBM Corp. (2015a, June 24). *IBM's commitment to people with disabilities*. Retrieved from http://www.ibm.com/able/product_accessibility/ibmcommitment.html

IBM Corp. (2015b). *Automated Accessibility Tester*. Retrieved from http://ibm.biz/bluemix-aat

IBM Corp. (2015c). *Digital Content Checker*. Retrieved from http://ibm.biz/bluemix-dcc

SSB Bart Group. (2015, Mar 6). *AMP for Mobile*. Retrieved from http://info.ssbbartgroup.com/AMP-forMobile.html

US Access Board. (2000, December 21). *Section 508 Standards for Electronic and Information Technology*. Retrieved from https://www.access-board.gov/guidelines-and-standards/communications-and-it/about-the-section-508-standards/section-508-standards

US Census Bureau. (2015, July 25). *Nearly 1 in 5 People Have a Disability in the U.S., Census Bureau Reports*. Retrieved from https://www.census.gov/newsroom/releases/archives/miscellaneous/cb12-134.html

W3C. (2008, December 11). *Web Content Accessibility Guidelines (WCAG) 2.0*. Retrieved from https://www.w3.org/TR/WCAG20/

World Health Organization and The World Bank. (2011). *World Report on Disability*. Retrieved from http://www.who.int/disabilities/world_report/2011/en/

Chapter 4
Combining Static Code Analysis and Machine Learning for Automatic Detection of Security Vulnerabilities in Mobile Apps

Marco Pistoia
IBM Corporation, USA

Omer Tripp
IBM T. J. Watson Research Center, USA

David Lubensky
IBM T. J. Watson Research Center, USA

ABSTRACT

Mobile devices have revolutionized many aspects of our lives. Without realizing it, we often run on them programs that access and transmit private information over the network. Integrity concerns arise when mobile applications use untrusted data as input to security-sensitive computations. Program-analysis tools for integrity and confidentiality enforcement have become a necessity. Static-analysis tools are particularly attractive because they do not require installing and executing the program, and have the potential of never missing any vulnerability. Nevertheless, such tools often have high false-positive rates. In order to reduce the number of false positives, static analysis has to be very precise, but this is in conflict with the analysis' performance and scalability, requiring a more refined model of the application. This chapter proposes Phoenix, a novel solution that combines static analysis with machine learning to identify programs exhibiting suspicious operations. This approach has been widely applied to mobile applications obtaining impressive results.

DOI: 10.4018/978-1-5225-0945-5.ch004

INTRODUCTION

Mobile devices have revolutionized many aspects of our lives. We use smartphones, tablets and wearable devices as portable computers and, often without realizing it, we run various types of security-sensitive programs on them, such as personal and enterprise email and instant-messaging applications, as well as social, banking, insurance and retail programs. These applications access and transmit over the network numerous pieces of private information, including our geographical location, device ID, contacts, calendar events, passwords, and health records, as well as credit-card, social-security, and bank-account numbers. Guaranteeing that no private information is exposed to unauthorized observers is very challenging, given the level of complexity that these applications have reached. Integrity concerns arise when mobile applications take untrusted user data as input to security-sensitive computations. In such cases, in order to avoid integrity violations, it is necessary to verify that the appropriate validation and/or sanitization routines are invoked. Program analysis tools for integrity and confidentiality enforcement have become a necessity, especially for mobile applications, which are updated very often and require high security assurance. Roughly speaking, such tools can be either "dynamic" or "static". Dynamic-analysis tools execute the program and infer security vulnerabilities based on the actual program executions. Such solutions can be quite time consuming because they require installing the program under analysis and then executing it multiple times, with different inputs and different execution choices on each run, in order to maximize the number of execution paths explored by the analysis. Often, however, there is no real guarantee that all the possible paths of execution have been explored. This means that some security vulnerabilities may remain undiscovered by the time the application is released. Conversely, a static-analysis tool does not require installing or executing the program. Rather, a formal model that over-approximates all feasible program executions is built, and the analysis reports results based on that model. Unlike dynamic analysis, static analysis has the potential of never missing any vulnerability, but it may report false positives; that is, it might signal security issues that are never exposed at run-time. Some analysis tools, even at the commercial level, have very high false-positive rates. Numerous user studies have proved that an analysis whose false-positive rate is too high becomes unusable because developers are forced to spend large amounts of time to filter out spurious issues, and when the false-positive rate is excessive, the effort becomes unbearable. Therefore, in order to reduce the number of false positives, a static analysis has to be very precise. Unfortunately, however, precision is in conflict with the analysis' performance and scalability, requiring a more refined model of the application and analysis thereof. A large body of research work has studied how to increase the precision of static-analysis tools without affecting their scalability. At the same time, applications have become increasingly more complex, with the addition of frameworks and dynamically loaded components, thereby making the compromise between precision and scalability even more difficult to reach. This chapter proposes Phoenix, a novel solution that combines static program analysis with machine learning. The idea behind Phoenix is to use relatively scalable static analysis to approximate possible program behaviors, and to then apply machine learning in order to identify programs exhibiting suspicious sequences of operations. This solution has been widely applied to mobile applications obtaining impressive results, with low false-positive and false-negative rates.

Phoenix comprises two components: a novel static analyzer, which performs demand-driven pointer analysis and lends itself to modular reasoning, and a machine-learning engine, which acts on the results of the static analyzer and filters out the majority of the false positives, while retaining almost the totality of the true positives. These two components are described in the remainder of this chapter.

STATIC PROGRAM ANALYSIS

This section presents the static program analysis component of Phoenix. It first presents an overview of the problem and a motivating example, and then goes into the details of the static-analysis engine, that is modular and allows for analysis of frameworks.

Software security is an ever-growing concern, especially when it comes to mobile applications. By design, mobile applications feed on inputs whose source is untrusted, perform numerous security-sensitive operations (such as database accesses and transfers of Web content to remote machines), and expose private data to potentially malicious observers. Mobile devices have access to numerous sources of private data, including the device's geographical location, the device identifier, the user's photographs, videos, audio files, calendar events, and contacts. Other pieces of private information are application-specific, and may include the user's authentication credential, healthcare records, social-security number, as well as credit-card and bank-account numbers. It is not surprising, then, that six out of the ten most critical mobile-application vulnerabilities according to the Open Web Application Security Project (OWASP) are information-flow-security violations, which can break *integrity* (whereby untrusted inputs flow into security-sensitive computations) or *confidentiality* (whereby private information is revealed to public observers). During the last decade, there has been intensive research on methods and algorithms for automatically detecting information-flow violations in Web applications. However, many of the published approaches are not readily applicable to industrial Web and mobile applications. Solutions based on type systems tend to be overly complex and conservative (Volpano, Irvine, & Smith, 1996; Myers, 1999; Shankar, Talwar, Foster, & Wagner, 2001), and are therefore unlikely to enjoy broad adoption, whereas those based on program slicing are often unsound (Tripp, Pistoia, Fink, Sridharan, & Weisman, 2009) or limited in scalability (Hammer, Krinke, &Snelting, 2006; Snelting, Robschink, & Krinke, 2006).

Phoenix includes a sound and highly accurate static security scanner, which also scales to large code bases, and particularly to Web and mobile applications, being designed for commercial needs as part of a product offering, IBM Security AppScan Source. This scanner performs a form of abstract interpretation (Cousot & Cousot, 1977) known as taint analysis (Sabelfeld & Myers, 2003): It statically detects data flows wherein information returned by a "source" reaches the parameters of a "sink" without being properly endorsed by a "downgrader". Depending on whether the problem being solved is related to integrity or confidentiality, a *source* is a method that injects untrusted or secret input into a program, a *sink* is a method that performs a security-sensitive computation or exposes information to public observers, and a *downgrader* is a method that sanitizes untrusted data or declassifies confidential data, respectively. The static security scanner included in Phoenix is equipped with a thorough configuration of triples of sources, sinks and downgraders for all known integrity and confidentiality problems, partitioned into security rules, such as Cross-site Scripting (XSS) and Structured Query Language injection (SQLi).

The key idea behind this static analysis is to track vulnerable information flows (emanating from sources) in a demand-driven manner, without eagerly building any complete representation of the subject application. Specifically, this static analyzer builds a call-graph representation of the program based on intraprocedural type inference. Furthermore, when there is a need to compute an aliasing relationship, stemming from flow of vulnerable information into the heap, the analyzer issues a granular aliasing query focused on the flow at hand, thereby obviating the need for whole-program pointer analysis. This enables sound and efficient scanning of large applications, where typically only a small portion of the application requires modeling. This characteristic is enabled by the fact that the static analyzer does not need to build any form of whole-program representation. In another view, the static analyzer presented in this chapter can be thought of as an extended type system, where a fully automated context-sensitive, interprocedural inference engine automatically attaches security annotations to program locations and propagates them. The static analyzer presented in this chapter enforces the following two properties:

1. The inference process is fully automated, and thus no complex, non-standard type system is forced on the developer.
2. The analysis is infinitely context sensitive (up to recursion), and consequently, it does not produce overly conservative results.

These properties lift the two most significant barriers that have so far prevented type systems from enjoying broad industrial adoption. To our knowledge, the static analyzer proposed in this chapter is the first taint-analysis algorithm that performs demand-driven analysis from the bottom up, including representing the program's type hierarchy, call graph and data-flow propagation graph. This is the key to achieving both accuracy and scalability without sacrificing soundness.

In essence, the static analyzer included in Phoenix makes the following specific contributions:

- **Demand-Driven Taint Analysis:** Phoenix's static analysis is based a demand-driven security analysis algorithm that is sound (even in the presence of multi threading), accurate and scalable. This section illustrates the design of the entire analysis stack in support of this feature.
- **Framework and Library Support:** Beyond the core analysis, this section presents novel extensions enabling effective modeling of framework and library code. These extensions are important for an analysis targeting real-world Web and mobile applications, which are built atop reusable frameworks.
- **Implementation:** The algorithm behind Phoenix has been fully implemented. It supports Java, .NET and JavaScript programs, and is currently used in a commercial product.

Motivation and Overview

To illustrate some of the unique features of the static analyzer embedded in Phoenix, we use the Aliasing5 benchmark from the Stanford SecuriBench Micro suite, shown in Listing 1. Designed for expository purposes, this example shows a Java application reading untrusted data from its parameters. Specifically, this example highlights the importance of tracking aliasing relationships between program variables and fields for sound security analysis, with buf flowing into two formal arguments of method foo (line 6).

The flow of the entire program is as follows: The doGet handler of the Aliasing5 servlet first initializes a fresh StringBuffer object, buf, with the string "abc" (line 5). It then invokes method foo, such that its first two formal arguments (buf and buf2) are aliased. Next, foo assigns the content of an untrusted parameter, "name", to variable name, in the source statement at line 10. This untrusted value subsequently taints the buffer pointed-to by buf (line 11). Because of the aliasing relationship between buf and buf2, the security-sensitive operation at line 13, which renders the content of buf2 to the response HTML, becomes vulnerable.

To detect the vulnerability in this program, the security scanner must account for the aliasing between buf and buf2 in foo. Existing approaches have all addressed this requirement by applying a preliminary whole-program pointer analysis, such as Andersen's flow-insensitive analysis (Andersen, 1994), to eagerly

Listing 1. The Aliasing5 benchmark from the SecuriBench Micro Suite

```
1: public class Aliasing5 extends HttpServlet {
2:   protected void doGet(HttpServletRequest req,
3:          HttpServletResponse resp)
4:          throws ServletException, IOException {
5:       StringBuffer buf = new StringBuffer("abc");
6:       foo(buf, buf, resp, req);
7:   }
8:   void foo(StringBuffer buf,
9:          StringBuffer buf2, ServletResponse resp,
10:          ServletRequest req) throws IOException {
11:       String name = req.getParameter("name");
12:       buf.append(name);
13:       PrintWriter writer = resp.getWriter();
14:       writer.println(buf2.toString()); /* BAD */
15:   }
16: }
```

compute an aliasing solution before starting the security analysis (Tripp et al., 2009). Performing a global aliasing analysis places a significant limitation on the scalability of the client security analysis, which is mitigated (but not lifted) if the aliasing analysis is coarse (i.e., context insensitive, flow insensitive, etc.). In that case, however, the ensuing security analysis becomes imprecise, often yielding an excess of false reports due to spurious data flows.

Phoenix, instead, performs on-demand alias resolution. It tracks symbolic representations of security facts, known as "access paths", and augments the set of tracked representations to account for aliases of tracked objects. Loosely speaking, an access path is a sequence of field identifiers, rooted at a local variable, such as x.f.g. This access path evaluates to the object o reached by dereferencing field f of the object pointed-to by x, and then dereferencing field g of o (or ⊥ if no such object exists).

Phoenix starts by modeling the effect of the source statement at line 10 as the seeding data-flow fact name.*. The * notation simply represents the fact that all objects reachable through variable name are to be considered untrusted. Then, the flow at line 11 leads the analysis to track both name.* and buf.content.*. However, because there is a flow into the heap at line 11, the analysis further issues an on-demand interprocedural aliasing query, which establishes that buf.content is aliased with buf2.content. Therefore, the analysis additionally tracks buf2.content.*. This exposes the vulnerability at line 13, where the toString call renders buf2.content to the response HTML.

Core Taint Analysis

The Phoenix static-analysis algorithm takes as input a Web or mobile application, along with its set of supporting libraries, and validates it with respect to a specification in the form of a set of "security rules". A *security rule* is a triple ⟨*Src*, *Dwn*, *Snk*⟩, where *Src*, *Dwn* and *Snk* are patterns for matching sources, downgraders and sinks in the subject program, respectively. A pattern match is either a method call or a field dereference. A vulnerability is reported for flows extending between a source and a sink belonging to the same rule, without a downgrader from the rule's *Dwn* set mediating the flow.

Phoenix interleaves call-graph construction with tracking of vulnerable information flows. This is to avoid building eager whole-program representations. Both the call graph and the data-flow solution computed atop the call graph are expanded on demand, ensuring scalability while retaining a high degree of accuracy.

Type-Hierarchy and Call-Graph Construction

As mentioned earlier, Phoenix refrains from building global program representations. Instead, it computes its supporting type hierarchy on demand. For this, Phoenix utilizes lazy data structures, which provide sophisticated mechanisms for caching and demand evaluation of type information at the granularity of individual methods and class fields.

Call-graph construction is also performed lazily. The call graph is built based on local reasoning, by resolving virtual calls according to an intra-procedural type-inference algorithm (Bacon and Sweeney, 1996). Call sites are not necessarily expanded eagerly (that is, before the data-flow analysis stage). Rather, an oracle is used to determine whether any given call site may lead to the discovery of source statements. Our oracle is sound, and is based on control-flow reachability between the calling method and source methods within the type-hierarchy graph (Dean, Grove, & Chambers, 1995).

Data-Flow Analysis

For a formal description of Phoenix's data-flow analysis algorithm, we use a standard description of the program's state, based on the domains shown in Table 1.

A program state, $\sigma = \langle \mathbf{E}, \mathbf{H} \rangle \in \mathit{States} = \mathit{Env} \times \mathit{Heap}$, maintains the pointing from variables to their values, as well as from object fields to their values. To describe the algorithm we use the syntactic structures shown in Table 2.

Table 1.

VarId	Program variables	$\mathit{Val} = \mathit{Loc} \cup \{\mathit{null}\}$	Values
FldId	Field identifiers	$\mathit{Env}: \mathit{VarId} \rightarrow \mathit{Val}$	Environments
Loc	Unbounded set of objects	$\mathit{Heap}: \mathit{Loc} \times \mathit{FldId} \rightarrow \mathit{Val}$	Heap

Table 2.

Statement	Meaning
x = new Object()	$[[\text{x = new Object()}]]\sigma = \sigma[[\text{x} \rightarrow o \in \mathit{Loc}.\ o \text{ is fresh}]]$
x = y	$[[\text{x = y}]]\sigma = \sigma[[\mathbf{E}(\text{x}) \rightarrow \mathbf{E}(\text{y})]]$
x.f = y	$[[\text{x.f = y}]]\sigma = \sigma[[\mathbf{H}(\langle \mathbf{E}(\text{x}), \text{f}\rangle) \rightarrow \mathbf{E}(\text{y})]]$
x = y.f	$[[\text{x = y.f}]]\sigma = \sigma[[\mathbf{E}(\text{x}) \rightarrow \mathbf{H}(\langle \mathbf{E}(\text{y}), \text{f}\rangle)]]$

These are kept to a minimum to simplify the description of the analysis. Extending the core language to contain procedure calls is straightforward (Cheng & Hwu, 2000).

Instrumented Semantics

To track security facts, we instrument the concrete semantics to further maintain untrusted (or tainted) access paths. Informally, an access path is a symbolic representation of a heap location. For example, access path x.g denotes the heap location pointed-to by field g of the object pointed-to by variable x. Security analysis over access paths tracks the set of paths evaluating to untrusted values.

More formally, an access path is a (possibly empty) sequence of field identifiers rooted at a local variable; i.e., an element in $VarId \times \left(.FldId\right)^{*}$. The meaning of access path $x.f_1...f_n$ is the unique value $o \in Val$ reached by first dereferencing x using **E**, and then following the references through $f_1...f_n$ in **H**, or $\perp$ if there are intermediate null dereferences in the path. This is defined inductively as follows:

$$\left[\!\left[x \cdot e\right]\!\right]\sigma = \begin{cases} \boldsymbol{E}\left(x\right), & x \in dom\left(\boldsymbol{E}\right) \\ \perp, & otherwise \end{cases}$$

$$\left[\!\left[x \cdot f_1 ... f_n\right]\!\right]\sigma = \begin{cases} \boldsymbol{H}\left(\left\langle \left[\!\left[x \cdot f_1 ... f_{n-1}\right]\!\right], f_n \right\rangle\right), & \left[\!\left[x \cdot f_1 ... f_{n-1}\right]\!\right]\sigma \neq \perp \wedge \\ & \left\langle \left[\!\left[x \cdot f_1 ... f_{n-1}\right]\!\right], f_n \right\rangle \in dom\left(\boldsymbol{H}\right) \\ \perp, & otherwise \end{cases}$$

An instrumented concrete state is a triple, $\sigma = \langle \mathbf{E}, \mathbf{H}, \mathbf{T} \rangle$, where **T** is a set of tainted access paths. We assume a security specification, *S*, which seeds the set **T** when evaluating certain assignment and field-read statements (according to the *Src* set of the provided security rules). The semantic rules for updating **T** appear in Figure 1.

Operator **A** in Figure 1 is defined by the semantic rules of Figure 2, which define the backward data-flow equations employed by Phoenix to perform demand-driven pointer analysis:

Figure 1. Forward data-flow equations

$$\mathbf{T} \xrightarrow{\texttt{x = new...}} \mathbf{T}$$

$$\mathbf{T} \xrightarrow{\texttt{x = y}} \mathbf{T} \cup \{\texttt{x.f}_1 \ldots \texttt{f}_n : \texttt{y.f}_1 \ldots \texttt{f}_n \in \mathbf{T}\}$$

$$\mathbf{T} \xrightarrow{\texttt{x = y.f}} \mathbf{T} \cup \{\texttt{x.f}_1 \ldots \texttt{f}_n : \texttt{y.f.f}_1 \ldots \texttt{f}_n \in \mathbf{T}\}$$

$$\mathbf{T} \xrightarrow{\texttt{x.f = y}} \mathbf{T} \cup \{\mathrm{A}(\texttt{x})\texttt{.f.f}_1 \ldots \texttt{f}_n : \texttt{y.f}_1 \ldots \texttt{f}_n \in \mathbf{T}\}$$

Figure 2. Backward data-flow equations

$$\mathbf{A} \xrightarrow{\texttt{x = new...}} \mathbf{A}$$
$$\mathbf{A} \xrightarrow{\texttt{x = y}} \mathbf{A} \cup \{\texttt{y.f}_1 \ldots \texttt{f}_n : \texttt{x.f}_1 \ldots \texttt{f}_n \in \mathbf{A}\}$$
$$\mathbf{A} \xrightarrow{\texttt{x = y.f}} \mathbf{A} \cup \{\texttt{y.f.f}_1 \ldots \texttt{f}_n : \texttt{x.f}_1 \ldots \texttt{f}_n \in \mathbf{A}\} \cup \{\texttt{x.f}_1 \ldots \texttt{f}_n : \texttt{y.f.f}_1 \ldots \texttt{f}_n \in \mathbf{A}\}$$
$$\mathbf{A} \xrightarrow{\texttt{x.f = y}} \mathbf{A} \cup \{\texttt{y.f}_1 \ldots \texttt{f}_n : \texttt{x.f.f}_1 \ldots \texttt{f}_n \in \mathbf{A}\} \cup \{\texttt{x.f.f}_1 \ldots \texttt{f}_n : \texttt{y.f}_1 \ldots \texttt{f}_n \in \mathbf{A}\}$$

Access-Path Widening

The key difficulty in using the symbolic access-path representation for static security analysis is that this representation of the heap, which is known as storeless (Deutsch, 1992), is unbounded. This problem manifests itself when dealing with recursive data structures, such as linked lists. To deal with this problem, we apply widening by introducing a special symbol, *. An access path now has either the concrete form $x.f_1...f_n$, or the widened form $x.f_1...f_n.*$ where

$$[\![x \cdot f_1 ... f_n \cdot *]\!]\sigma = \left\{o : \exists f_{n+1} ... f_k \in \left(FldId\right)* \cdot o = [\![x \cdot f_1 ... f_n \cdot f_{n+1} ... f_k]\!]\sigma\right\}$$

That is, a widened access path potentially points to more than one object.

In this way, the analysis can track a bounded number of access paths in a sound manner by restricting the length of an access path to some constant c, and allowing for insertion of $*$ at the end of a path of length c instead of extending it when accounting for the effect of a field-assignment statement.

On-Demand Aliasing

As mentioned earlier, Phoenix features the ability to soundly track symbolic security facts. The key idea is to perform alias analysis on demand, when an untrusted value flows into an object field (i.e., untrusted data flows into the heap). We first illustrate this situation through a simple example, where we assume that initially there is a single taint fact, **T** = {z.g}, and the last statement—assigning a value to o.sinkfld—is a sink, and as such must not be assigned an untrusted value (Box 1).

We highlight in red the access paths that would be missed by a forward data-flow analysis without on-demand alias-analysis capabilities, such as the Interprocedural, Finite, Distributive Subset (IFDS) framework (Reps, Horwitz, & Sagiv, 1995). Such an analysis would ignore the assignment x = y.f because it is not affected by **T**, thereby missing the aliasing relation between x.h and y.f.h at the point when it becomes relevant, which is the following two statements: The first, x.h = z.g, contaminates x.h, and thus also y.f.h, and the second dereferences y.f.h into w.

In contrast, Phoenix's static analysis is fully sound, as can be easily proven. Phoenix handles cases, such as the one above, by performing on-demand alias analysis. Upon encountering the field-assignment statement x.h = z.g, Phoenix traverses the control-flow graph backwards seeking aliases of x.h. It then finds that y.f.h is an alias of x.h, and propagates this additional security fact forward, which ensures that the security vulnerability is discovered. The Phoenix propagation steps are visualized above using labeled edges, the label consisting of the step index (in square brackets) followed by the learned taint fact.

Box 1.

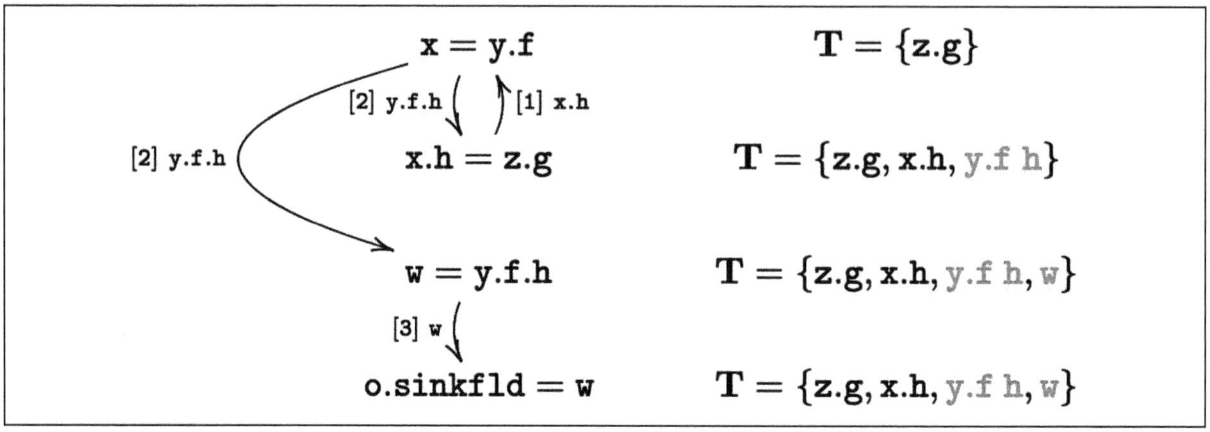

Formally, Phoenix computes a fixed-point solution for the equations in Figure 2 while traversing the control flow backwards from the statement performing the heap update. The seeding value for **A** in our example is the singleton set {x}.

Extensions: Library and Framework Modeling

Modern Web and mobile applications are often built atop one or more frameworks, such as Struts, Spring, JavaServer Faces (JSF) and jQuery (Sridharan, Artzi, Pistoia, Guarnieri, Tripp, & Berg, 2011; Vosloo & Kourie, 2008). Frameworks typically invoke application code using reflective constructs, based on information provided in external configuration files, which complicates static analysis of Web applications.

To address this concern, Phoenix is fully integrated with Framework For Frameworks (F4F), a recent solution augmenting taint-analysis engines with precise framework support (Sridharan et al., 2011). F4F automatically generates static-analysis artifacts, which can be integrated into a taint-analysis engine to ensure that the interaction of an application with the frameworks it uses is modeled soundly and accurately.

Phoenix's integration with F4F exploits the fact that static analysis can operate on non-executable yet legal code. Phoenix transforms the F4F output into synthetic code that soundly models data flows involving framework code. This choice has several advantages compared to direct modeling of frameworks within the Phoenix static-analysis engine, being:

- More lightweight: no need to directly generate Intermediate-Representation (IR) code.
- More portable and reusable: the synthetic code generated by F4F can be plugged into any existing analysis.
- More intelligible to the developer, who is presented with simple code instead of IR code.

Before statically analyzing an application, Phoenix takes the output of F4F and transforms it as follows. Each call replacement has a synthetic method associated with it. This is the method that Phoenix should consider in place of the one specified in the application source code. For every synthetic method, Phoenix creates code corresponding to the instructions for that synthetic method that are specified in the output of F4F. In most cases, this can be done straightforwardly. However, there were several interesting problems that need to be addressed.

One case is simulating method invocations from synthetic methods. Such invocations are on uninitialized variables, which causes Phoenix's intraprocedural type inference to ignore them. Solving this by initializing the variables is problematic: Some declared types are abstract, and some do not have a default constructor. Phoenix solves this problem by adding a level of indirection via a method call that returns null. Since the assignment to null is performed in a different procedure, Phoenix's type inference accepts the call as valid, with a result sufficient to model taint propagation faithfully.

Another problem arises when synthetic methods invoke default-scope or protected methods in a class of another package. Since these methods can only be invoked from classes in the same package, Phoenix extends that package with an additional public synthetic class containing a public synthetic method that calls the default-scope or protected method, and returns its return value. Being public and in the same package as the restricted method, this synthetic method can be invoked without exceptions.

MACHINE LEARNING

The scale and complexity of modern software systems often lead to missed security vulnerabilities. Static analysis has emerged as a promising solution for automated security auditing, which can scale to millions of lines of code while accounting for nontrivial program behaviors, resulting in the detection of severe and sometimes also subtle vulnerabilities (Tripp et al., 2009; Guarnieri et al., 2011; Tripp et al., 2013; Guha, Krishnamurthi, & Jim, 2009).

In particular, static analysis has shown great value when applied to information-flow vulnerabilities (Sabelfeld & Myers, 2006; Denning & Denning, 1977). These are the most serious and prevalent forms of vulnerability in today's landscape of Web and mobile applications.

A popular method of detecting information-flow vulnerabilities using static analysis, as we saw in the previous section. The analysis checks for reachability between source and sink statements via a downgrader-free path. Such paths, which we refer to as *witnesses* or *counterexamples*, are reported as potentially vulnerable. As an illustration, we provide in Figure 3 an exemplary witness reported by a commercial taint analysis for client-side JavaScript.

Figure 3. Security witness reported by a commercial static security checker for JavaScript

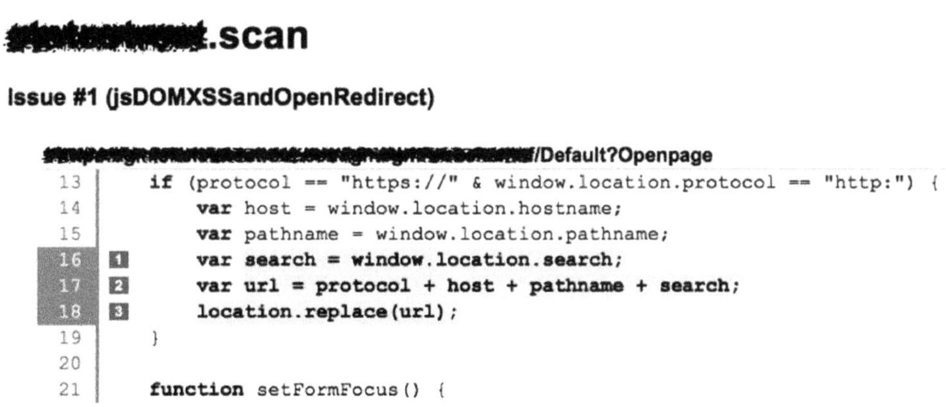

The reported vulnerability in this case is *open redirect*, which occurs when the user is able to influence the target URL of a redirection operation. The analysis bases this warning on information flow between a statement that obtains the URL query string (stored as location.search) and a statement performing redirection (location.replace()). The intermediate statement concatenates the return value from the source (pointed-to by variable search) with other strings, and thus the taint tag is carried across to the resulting url string.

Generalizations and extensions of taint analysis include features such as string sensitivity, whereby the analysis explicitly tracks string values and their structure for increased precision (Tateishi et al., 2011); typestate-based tracking rules to refine the security specification (Livshits & Lam, 2005); as well as quantitative notions of information flow for more accurate and informative warnings (McCamant & Ernst, 2008). In all of these cases, however, the key concept of verifying disjointness between sources and sinks modulo permitted exceptions remains the same.

Static security verification, as presented in the previous section, is not a silver bullet. To handle industry- scale applications, the analysis must apply aggressive approximations. Notable dimensions of precision loss, suffered also by Phoenix's static analyzer, include *flow insensitivity*, whereby the analysis does not track the order in which memory updates occur and instead conservatively accounts for all possible update orders; *path insensitivity*, whereby the analysis ignores path conditions, thereby traversing infeasible execution paths; and *context insensitivity*, whereby the analysis refrains from modeling the calling context of a method, thereby analyzing infeasible invocation scenarios.

These (and other) sources of inaccuracy shrink the analysis' state space (also known as the *abstract state*) by an exponential factor, and hence contribute significantly to the scalability of the analysis. As an example, if path conditions are accounted for, then the analysis has to track two different runtime program states when reaching a branching condition. Path insensitivity saves the analysis from this type of state-space blowup.

Approximations applied by the analysis can result in false alarms. These may be due to various reasons, including, for example, *infeasible control flow*, if the witness contains invalid branching decisions or call sites that are resolved incorrectly; *ignoring of downgrading operations*, if a proprietary downgrader is applied or downgrading occurs inline; and *imprecise tracking of data flow*, e.g. due to coarse modeling of string operations and/or aliasing between variables.

Indeed, while approximation enables scalability, it often also results in an excess of false alarms. These plague the reports by static analysis tools (Muske, Baid, & Sanas, 2013), thereby hindering their usability. According to interviews with professional developers, false alarms are the most significant barrier to adoption of tools based on static program analysis by developers (Johnson, Song, Murphy-Hill, & Bowdidge, 2013). As things now stand, developers prefer to release and deploy insecure software rather than find and fix latent vulnerabilities using off-the-shelf static security checkers. This is unfortunate.

The goal of Phoenix is to improve the usability of static security checkers by cleansing their output, particularly of the static security analyzer embedded in Phoenix. We put forward two basic requirements:

1. **Generality:** Although our favorite target for cleaning the output is the static security analyzer of Phoenix, the cleansing technique should not be specific to a given tool or analysis technique. It should rather treat the analysis algorithm as opaque for wide applicability and ease of integration into legacy as well as new analysis tools.

2. **Customizability:** Different users have different preferences when reviewing security warnings. Some prefer to aggressively suppress false alarms, even at the expense of eliminating certain true issues, whereas others may opt for completeness at the price of more false warnings.

Driven by these requirements, we have developed a method for filtering false warnings that combines lightweight user interaction with heavyweight automation. In our approach, the user classifies a small portion of the raw warnings output by the analysis. The user also specifies a tradeoff between elimination of false alarms and preservation of true findings. These two inputs are fed into a statistical learning engine, which abstracts the warnings into feature vectors that it uses to automatically build a filter. The filter, which is instantiated according to the user-specified policy, is next applied to the (vast majority of) remaining warnings, resulting in (many) less warnings than those initially reported by the security checker. Importantly, our learning-based approach is guided solely by the warnings themselves. It does not make any assumptions about, or access to, the internals of the analysis tool.

To build the filter, Phoenix searches through a library of classification algorithms. It computes a competency score for each of these algorithms based on the user-classified warnings. The most favorable candidate is then applied to the remaining warnings.

To evaluate the cleansing technique of Phoenix, we ran Phoenix's security checker on a set of 1,700 JavaScript programs, taken from a diversified set of 675 top-popular Web and mobile applications, which resulted in a total of 3,758 warnings. A security specialist then classified these as either true or false warnings. The results are highly encouraging. As an example, for a policy biased toward preservation of true positives, given only 200 classified warnings, Phoenix is able to boost precision by a factor of 2.868 while reducing recall by a negligible factor (×1.006). Conversely, if the user is more biased toward elimination of false alarms, still based on only 200 classified warnings, Phoenix achieves a recall-degradation factor of only 2.212 and a precision-improvement factor of 9.014. In all cases, and across all policies, Phoenix is able to improve precision by a factor ranging between ×2.868 and ×16.556 in return to user classification of only 200 warnings.

From the point of view of machine learning, Phoenix makes the following principal contributions:

1. **Boosting Usability via Learning:** We propose a novel and general technique to boost the usability of static security checkers. In the Phoenix approach, users invest tolerable effort in classifying a small portion of the warnings, and in return a statistical learning engine computes a filter over the remaining warnings. The filter is parameterized by the user's preference regarding the tradeoff between true and false alarms.
2. **Characterization of Security Warnings:** We characterize different features of static security warnings, and draw general conclusions about their correlation with the correctness of a reported warning. This insight is of independent value, and can serve for other studies as well. We also discuss and analyze the relative merits and weaknesses of different classification algorithms, which is again of general applicability.

Motivation and Overview

In this section, we motivate the need for the Phoenix machine-learning system and provide a high-level description of its main components and properties.

Limitations of Static Analysis

As highlighted above, static program analysis has inherent limitations in precisely modeling run-time behaviors of the subject program. Added to these, the analysis often deliberately sacrifices precision (even when a precise model is possible) in favor of scalability. Following are design choices that are often made for the analysis to scale to large codes:

- **Flow Insensitivity:** The analysis does not track the order in which memory updates occur, and instead conservatively accounts for all possible update orders. A simple example is the following:

```
x.f = read();
x.f = "";
write(x.f);
```

Even though the value of x.f is benign when it flows into the write() sink, the analysis abstracts away the order of updates to x.f and simply records that at some point that field was assigned an untrusted value.

- **Path Insensitivity:** Path conditions are ignored. Instead, the analysis assumes that all paths through the control-flow graph (CFG) are feasible. Here is an example:

```
x.f = "";
if (b) {
   x.f = read();
}
if (!b) {
   write(x.f);
}
```

The above code is not vulnerable, as the conditions governing the source and sink statements are mutually exclusive. Yet, a path-insensitive analysis would report the infeasible trace through both the source and the sink as a warning.

- **Context Insensitivity:** The analysis abstracts away the context governing the invocation of a method, thereby merging together different execution contexts of the same method, as this example illustrates:

```
y1 = id(x);
y2 = id(read());
write(y1);
```

Here, id() is simply the identity method, which echoes back its input. In the first invocation, the input is trusted, while the second invocation is with an untrusted argument. Merging together the two contexts, the analysis conservatively judges that id() may return an untrusted value, and therefore y1 is treated as untrusted causing a vulnerability to be reported when write() is called.

Design choices like the ones above, which are pertinent for performance and scalability, each contribute to the analysis's imprecision. Worse yet, there are significant interaction effects between the different sources of imprecision. This defines the need for complementary machinery to cleanse the analysis's output. In our approach, this step is carried out interactively, with help from the user, by casting the warnings reported by the analysis into a statistical learning framework.

System Architecture

We describe here the high-level architecture of Phoenix. The input to Phoenix is a set of raw warnings $\{w_1, ..., w_n\}$ output by the static security checker. Next, the user is asked to classify a subset of the warnings. This subset is selected at random to avoid biases. The result is a classified subset $\{(w_{i1},b_{i1}),...,(w_{ik},b_{ik})\}$ of the warnings, where the labels b_{ij} are boolean values (indicating whether the warning is true of false). Naturally, the accuracy of the filter computed by Phoenix is proportionate to the number of warnings reviewed by the user. However, as we demonstrate experimentally in Section 5, even a relatively small sample of 100 warnings suffices for Phoenix to construct a highly precise filter.

To cast security warnings into a statistical setting, we need to derive simple-structured features from the warnings, which are complex objects that cannot tractably be learned directly. This part of the flow is what we call *feature mapping*. A given warning is abstracted as a set of attributes, including e.g., the respective line numbers of the source and sink statements, the time required by the analysis to compute the witness, the number of flow steps along the witness, etc.

The feature vectors, combined with the user-provided true/false tags and policy, provide the necessary data to learn and evaluate filters. Given training data of the following form:

```
[length = 14, time = 2.5, srcline = 10, . . .]    →    false
[length = 6, time = 1.1, srcline = 38, . . .]     →    true
[length = 18, time = 3.6, srcline = 26, . . .]    →    false
...
```

the Phoenix system partitions the data into training and testing sets of equal cardinality. Phoenix then generates a set F of candidate filters by training different classification algorithms on the training set. Phoenix also converts the policy into a scoring function SCORE. Next, each of the candidate filters is applied to the testing set, and the resulting classifications are reduced to a score via the SCORE function based on the rate of true positives, false positives and false negatives. Finally, the filter that achieves the highest score is applied to the remaining warnings. The user is presented with the findings surviving the filter.

Features and Learning Algorithms

Naturally, the efficacy of Phoenix is dependent on the available features and learning algorithms. We describe both in detail in Sections 3 and 4, respectively. Here we give a brief informal description of both.

Features

As explained earlier, features are an abstraction of witnesses reported by the analysis tool. Most of the features that we have defined reflect basic characteristics of the witness, such as:

1. Its length (i.e., the number of statements along the path),
2. The syntactic location of the source and sink statements, and
3. The context manipulated by the JavaScript code (plain DOM elements, Flash, JavaScript, etc).

We have also defined general features that are not derivable directly from the witness itself, but rather reflect metadata associated with the witness, such as the time required by the analysis to detect the respective violation.

We note, importantly, that simply defining a large set of arbitrary features may lead the learning algorithm to discover false correlations. For this reason, we have made a careful selection of features. Behind each of the features is a justification for why it may correlate with true/false alarms. As an example, many findings involve imported code, such as the swfobject library for embedding Flash content in HTML files. These often invoke sources and/or sinks (e.g., reading or writing of the document's URL). The developer may deem flows emanating from, passing through or arriving at such libraries as spurious. This may happen, e.g., if the library applies inline sanitization that the analysis misses. Thus, the source/sink location may become an important feature, which correlates well with user classifications.

Learning Algorithms

There is a wide spectrum of classification techniques (Witten & Frank, 2005; Hastie, Tibshirani, & Friedman, 2009). These range between tree-based classification, such as decision trees (Quinlan, 1993); rule-based algorithms (Nevill-Manning, Holmes, & Witten, 1995); functional methods, which compute a geometrical boundary between the instances (Pontil & Verri, 1998); Bayesian techniques, which compute the probability of events (actual vs. spurious vulnerabilities in our case) as a function of feature attributes (Ng & Jordan, 2002); and even (unsupervised) clustering techniques like K-means that form clusters and tag new instances according to their assigned cluster (Cleary & Trigg, 1995).

An important property of Phoenix is that it lets the user specify the filtering policy. We validated experimentally that different policies are best served by different classification algorithms. As an example, if the user is strongly biased toward elimination of false alarms, then a rule-based classifier may identify a feature f and a threshold value v, such that (almost) any alarm for which $f \geq v$ is false. This would satisfy the user's preference perfectly, but at the same the classifier may also eliminate true vulnerabilities. Striking a non-trivial balance between precision and recall, on the other hand, may leads toward a more sophisticated classifier—such as a decision tree or a kernel Support Vector Machine (SVM). In light of this observation, we have linked into Phoenix eight popular classifiers that subsume all the algorithmic categories above (functional, rule based, tree based, etc.).

Learning Features

As we highlighted earlier, the choice of which features to map a security witness to has important bearing on the overall quality of the filtering algorithm. The main concern with introducing arbitrary features is that the algorithm could be led to discover false correlations between such features and witness correctness. We dedicate this section to explaining the rationale behind the features we have selected. We divide the discussion according to feature categories.

Lexical Features

A natural category of features are those recording syntactic properties of the witness. Specifically, we have defined seven such features, as follows:

1. **Source Identifier (srcid):** The name of the field or function evaluated in the source statement (for example, document.location).
2. **Sink Identifier (srcid):** The name of the field or function evaluated in the sink statement (for example, window.open()).
3. **Source Line Number (srcline):** The line number of the source statement.
4. **Sink Line Number (srcline):** The line number of the sink statement.
5. **Source URL (srcurl):** The URL of the JavaScript function containing the source statement.
6. **Sink URL (sinkurl):** The URL of the JavaScript function containing the sink statement.
7. **External Objects (extobjs):** Flags indicating whether the witness is performing mailto or embed functionality (for example, Flash).

Syntactic features are effective in uncovering patterns due to third-party libraries or usage of frameworks. They also assist in localizing noisy sources and sinks. Indeed, our reason for recording source and sink line numbers as well as the URLs of the files enclosing the source and sink statements is to capture instances where the flow either emanates from, or arrives at, third-party libraries. Another meaningful characterization of the flow is the context(s) it manipulates via mailto and embed statements. Certain attack vectors fail in such special contexts, which could be a source of false alarms.

Finally, for source and sink identifiers, the motivation is to detect "noisy" operations, such as document.location.url, which can cause an open-redirect attack if assigned an untrusted value, although this happens very rarely (Tripp, Ferrara, & Pistoia, 2014).

Quantitative Features

A second category of features includes those that record quantitative measures of the witness. These are not to be confused with numerical yet non-quantitative features, such as line numbers. The latter are meaningful only inasmuch as equality checking is concerned, whereas quantitative measures can meaningfully be subjected to less-than and greater-than comparisons. We define the following five quantitative features:

1. **Total Results on (Results):** The overall number of findings reported on the file containing the sink statement.
2. **Number of Steps (Steps):** The number of flow milestones comprising the witness path.
3. **Time (Time):** The total time spent by the analysis on the scope containing the witness.
4. **Number of Path Conditions (Conditions):** The number of branching statements (either loops or conditions) along the witness path.
5. **Number of Functions (Functions):** The number of functions enclosing statements along the witness path.

Intuitively, all five of these features satisfy the following property: the greater their value is, the less likely it is for the witness to be correct. First, for overall number of results, it is unlikely (albeit not impossible) for a single file to contain a large number of vulnerabilities. A more likely hypothesis is that the functions in the file are complicated, leading to their conservative and thus imprecise analysis. The time feature captures a similar pathology. Often imprecision leads the analysis to explore dead code or infeasible execution paths, which in turn lead to further imprecision, and so on. Thus, if the analysis has spent a long time on a given scope, then that is likely a symptom indicating that the model it has created is overly conservative. Finally, for methods, steps and path conditions, the higher these counts are, the more the analysis is exposed to errors due to over-approximations, such as infeasible branching and incorrect resolution of call sites.

Security-Specific Features

The third and final category relates to features that are intrinsic to the security domain. We have identified two such features:

1. **Rule Name (rname):** The name of the violated security rule.
2. **Severity (Severity):** The severity of the violation as determined by the analysis tool.

There are various other security properties that are reported by industrial security checkers, but these change across tools, and so for generality we avoided from including them.

Similarly to the source and sink identifiers, the involved security rule may prove "noisy." This could be either because:

1. The rule is perceived as less relevant by the user, or
2. The sources and sinks defined by the rules are noisy, or
3. There are defense measures (such as framework-level sanitizers) that suppress the given type of vulnerabilities, which the analysis is not aware of.

The severity of a finding also hints toward its correctness. In specific, low-severity witnesses are less likely to be accepted by the user as actionable.

Learning Algorithms

Our approach to static security analysis reduces the problem to binary classification, either in the online or offline setting. For a set of reports, we have feature data (given in Section 3) along with user-generated labels. Binary classification is a classic problem in machine learning, and a variety of methods from the field can be applied; for a survey, see Witten and Frank (2005), Bishop (2006), and Hastie et al. (2009). In this section, we give a brief overview of methods we use to evaluate the approach in Section 5, and discuss their suitability for static security analysis. We discuss four categories of methods: functional, clustering, tree/rule based, and Bayesian. Methods in different classes can be combined, and some state of the art methods borrow ideas from several categories. For example, the NB-Tree method builds a decision tree with Naive Bayes' classifiers at the leaves. We provide a general overview, rather than delving too deeply into the details of the methods we compared. We close with a comparative discussion of the methods, identifying potential advantages of some categories, and presenting strategies for comparison between methods useful from the user's perspective.

Functional Methods

Functional methods include logistic regression (Bishop, 2006), linear support vector machines (Pontil & Verri, 1998; Schölkopf, Smola, Williamson, & Bartlett, 2000), and generalizations, such as neural nets and kernel SVMs (Hastie et al., 2009). For convenience, we refer to these models as functional classification. These methods classify by learning a boundary either in feature space, or in a derived space related to features space by particular mappings. Once trained, the model can be used for prediction by noting where in the decision space an incoming feature would fall.

For example, SVMs search for a hyperplane in feature space that best separates labeled data (according to a maximum margin condition). Given a set of features x_i with labels $y_i \in \{-1,1\}$, SVMs solve for a hyperplane *w* that solves a strictly convex optimization problem:

$$\min_{w,\gamma} \frac{1}{2} w^2 + \lambda \sum_i \max\left(0, 1 - y_i\left(x_i^T w - \gamma\right)\right)$$

By inspection, the method weighs a regularization term, w^2, against a robust classification measure, which is 0 for example *i* when the predicted label $x_i^T w - \gamma$ has the correct sign, and grows linearly if the sign is incorrect. The problem is strictly convex, so always guaranteed to have a unique solution, which is easily found using iterative methods.

One weakness of SVMs, and other linear methods (e.g., logistic regression) is the richness of the model space—there are limits to how well a linear classifier can perform. To address this issue, kernel SVMs neural net models train multiple layers of derived features from the data, while kernel SVMs make use of a kernel function Φ that maps the features to a different space, yet allows simple evaluation of inner products $\Phi\left(x_i\right)^T \Phi\left(x_j\right)$. Kernels and their applications have a rich literature (Aronszajn, 1950; Saitoh, 1998).

Instance-Based Classification

Instance-based learning requires a distance function to measure the distance of an incoming measurement to existing instances. Given such a function, one can simply find the nearest labeled instance to an unknown data point, and use that label to predict the class of the input.

For example, the Kstar algorithm uses a generalized distance function that models the distance between two instances through a series of possible transformations. Considering the entire set of possible transformations gives a probabilistic measure that takes into account influence from several labeled points (Cleary & Trigg, 1995).

Tree- and Rule-Based Methods

Tree- and rule-based methods are divide-and-conquer algorithms that try to efficiently partition data instances according to labels.

For example, decision trees partition data into branches according to attribute values and labels. In order to build the tree, a feature attribute must be chosen at every branch point. These choices are made in a way that maximizes the so-called "information gain", which is a data-dependent measure that can be quickly computed. Decision trees work top-down, finding an attribute to split on at each point, and continuing recursively into the branches (Quinlan, 1993).

Rule-based methods attempt to find covering rules that describe each class, excluding others. For discrete attributes, rules can include membership in subsets of domain values (e.g., use of frameworks is binary and can be used in a split or a rule). For continuous or ordinal variables, rules can include thresholds (e.g., time taken to complete analysis larger than a set value). For example, the 1R classifier generates a one-level decision tree, with each rule in the set testing only one particular attribute (Nevill-Manning et al., 1995).

Bayesian Methods

Bayesian methods include Naive Bayes and Bayesian Networks, and directly model the probability of class events as a function of feature attributes. Naive Bayes assumes independence of attributes, and uses Bayes' rule to compute the probability of class given feature as:

$$P\left(C=c \mid X=x\right)=\frac{P\left(X=x \mid C=c\right)P\left(C=c\right)}{P\left(X=x\right)}$$

where the probabilities on the right hand side are learned from the data (Ng & Jordan, 2002). It is important to note that Naive Bayes also assumes quantitative features have a Gaussian distribution, but there are generalizations that relax this assumption. More importantly, independence of attributes can be relaxed as well, and more sophisticated methods such as Bayesian Networks are able to learn covariance structure from the data, e.g., by maximum likelihood estimation (Heckerman, Geiger, & Chickering, 1995).

Performance Measurement

The performance of classification methods can be understood through a range of statistics, such as precision, recall, and accuracy. Every method may have a whole range of values associated to it. For example, precision is the ratio of positives correctly identified to all the instances labeled positive, while recall is the proportion of positive instances correctly identified. There is a natural tradeoff between recall and false positives. By trying to catch a higher portion of positive instances, one will necessarily mislabel a greater portion of negative instances as well. Thus, increasing recall leads to decreased precision.

This suggests two ways of comparing binary classification algorithms:

1. One way is to compute an aggregate measure of quality across a range of possible policies. This is typically done using Receiver Operator Characteristic (ROC) curves (Pepe, 2003). ROC curves plot the true positive rate (recall) as a function of the false positive rate for each classifier. The area under and ROC curve, called AUC, serves as an aggregate measure of quality; the higher the AUC, the better the classifier (overall). An AUC of 1 means that the classifier can perfectly separate the two classes; an AUC of 0.5 means that the classifier is essentially no better than a random guess. Two classifiers with the same AUC may perform differently in different regions of the 2D space defined by true positives and false positives.
2. A second way is to prescribe a policy choice, for example, to require high recall. This is appropriate for static security analysis, since we want to miss as few real issues as possible. Then, for high recall, one can compare classifiers using their false positive rates (the lower the better), or similar measures, such as precision (the higher the better). Different classifiers may be superior for different policy choices.

RELATED WORK

Previous work that relates to Phoenix's novel contributions falls into two areas of research: program analysis for security and machine learning applied to reduction of false positives. In this section, we concentrate on related work on these two areas.

Program Analysis for Security

There is a rich body of work on taint analysis. We here concentrate on static taint analysis, and refer the reader to Chang, Streiff, and Lin (2008) and Newsome and Song (2005) for a survey of dynamic taint-analysis techniques. A detailed overview of works on program slicing is given in Sridharan, Fink, and Bodík (2006) and references therein.

The notion of tainted variables became known with the Perl language. Typically, the data manipulated by a program can be tagged with security levels (Denning & Denning, 1977), which assume a poset structure. Under certain conditions, this poset is a lattice (Denning, 1976). Given a program, the principle of non-interference dictates that low-security behavior of the program be not affected by any high-security data, unless that high-security data has been previously downgraded (Goguen & Meseguer, 1982). Taint analysis is an information-flow problem in which high data is the untrusted output of a source, low-security operations are those performed by sinks, and untrusted data is downgraded by sanitizers.

Volpano et al. (1996) show a type-based algorithm that certifies implicit and explicit flows and also guarantees non-interference. Shankar et al. (2001) present a taint analysis for C using a constraint-based type-inference engine based on cqual. Similarly to the propagation graph built by Phoenix, a constraint graph is constructed for a cqual program, and paths from tainted nodes to untainted nodes are flagged.

Java Information Flow (Jif) (Myers, 1999) uses type-based static analysis to track information flow. Based on the Decentralized Label Model (Myers & Liskov, 1997), Jif considers all memory as a channel of information, which requires that every variable, field, and parameter used in the program be statically labeled. Labels can either be declared or inferred. Ashcraft and Engler (2002) also use taint analysis to detect software attacks due to tainted variables. Their approach provides user-defined sanity checks to untaint potentially tainted variables. Pistoia, Flynn, Koved and Sreedhar (2005) present a static analysis to detect tainted variables in privilege-asserting code in access-control systems based on stack inspection.

Snelting, Robschink, and Krinke (2006) make the observation that Program Dependence Graphs (PDGs) and non-interference are related. Based on this observation, Hammer, Krinke, and Snelting (2006) present an algorithm for verifying non-interference: For output statement s, *backslice*(*s*) must contain only statements whose security label is lower than *s*, where *backslice* is a function that maps each statement *s* to its static backward slice. Though promising, this approach has not been shown to scale.

Livshits & Lam (2005) analyze Java, Enterprise Edition (EE) applications by tracking taint through heap-allocated objects. Their solution requires prior computation of a flow-insensitive, context-sensitive may-points-to analysis, based on Binary Decision Diagrams (BDDs) (Whaley & Lam, 2004), which limits the scalability of the analysis (Lhoták & Hendren, 2006). The points-to relation is the same for the entire program ignoring control flow. By contrast, the PDG-based algorithm in Hammer et al. (2006) handles heap updates in a flow-sensitive manner, albeit at a much higher cost. Livshits and Lam's analysis requires programmer-supplied descriptors for sources, sinks and library methods dealing with taint carriers. Guarnieri, Pistoia, Tripp, Dolby, Teilhet and Berg (2011) present a taint analysis for JavaScript. Their work relies on the whole-program analysis by Andersen (1994). While being sound, the analysis has not been shown to scale to large programs.

Wassermann and Su extend Minamide's string-analysis algorithm (Minamide, 2005) to syntactically isolate tainted substrings from untainted substrings in PHP applications. They label non-terminals in a context-free grammar with annotations reflecting taintedness and untaintedness. Their expensive yet elegant mechanism is applied to detect both SQLi (Wassermann & Su, 2007) and XSS (Wassermann & Su, 2008) vulnerabilities. Subsequent work by Tateishi, Pistoia, & Tripp (2011) enhances taint-analysis precision through a string analysis that automatically detects and classifies downgraders in the application scope.

McCamant & Ernst (2008) take a quantitative approach to information flow: Instead of using taint analysis, they cast information-flow security to a network-flow-capacity problem, and describe a dynamic technique for measuring the amount of secret data that leaks to public observers.

Phoenix's scalability stems from its demand-driven analysis strategy. Demand-driven pointer analysis was originally introduced by Heintze and Tardieu (2001). Since then, there have been several works on demand-driven points-to analysis via context-free-language reachability (Sridharan & Bodík, 2006; Zheng and Rugina, 2008; Yan, Xu, & Rountev, 2011). For taint analysis, our empirical data suggests that only a small fraction of a large program is expected to be influenced by source statements. Fuhrer, Tip, Kieżun, Dolby, and Keller (2005) take a demand-driven approach in replacing raw references to generic library classes with parameterized references. At a high level, this analysis resembles the alias analysis performed by Phoenix, as constraints on type parameters are first propagated backwards to allocation sites and declarations, and from there they are propagated forward.

False-Positive Elimination

The challenge of eliminating as many false positives as possible from the report of a static analysis, without introducing an excessive number of false negatives, has been the subject of numerous research studies.

Junker, Huuck, Fehnker, and Knapp (2012) cast static-analysis clients as syntactic model-checking problems. Violations of the verification property result in a counterexample (or path), which is checked for feasibility using a Satisfiability Modulo Theories (SMT) solver. If the path is infeasible, then the causes for its infeasibility are extracted via path slicing and an observer automaton is constructed to exclude all paths sharing the same cause. Junker, et al. have integrated their approach into the Goanna tool for static analysis of C/C++ programs. In another study, Fehnker, Huuck, Seefried, and Tapp (2010) share their experience in applying Goanna to large software systems consisting of 106 and more lines of code, including the Firefox browser. They report on an excessive number of false alarms, and address this usability issue by refining the security rules of Goanna. In contrast with many other analyses, the Goanna rules are phrased as automata, which are translated into a Kripke structure for model checking. Unlike Phonenix, these works are strictly concerned with reducing the number of false positives, without giving the user of the analysis the flexibility to express a preference that reflects a bias over true versus false positives.

Muske, Baid, and Sanas (2013) tackle the problem of false warnings via a partitioning approach. They assume two partitioning phases: The first divides the warnings into equivalence classes, assigning a leader to each class, such that if the leader is a false warning, then all other warnings are also false. The second step is to partition the leader warnings. This is done based on the variables modified along the path and their modification points. The entire process is meant to facilitate manual review of the warnings generated by a static analysis tool. The authors report on 50%-60% reduction in review effort corresponding to 60% redundant warnings on average, which are more readily eliminated thanks to their technique. Just like for Phoenix, the purpose of this work is to reduce the burden on the analyst reviewing the results of a static analysis. The difference is that this work does that by partitioning the analysis results, in such a way that results with the same characteristics are grouped in the same equivalence class, whereas Phoenix is based on an algorithm that learns the characteristics of false positives and prevents from presenting to the analyst flows that are highly likely to lead to other false positives.

An orthogonal approach that Muske, Datar, Khanzode, and Madhukar (2013) propose is to first apply a scalable yet imprecise abstract interpretation, and then remove false warnings using a bounded model-checking technique, which features the opposite tradeoff. The main goal is the elimination of false reports. This approach consists of two distinct techniques to identify equivalence between assertions, thereby obviating the need to verify all but one of the assertions, and a third technique for skipping verification of assertions that do not influence any of the false alarms. However, this work does not allow the analyst to choose whether the filter applied to the result of an analysis should err more towards the false-positive or true-positive side.

Johnson et al. (2013) report on interviews with developers to learn the reasons why static analysis tools are underused. They conclude that the two main barriers to adoption of such tools are:

1. False positives, and
2. The way in which warnings are presented to the developer.

These are related: if the warning is hard to understand, then false alarms become harder to identify and eliminate.

EFindBugs, developed by Shen, Fang, and Zhao (2011), is an improved version of the popular FindBugs tool (Ayewah, Pugh, Morgenthaler, Penix, & Zhou, 2011) that addresses the excess of false positives commonly reported by FindBugs. This is achieved via a two-staged error ranking strategy. First, EFindBugs is applied to a sample program. The resulting warnings are classified manually, and an approximate defect likelihood is assigned to each bug category and bug kind. This determines the initial ranking of reported bugs. Defect likelihood is later tuned in a self-adaptively manner when EFindBugs is run on the user's application thanks to users' feedback on reported warnings. This optimization process is executed automatically and based on the correlations among error reports with the same bug pattern. Unlike Phoenix, this approach does not offer the analyst the ability to finely tune the results according to the specific needs of the analyst.

CONCLUSION

We have taken a step toward improving the quality of static security analysis of mobile code and then bridging the usability gap separating between developers and automated security checkers based on static program analysis. The strongest detractor for developers is the excess of false warnings and the difficulty to understand the findings reported by the analysis tool (often because they are spurious and represent infeasible program behaviors). Thus, in aiming for completeness, static security checkers end up discouraging the user to the point that the tool is not used at all even at the cost of missing actual vulnerabilities that the tool is capable of finding.

In our approach, the report from Phoenix is cleansed by combining user interaction with statistical learning techniques. The user specifies a policy, or preference, that reflects a bias over true versus false warnings. The user also contributes classifications for a small fragment of the overall warnings. These trigger the creation and evaluation of multiple candidate filters, which are each evaluated according to the user-provided policy. The best filter is then applied to the remaining warnings, leaving the user to review only a (small) subset of the raw warnings output by the tool. Experiments that we have performed over Phoenix are highly encouraging. For example, given only 200 classified warnings, if the user expresses a preference toward preservation of true positives, Phoenix is able to improve precision by a factor of 2.868 while reducing recall by a negligible factor (×1.006). Conversely, if the user favors elimination of false alarms, still based on only 200 classified warnings, Phoenix achieves a recall-degradation factor of only 2.212 and a precision-improvement factor of 9.014. Other policies are enforced in an equality effective manner. In all cases, and across all policies, Phoenix is able to improve precision by a factor ranging between ×2.868 and ×16.556 in return to user classification of only 200 warnings.

REFERENCES

Aronszajn, N. (1950). Theory of Reproducing Kernels. *Transactions of the American Mathematical Society*, 68.

Ashcraft & Engler. (2002). Using Programmer-Written Compiler Extensions to Catch Security Holes. *S&P*.

Ayewah, N., Pugh, W., Morgenthaler, J. D., Penix, J., & Zhou, Y. (2007). Using Findbugs on Production Software. In OOPSLA Companion. doi:10.1145/1297846.1297897

Bacon, D. F., & Sweeney, P. F. (1996). *Fast Static Analysis of C++ Virtual Function Calls*. OOPSLA. doi:10.1145/236337.236371

Bishop, C. M. (2006). *Pattern Recognition and Machine Learning* (Vol. 1). Springer.

Chang, W., Streiff, B., & Lin, C. (2008). *Efficient and Extensible Security Enforcement Using Dynamic Data Flow Analysis*. CCS. doi:10.1145/1455770.1455778

Cheng, B., & Hwu, W. W. (2000). Modular Interprocedural Pointer Analysis Using Access Paths: Design, Implementation, and Evaluation. In *Proceedings of the ACM SIGPLAN 2000 Conference on Programming language design and implementation*. doi:10.1145/349299.349311

Cleary, J. G., & Trigg, L. E. (1995). *K^*: An Instance-based Learner Using an Entropic Distance Measure*. ICML.

Cousot, P., & Cousot, R. (1977). Abstract Interpretation: A Unified Lattice Model for Static Analysis of Programs by Construction or Approximation of Fixpoints. In POPL. doi:10.1145/512950.512973

Dean, J., Grove, D., & Chambers, C. (1995). Optimization of Object-oriented Programs Using Static Class Hierarchy Analysis. In *Proceedings of the 9th European Conference on Object-Oriented Programming, ECOOP '95*. doi:10.1007/3-540-49538-X_5

Denning, D. E. (1976). A Lattice Model of Secure Information Flow. *Communications of the ACM*, *19*(5), 236–243. doi:10.1145/360051.360056

Denning, D. E., & Denning, P. J. (1977). Certification of Programs for Secure Information Flow. *Communications of the ACM*, *20*(7), 504–513. doi:10.1145/359636.359712

Deutsch, A. (1992). *A Storeless Model of Aliasing and Its Abstractions Using Finite Representations of Right regular Equivalence Relations*. ICCL. doi:10.1109/ICCL.1992.185463

Fehnker, Huuck, Seefried, & Tapp. (2010). Fadetogrey: Tuning Static Program Analysis. *ENTCS*, 266.

Fuhrer, R., Tip, F., Kieżun, A., Dolby, J., & Keller, M. (2005). *Efficiently Refactoring Java Applications to Use Generic Libraries*. ECOOP. doi:10.1007/11531142_4

Goguen, J. A., & Meseguer, J. (1982). Security Policies and Security Models.S&P. doi:10.1109/SP.1982.10014

Guarnieri, S., Pistoia, M., Tripp, O., Dolby, J., Teilhet, S., & Berg, R. (2011). *Saving the World Wide Web from Vulnerable JavaScript*. ISSTA. doi:10.1145/2001420.2001442

Guha, A., Krishnamurthi, S., & Jim, T. (2009). *Using Static Analysis for Ajax Intrusion Detection*. WWW. doi:10.1145/1526709.1526785

Hammer, C., Krinke, J., & Snelting, G. (2006). Information Flow Control for Java Based on Path Conditions in Dependence Graphs.S&P.

Hastie, T., Tibshirani, R., & Friedman, J. (2009). *The Elements of Statistical Learning* (Vol. 2). Springer. doi:10.1007/978-0-387-84858-7

Heckerman, D., Geiger, D., & Chickering, D. M. (1995). Learning Bayesian Networks: The Combination of Knowledge and Statistical Data. *Machine Learning*, *20*(3), 197–243. doi:10.1007/BF00994016

Heintze, N., & Tardieu, O. (2001). *Demand-Driven Pointer Analysis*. PLDI. doi:10.1145/378795.378802

Johnson, B., Song, Y., Murphy-Hill, E., & Bowdidge, R. (2013). Why Don't Software Developers Use Static Analysis Tools to Find Bugs? ICSE.

Junker, M., Huuck, R., Fehnker, A., & Knapp, A. (2012). *SMT-based False Positive Elimination in Static Program Analysis*. ICFEM. doi:10.1007/978-3-642-34281-3_23

Lars Ole Andersen. (1994). *Program Analysis and Specialization for the C Programming Language*. (PhD thesis). University of Copenhagen, Copenhagen, Denmark.

Lhoták, O., & Hendren, L. J. (2006). Context-Sensitive Points-to Analysis: Is It Worth It? CC.

Livshits, V. B., & Lam, M. S. (2005). Finding Security Vulnerabilities in Java Applications with Static Analysis. USENIX Security.

McCamant, S., & Ernst, M. D. (2008). *Quantitative Information Flow as Network Flow Capacity*. PLDI. doi:10.1145/1375581.1375606

Minamide, Y. (2005). *Static Approximation of Dynamically Generated Web Pages*. WWW. doi:10.1145/1060745.1060809

Muske, T. B., Baid, A., & Sanas, T. (2013). *Review Efforts Reduction by Partitioning of Static Analysis Warnings*. SCAM. doi:10.1109/SCAM.2013.6648191

Muske, T. B., Datar, A., Khanzode, M., & Madhukar, K. (2013). *Efficient Elimination of False Positives Using Bounded Model Checking*. VALID.

Myers, A. C. (1999). *JFlow: Practical Mostly-static Information Flow Control*. POPL. doi:10.1145/292540.292561

Myers, A. C., & Liskov, B. (1997). *A Decentralized Model for Information Flow Control*. SOSP. doi:10.1145/268998.266669

Nevill-Manning, C. G., Holmes, G., & Witten, I. H. (1995). The Development of Holte's 1R Classifier. In *Proceedings of the Second New Zealand International Two-Stream Conference on Artificial Neural Networks and Expert Systems*. IEEE. doi:10.1109/ANNES.1995.499480

Newsome, J., & Song, D. (2005). *Dynamic Taint Analysis for Automatic Detection, Analysis, and Signature Generation of Exploits on Commodity Software*. NDSS.

Ng, A. Y., & Jordan, M. I. (2002). On Discriminative vs. Generative Classifiers: A Comparison of Logistic Regression and Naive Bayes. *Advances in Neural Information Processing Systems*, 2.

Pepe, M. S. (2003). *The Statistical Evaluation of Medical Tests for Classification and Prediction*. Oxford University Press.

Pistoia, M., Flynn, R. J., Koved, L., & Sreedhar, V. C. (2005). *Interprocedural Analysis for Privileged Code Placement and Tainted Variable Detection*. ECOOP. doi:10.1007/11531142_16

Pontil, M., & Verri, A. (1998). Properties of Support Vector Machines. *Neural Computation*, 10. PMID:9573414

Quinlan, J. R. (1993). *C4.5: Programs for Machine Learning* (Vol. 1). Morgan Kaufmann.

Reps, T., Horwitz, S., & Sagiv, M. (1995). *Precise Interprocedural Dataflow Analysis via Graph Reachability*. POPL. doi:10.1145/199448.199462

Sabelfeld, A., & Myers, A. C. (2003). Language-based Information-flow Security. *IEEE Journal on Selected Areas in Communications*, *21*(1), 5–19. doi:10.1109/JSAC.2002.806121

Saitoh, S. (1988). *Theory of Reproducing Kernels and Its Applications*. Longman.

Scholkopf, B., Smola, A. J., Williamson, R. C., & Bartlett, P. L. (2000). New Support Vector Algorithms. *Neural Computation*, 12.

Shankar, U., Talwar, K., Foster, J. S., & Wagner, D. (2001). Detecting Format String Vulnerabilities with Type Qualifiers. USENIX Security.

Shen, H., Fang, J., & Zhao, J. (2011). *EFindBugs: Effective Error Ranking for FindBugs*. ICST.

Snelting, Robschink, & Krinke. (2006). Efficent Path Conditions in Dependence Graphs for Software Safety Analysis. *TOSEM*, *15*(4).

Sridharan, M., Artzi, S., Pistoia, M., Guarnieri, S., Tripp, O., & Berg, R. (2011). *F4F: Taint Analysis of Framework-based Web Applications*. OOPSLA. doi:10.1145/2048066.2048145

Sridharan, M., & Bodík, R. (2006). Refinement-based Context-sensitive Points-to Analysis for Java. In *ACM SIGPLAN Conference on Programming Language Design and Implementation (PLDI 2006)*. doi:10.1145/1133981.1134027

Sridharan, M., Fink, S. J., & Bodík, R. (2007). *Thin Slicing*. PLDI. doi:10.1145/1250734.1250748

Tateishi, T., Pistoia, M., & Tripp, O. (2011). *Path- and Index-sensitive String Analysis Based on Monadic Second-order Logic*. ISSTA. doi:10.1145/2001420.2001441

Tripp, O., Ferrara, P., & Pistoia, M. (2014). *Hybrid Security Analysis of Web JavaScript Code via Dynamic Partial Evaluation*. ISSTA. doi:10.1145/2610384.2610385

Tripp, O., Pistoia, M., Cousot, P., Cousot, R., & Guarnieri, S. (2013). Andromeda: Accurate and Scalable Security Analysis of Web Applications. FASE.

Tripp, O., Pistoia, M., Fink, S. J., Sridharan, M., & Weisman, O. (2009). TAJ: Effective Taint Analysis of Web Applications. PLDI.

Volpano, Irvine, & Smith. (1996). A Sound Type System for Secure Flow Analysis. *JCS*, *4*(2-3).

Vosloo & Kourie. (2008). Server-centric Web Frameworks: An Overview. *ACM Comput. Surv.*, *40*(2), 4:1–4:33.

Wassermann, G., & Su, Z. (2007). *Sound and Precise Analysis of Web Applications for Injection Vulnerabilities*. PLDI. doi:10.1145/1250734.1250739

Wassermann, G., & Su, Z. (2008). *Static Detection of Cross-site Scripting Vulnerabilities*. ICSE. doi:10.1145/1368088.1368112

Whaley, J., & Lam, M. S. (2004). *Cloning Based Context-Sensitive Pointer Alias Analysis Using Binary Decision Diagrams*. PLDI. doi:10.1145/996841.996859

Witten, I. H., & Frank, E. (2005). *Data Mining: Practical Machine Learning Tools and Techniques*. Morgan Kaufmann.

Yan, D., Xu, G., & Rountev, A. (2011). Demand-driven context-sensitive alias analysis for java. In *Proceedings of the 2011 International Symposium on Software Testing and Analysis*. doi:10.1145/2001420.2001440

Zheng, X., & Rugina, R. (2008). Demand-driven alias analysis for c. In *Proceedings of the 35th annual ACM SIGPLAN-SIGACT Symposium on Principles of Programming Languages*. doi:10.1145/1328438.1328464

Chapter 5
Trust Profiling to Enable Adaptive Trust Negotiation in Mobile Devices

Eugene Sanzi
University of Connecticut, USA

Thomas P. Agresta
University of Connecticut Health Center, USA

Steven A. Demurjian
University of Connecticut, USA

Amanda Murphy
Canisius College, USA

ABSTRACT

In order to secure mobile devices, there has been movement to trust negotiation where two entities are able to establish a measure of mutual trust, even if no prior contact between either entity has existed in the past. This chapter explores adaptive trust negotiation in a mobile environment as a means to dynamically adjust security parameters based on the level of trust established during the negotiation process thereby enhancing mobile security. To accomplish this, the chapter proposes a trust profile that contains a proof of history of successful access to sensitive data to facilitate identification and authentication for adaptive trust negotiation. The trust profile consists of a set of X.509 identity and attribute certificates, where a certificate is added whenever a user via a mobile application makes a successful attempt to request data from a server where no relationship between the user and server has previously existed as a result of trust negotiation. Our approach allows the user to collect an ever-growing amount of profile data for future adaptive trust negotiation.

INTRODUCTION

As the shift towards mobile device and application usage over traditional PCs as a dominant computing platforms occurs (Gartner, 2015), criminals are increasingly focusing on mobile devices as a means to steal data from unsuspecting users (Montopoli, 2013). Despite the surge in mobile device attacks, several industries are increasingly relying on mobile devices (West, 2012). There has been an emphasis on securing banking and financial platforms (Herzberg, 2003) with users adapting payments via mobile devices, as evidenced by Apple Pay, Google Wallet, and Samsung Pay. The ubiquity of mobile devices

DOI: 10.4018/978-1-5225-0945-5.ch005

in our daily lives has been led by the fitness and healthcare industries both for individuals monitoring their fitness activities and medical conditions such as: family members, care givers, etc.; and primary physicians, psychiatrists, on-call physicians, nurses, therapists, specialists, pharmacists, etc., seeking to access patient-collected fitness/health data in their daily activities. The healthcare industry is increasingly relying on mobile devices for quick and easy access to patient records via mHealth (Himiss, 2014) apps during treatment (Ventola, 2014) with an estimate that 80% of doctors rely on mobile devices in a report (Lewis, 2011) to access an electronic medical record (EMR) (Conn, 2014). In fact, a recent report (Aitken, n.d.) highlights 43,700+ medical apps in the Apple app store, with 69% apps targeting consumers/patients and 31% for use by medical providers. Apple has a separate category for Medical apps (iTunes, n.d.) and there has been a study comparing medical apps for both iOS and Android platforms (Seabrook, et al., 2014). Healthcare/medical apps for consumers and medical providers require a high degree of security due to the presence of protected health information (PHI) and personally identifiable information (PII).

To secure mobile devices, there has been increasing focus on *trust negotiation* (van der Horst T. W., Sundelin, Seamons, & Knutson, 2004), a procedure whereby two entities are able to establish a measure of mutual trust, even if no prior contact between either entity has existed in the past. *Adaptive trust negotiation* refers to the ability to dynamically adjust security parameters based on the level of trust established during the negotiation process. When a user via a mobile device attempts to access a server, a series of agreed upon credentials (e.g. attribute certificates) are exchanged to establish trust. The server vets the certificate, then determines if the user is trustworthy and the level of access to be allowed. Work by (Ryutov, Zhou, Neuman, Leithead, & Seamons, 2005) presents a framework for the adaptive trust negotiation process using a combination of TrustBuilder and the GAA-API (n.d.) for users to establish trust with online businesses based on the number and value of past purchases, to allow the user to make larger purchases of increasing value.

The usage of trust negotiation in healthcare information technology (HIT) systems was introduced by (Vawdrey, Sundelin, Seamons, & Knutson, 2003) and augmented by including additional assurance when accessing the EMR of a hospital (Elkhodr, Shahrestani, & Cheung, 2011) or employing trust negotiation to confirm the requestor's status as a licensed physician (Vawdrey, Sundelin, Seamons, & Knutson, 2003). One objective of this chapter is to explore the feasibility and utility of adaptive trust negotiation and its suitability for the healthcare domain, particularly for mHealth apps. Specifically, we expand existing capabilities in adaptive trust negotiation's ability to authorize users by increasing the granularity of security measures that can be utilized in an HIT system. For example, the remote server will be able to access portions of the medical provider's health record access history (i.e., a trust profile) to EMRs or other HIT systems that are exposed by the provider in the presented credentials. If the remote server grants access, the medical provider receives new identity and attribute certificates to augment the existing credentials that can be utilized as proof/history of successful access to PHI and PII for a future trust negotiation.

The adaptive trust negotiation process incorporating the trust profile in this chapter requires the user to present his/her authorizations (vetted set of credentials) to sensitive data from different systems that he/she has been successfully accessing over time. This history of user access is passed as a credential during the trust negotiation process, allowing past secure access to inform future access. A *Trust Profile* is created and modified over time to assemble a history of the successful access to serve as proof of past access to sensitive data. In support of the Trust Profile, the user has a *digital wallet* containing proof and history via new identity and attribute certificates detailing access by the user to sensitive data. A *Trust*

Profile is a subset of the user's digital wallet that can change based on a user's location and the type of data the user is attempting to access. The Trust Profile is independent of any particular mobile device and travels with the user and changes in response to the user's attempts to access sensitive data on multiple systems at different locations over time. From a practical perspective, the Trust Profile consists of a set of identity and attribute certificates, where a certificate is added to the trust profile whenever a user via a mobile application makes a successful attempt to request data from a server where no relationship between the user and server has previously existed independent of the domain. These certificates adhere to the X.509 (n.d.) standard for identity and attribute certificates. Consider a physician utilizing a mHealth app for accessing patient data: from an EMR at the physician's primary practice, from an EMR that the physician utilizes when he spends one day at a local city clinic, from an EMR that is in a hospital where a physician sees his patients, etc. Since all of these various accesses to patient data are on different EMRs, the physician's Trust Profile is constantly updated to record a history of the successful PHI accesses.

The Trust Profile is stored in a form that is presentable on behalf of the user to other, unrelated systems with similar sensitive data that the user is interested in gaining access (to which he/she has not been previously explicitly authorized to). The Trust Profile is compatible with mobile devices and allows a user to make requests to new, previously unknown systems. Additionally, the Trust Profile submitted by a user (the requestor) must be supported by an adaptive trust infrastructure via a set of interacting components including: a component to verify the structure and content of a Trust Profile; a component to determine the authenticity of the Trust Profile with respect to the user/credentials; a component to match the Trust Profile against a defined security policy of the receiving system; a component to deliver the sensitive data from the source to the requestor; and, a component to generate/add a record of the transaction to the Trust Profile. A presented Trust Profile contains credentials and the degree of access (create, read, update, and delete) which will be allowed to the requestor. To demonstrate the feasibility of our work, we utilize the Connecticut Concussion Tracker (CT^2) mHealth app, a joint effort between the Departments of Physiology and Neurobiology, and Computer Science & Engineering at the University of Connecticut, in collaboration with faculty in the Schools of Nursing and Medicine. CT^2 was developed in support of a newly passed law on concussions to be tracked for kindergarten through high school in Connecticut (State of Connecticut, n.d.).

This chapter contains 5 sections. The *Background* section discusses the healthcare domain and adaptive trust negotiation in conventional and mobile computing. The *Trust Profiling for Adaptive Trust Negotiation* section defines and explains our approach to Trust Profiles that extends adaptive trust negotiation for supporting mobile devices/applications. Next, the *Design and Prototyping of Trust Profiles* section implements the capabilities of Trust Profiles through an extension to the mHealth CT^2 app in support of an adaptive trust negotiation process that has been added to the CT^2 server. Then, the *Future Trends* section explores the areas of single sign-on (SSO) (Yu, Wang, & Mu, 2012), biometrics (Biometrics, n.d.), spatio-temporal access control (Ray & Toahchoodee, 2007), and their impact on securing mobile authentication procedures. Finally, the *Conclusion* section highlights the chapter contributions.

BACKGROUND

Background for the chapter is divided into five areas via examples in healthcare: *role-based access control (RBAC)* to identify the user by role and the nature of accessible sensitive data; *identity certificates* as a set of credentials he/she has accumulated from accessing sensitive data; *attribute certificates* that encode user credentials in a verifiable and claimable format; *trust agents* that offload computationally intensive operations from the mobile device to an external server; and, related trust models.

RBAC provides a set of permissions, roles, and users (Ferraiolo, Sandhu, Gavrila, Kuhn, & Chandramou, 2001). Permissions are a set of actions that one may take in regards to objects (data) with operations to create, read, update, and/or delete data. In healthcare, these permissions involve reading a patient's medical data, inserting new records into a patient's history, or reading a patient's insurance information with roles for nurses, physicians, billing staff, or secretaries. Each role contains only the permissions necessary to perform the associated job, e.g., an employee attempting to access billing data under the doctor role would not have the proper permission and would be denied access. Each user of the system is assigned one or many roles, but is limited to one role for any session. One major extension to RBAC provides functionality for role delegation (Na & Cheon, 2000), where the owner of a role may receive the ability to permit another user to act in their stead with respect to a subset of their permissions. RBAC has been a popular choice for access control within HIT systems (Fernández-Alemán, Señor, Lozoya, & Toval, 2013).

Identity certificates (Housley, Polk, Ford, & Solo, 2002) uniquely identify the certificate owner through cryptographic means in a public key infrastructure (PKI) using the X.509 standard. In PKI, a certificate authority (CA) disseminates an identity certificate to a user after he/she first proves their identity through traditional means (e.g., driver's license, birth certificate, passport, email from administrator of owned domain, etc.). The CA provides a cryptographic signature on the certificate that indicates that they endorse the user's claim to that identity and that the contents of the certificate have not been altered since the signature was created. If the certificate is verified, the system performing verification accepts the identity certificate if the system trusts the CA that signed it. The user's ownership of the certificate is proved via public/private key cryptography. The user's public key is listed within the certificate while the private key is kept secret by the user. To provide proof of ownership, the user can decrypt messages encrypted with his/her public key and provide responses encrypted with the private key that the associated public key is able to decrypt. In this chapter, the identity certificate can uniquely identify unknown entities and confirm that the unknown holder is in fact a member of the medical community. This virtual identity is utilized as an anchor point for a verifiable medical record access record history as credentials in the trust negotiation process.

Attribute certificates (Farrell & Housley, 2002) store data in a key-value pair format and are associated with an identity certificate through its serial number, which is unique inside the signing organization (the issuer), and the issuer. The attribute certificate is signed by an attribute authority (AA) in a manner similar to an identity certificate. An identity certificate may have one or more attribute certificates associated with it, but each attribute certificate is associated with one identity certificate. Similar to the identity certificate, the information within an attribute certificate contains a digital signature computed at the time of creation by the AA. During certificate verification, if the signature on the attribute certificate is found to be valid, the information within is trusted if the AA is trusted. The separation between the identity certificate and attribute certificate facilitates the addition of information that augments the identity of the holder without requiring reverification of the holder's identity. A more specialized, short-lived version of the attribute certificate referred to as the rule certificate may be generated that records the user's actual permissions on the HIT system for the current session (Mavridis, Georgiadis, Pangalos, & Khair, 2001). In healthcare, an attribute certificate might contain: role (e.g., primary physician, nurse, pharmacist, etc.), permissions (e.g., whether the holder is allowed to delegate responsibilities), or authorization (e.g., when the holder is allowed to access sensitive data).

Trust agents (van der Horst T. W., Sundelin, Seamons, & Knutson, 2005) are software components that are able to perform the trust negotiation process for others. A *local agent* runs on the device initiating or receiving a request to begin the trust negotiation process. A *remote agent* performs the same task but runs on a different device, performing the trust negotiation process on behalf of the device that desires trust. The trust negotiation process requires a substantial amount of computational power for the involved cryptographic processes. While a traditional PC's only bottleneck is the PC's ability to complete many cryptographic calculations quickly, mobile devices must be also be able to compute the calculations while maximizing battery life using less powerful CPUs. To address this issue, mobile devices can leverage surrogate trust negotiation (Sundelin, July 2003) where a trusted base station performs trust negotiation. In healthcare, a trust agent could operate as a software module running on a trust negotiation server owned by a healthcare organization that the mobile device contacts to perform the trust negotiation phase. *Generic software agents* may be used by medical servers receiving trust negotiation requests to generate new certificate-based credentials, offloading the responsibility of guarding the private key and signing certificates to a trusted third party.

Trust in this chapter represents the ability for two entities to: believe the authenticity of one another's credentials, utilize those credentials to ascertain whether each individual is entitled to privileged information, and believe that each will handle sensitive data appropriately once exchanged. Many different trust models have been proposed. (Artz & Gil, 2007) surveys a wide range of trust and trust distribution techniques, including trust in the accuracy of the information released (e.g. search engine results). The work also explores policy-based trust (users possess credentials that must be matched to security policies) and reputation-based trust (user behavior is inferred from past actions). (Sabater & Sierra, 2005) reviews different models of trust and classifies by: the conceptual model (cognitive vs. game theory); the information source (direct, witness, sociological, prejudice); the visibility type (global vs. subjective); the granularity (single context vs. multiple context); agent behavior (cheating not considered, agents can hide or bias information, agents can lie); the type of exchanged information; and the trust value. This work notes that the diversity of trust models creating trust in different domains makes it difficult to classify each model according to this criteria. As an example, our approach utilizes information sources from direct experience information, which is the access history the user presents as proof of past data access, and prejudice information, which is the role the user chooses to initiate the request. Direct experience is defined as trust values that are provided directly from the entity the user is initiating the request to, or other members of the community (other healthcare organizations). Prejudicial information is inferred based on the user's "group". For example, a hospital employee with an X-ray technician role cannot access the patient's billing data, but it can be inferred that the release of past X-ray data may be warranted if requested.

TRUST PROFILING FOR ADAPTIVE TRUST NEGOTIATION

In this section, we describe the trust profile, its usage, and the architecture required to enable adaptive trust negotiation in systems, demonstrated via the healthcare domain. In this context, trust is the ability of the two entities to believe one another, and that each will take proper responsibility in the handling of sensitive data. Companies that improperly disclose medical data stand to lose money and customers' trust. However, the proper dissemination of medical data is paramount in patient care/treatment and medical research. A *trust profile* is the entity that is constructed to support the adaptive trust negotiation process by providing a set of access history-based credentials that a data requestor and data holder exchange to establish trust. In the approach detailed in this chapter, a healthcare stakeholder builds a set of credentials into a trust profile over the course of his/her medical career, allowing him/her to build trust with the various HIT systems containing the data he/she needs.

The remainder of this section introduces and explains trust profiling for adaptive trust negotiation in four parts. In part one, we overview the trust profile and the negotiation process. Next, in part two, we examine the physical structure of the trust profile and the supporting network architecture. Part three discusses the trust profile processing which is decomposed into three components (validation, security policy, and data collection and delivery) that reads the trust profile and decides whether data will be disseminated. Lastly, part four has a comprehensive healthcare example of trust profile utilization leveraging the concepts presented in the first three parts.

Trust Profile and Negotiation Process

In part one of this section of the chapter, we present the trust profile and negotiation process that requires a set of credentials that are passed between the two entities attempting to establish trust. In previous works (Elkhodr, Shahrestani, & Cheung, 2011) (Vawdrey, Sundelin, Seamons, & Knutson, 2003), the credentials are based on what the user is (e.g., physician, billing agent) whereas this work allows for more fine grained control that tunes user access by adding credentials that detail actions that the user has been allowed to take in the past (e.g., access patient A's complete medical history, access a hospital's available public health data). This allows implementations of security policies that have more options with PHI disclosure. Based on the user's access history, the system may decide to: deny access if the user does not meet basic requirements for access (such as a physician attempting to access protected mental health data); allow access but trigger an extra layer of auditing (such as an alert being issued to an auditor when an unknown E.R. doctor requests medical data regarding a patient he/she has never treated); or allow access to the data (such as in the event a doctor who is an employee of the institution is treating a patient he/she has already treated, but is currently working in a remote location).

The attribute certificate, as introduced in the background section, is a container for records of user access while the identity certificate is a unique virtual identity that the user can claim ownership of. A user presents identity and attribute certificates that encode the trust profile, readable by the trust negotiation server of the HIT system (e.g., EMR), along with a request for the exact data needed. The HIT system's trust negotiation server determines certificate authenticity and ownership using PKI, then extracts the medical record access history of the user from the attribute certificate. The HIT system's trust negotiation server decides the level of access the user is allowed and generates new certificates that detail which records the user is allowed to access. The security policy reacts to the request dynamically and adjusts which credentials are required. For instance, a family physician requesting updates on his/her patient's medical record from other medical stakeholders (e.g., OB/GYN, podiatrist, dentist, cardiologist, etc.) that the patient has recently seen for treatment would be expected to present a medical record access history indicating that he/she has successfully authenticated and been granted access to the patient's medical history in the past.

Each stakeholder (e.g., primary physicians, psychiatrists, on-call physicians, nurses, therapists, specialists, pharmacists, etc.) in the healthcare domain that is expected to require access to secure HIT systems is granted the ability to build and maintain a trust profile. An initial trust profile is granted by the healthcare institution that employs the stakeholder, thus endorsing the professional's status as a trusted member of the medical community. Should the stakeholder leave the healthcare institution, his/her trust profile remains valid and moves with him/her as a permanent record of the access afforded to him/her by the institution. When joining a new healthcare institution with its own EMR, the stakeholder is granted additions to the trust profile indicating that this new institution also endorses his/her trustworthiness and begins recording requests for access into the trust profile. Additionally, when requesting medical data from HIT systems of healthcare institutions where there is no preexisting relationship between the owner of the system and the person requesting access, in the event that the trust negotiation phase is successful, the HIT system adds its own entries to the user's trust profile. This behavior allows practicing stakeholders to gradually build a permanent trust profile over the course of his/her medical career with trust endorsements from many different institutions that demonstrates a history of successful access of sensitive data in varied HIT systems. In the event that the physician requires access to medical data from an unknown healthcare institution, the physician can use this trust profile to obtain assurance that he/she is trusted by trustworthy entities.

Trust Profile Structure

In part two of this section of the chapter, we describe the trust profile's structure. The structure of the trust profile is a series of identity and attribute certificates which together form the physician's digital wallet. When a user is successful in trust negotiation with an unknown HIT system, the system requests a new public key from the user and generates an identity certificate that is utilized in future communication with the server. The server also generates and signs an attribute certificate containing records of access that is attached to the identity certificate. Thus, the user has verifiable proof of access to these records that can be utilized as credentials in attempts to access healthcare data residing at other healthcare organizations. In the case of mobile devices, the certificates may be stored locally on the device, or stored with a remote agent that can perform the trust negotiation procedure and generate the necessary public-private key pairs on behalf of the mobile device. A *medical authority*, similar to the certificate authority described in the *Background* section, is responsible for verifying that the HIT systems are certified, maintained by licensed healthcare providers, and proper security procedures are followed on the certificate processing and signing servers. Medical authorities establish trust between the healthcare organizations' HIT systems that endorse the trust profiles of those who have been allowed access to patient records as shown in Figure 2. Mutual trust must be established between the HIT systems through medical authorities to enable the trust negotiation process; in order for the HIT systems to trust the authenticity of the user's trust profile, the HIT system that signed it must be trusted.

While a later section of the chapter provides a detailed real-world example, for the reader to be able to understand the concepts in the rest of this section, a brief example is provided. To begin, a sample Physician Trust Profile is shown in Figure 1. The Physician has multiple roles (Physician that sees patients at the Family Health Center, Researcher and Professor at the UConn Health Center (UCHC) that includes the medical school, and Radiologist at St. Francis hospital that assesses imaging tests) that generate appropriate attribute certificates that are associated with X.509 certificates issued by the aforementioned health organizations. Note that each X.509 certificate in Figure 1 has one or more attribute certificates that represent the role of the user within the organization (e.g., UCHC has two attribute certificates for the roles Research and Professor). The Physician presents his trust profile containing the multiple certificates to a new health organization (Hartford Hospital) that he needs to have access to for treating one of his patients at St. Francis Hospital. Now suppose that a physician attempts to access a patient's health

Figure 1. A sample physician trust profile

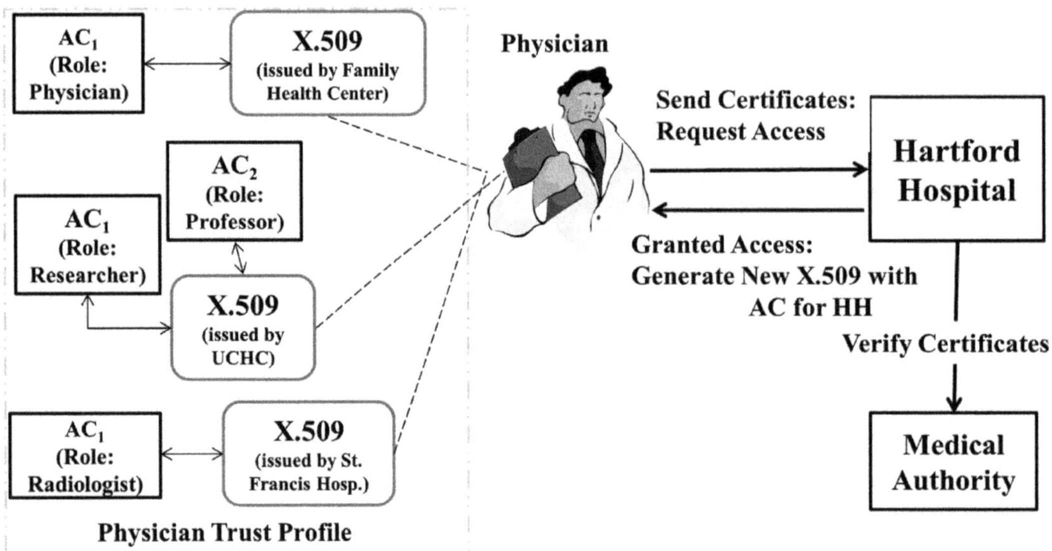

Figure 2. Trust profile process

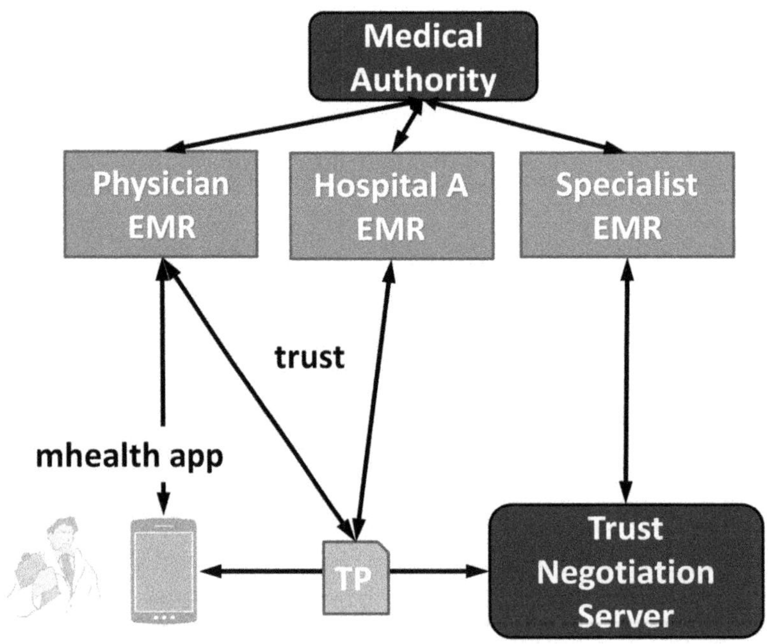

data in the EMR at his/her local practice using a mobile health (mHealth) application on his/her mobile device and discovers that the patient has recently visited an unknown specialist for a related condition; this is shown in Figure 2. In such a situation, the mHealth application used by the physician will present his/her trust profile to the specialist's EMR through a remote agent running on a trust negotiation server maintained by his/her hospital. The physician's trust profile details: previously successful attempts of the physician accessing patient data at his/her local EMR; and, patient data the physician has previously accessed from other specialists. The trust negotiation server for the specialist's EMR will read the presented trust profile, determine its legitimacy, decide the level of access to be authorized to the physician, and determine which additional actions the system must execute to ensure data integrity and security. If the attempt at access is successful, the specialist's server automatically returns to the physician a new set of digital credentials that the physician can add to his/her Trust Profile.

Trust Profile Process

In part three of this section of the chapter, the *trust profile process* is presented by explaining its three components as shown in Figure 3: *validation* to ensure that the credentials within the user's certificates are correct and have not been modified; *security policy* to enforce a set of requirements on the presented credentials; and, *data collection and delivery* to retrieve the PHI data and transfer it securely to the user. When a physician is attempting to access information from multiple HIT systems (EMRs), in order to identify the correct location(s), a medical record discovery service such as a master patient index (MPI) is utilized. MPI is a uniform index that is able to cross reference a patient's medical data that is stored in multiple HIT systems (EMRs) in support of health information exchange (HIE) (HIS, n.d.). The physician sends a request for healthcare data and an appropriate subset of his/her trust profile. Recall from Figure 2, the physician is attempting to obtain a copy of his/her patient's medical records at a specialist's office EMR, where his/her trust profile provides proof that he/she has treated this patient before by showcasing successful authorizations to the patient's data in the Family Medical Center and St. Francis Hospital EMRs. The HIT system (specialist's office EMR) receives the trust profile and passes it to the healthcare organization's trust negotiation server, which completes trust profile processing and adaptive trust negotiation on its behalf.

Figure 3. Trust components and their interdependencies

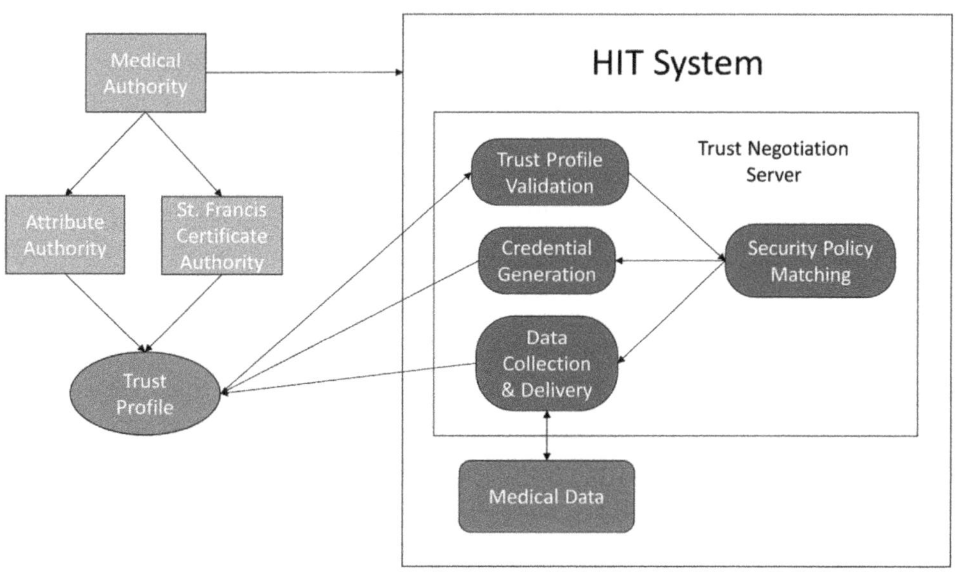

The trust negotiation server's *validation component* as shown in the upper middle of Figure 3 is responsible for performing analysis on the presented trust profile and determining its authenticity. The *validation component* begins by checking the user's identity certificates for validity; verifying that the certificate has not been altered and checking that the signer is trusted. A challenge is sent to the physician's trust agent utilizing public key cryptography to prove that the physician is the rightful owner of the record. A successful response indicates that the physician is the rightful owner of the identity certificate and thus the trust profile specified in the associated attribute certificates. The associated attribute certificates are checked to ensure that they have not been altered, and that the associated attribute authority (AA) is trusted. The trust negotiation component now knows that: the trust profile is valid, the information contained within is trustworthy, and the user responsible for initiating the connection is the rightful claimant to the presented trust profile. The healthcare data request (to an EMR) and the information within the attribute certificate are extracted and sent to another subcomponent of the trust negotiation component that matches the trust profile to a security policy.

The *security policy component* as shown on the right of Figure 3 receives the user's request and the extracted attributes from the validation component. The security policy contains all rules that govern the security policy component's responses to the requestor. The security policy component matches the trust profile against the policy and decides the requestor's level of access to the data and which other necessary actions the HIT system will undertake to ensure data security. The enacted policy differs depending on the nature of the request and other attributes present in the trust profile. For example, a physician working in the E.R. requesting data to treat a patient that arrived from an automobile accident would cause the HIT system to enact a policy that requires indications that the physician has accessed data under the role of an E.R. physician, but the requirement that the physician has treated the patient before is relaxed (since it is likely the patient has not been treated by that E.R. physician previously). Since there is no indication in the trust profile that the physician has treated this patient previously, the security policy component would dispatch an audit notification to an auditor for later verification. Once the security policy component has completed this process, the request is passed to the data collection and delivery component in the bottom of Figure 3. However, if the user's credentials do not match the security policy, the trust negotiation server sends a message to the user stating that the request is denied.

The *data collection and delivery component* shown in the bottom middle of Figure 3 is only enabled in the case when a user is successful in the trust negotiation process. The data collection and delivery component is responsible for: creating a secure record of the transaction for the user to add to his/her digital wallet, collecting the requested medical data, performing any additional actions required by the security policy, and securely delivering the requested data to the user as shown in Figure 3. If the user does not possess an identity certificate from this institution, the component requests a public key from the user. The user generates a public-private key pair, using a remote agent, from his/her mobile device and sends the public key to the server. The trust negotiation server creates a certificate signing request and forwards it to the institution's certificate authority (CA), as shown in Figure 2. The data collection and delivery component utilizes the institution's AA to create an attribute certificate that encodes records of access for the data that is to be sent the user, e.g., this would be creating the attribute certificate for Hartford Hospital from Figure 1 or for the request to the Specialist EMR in Figure 2. This process is represented by the credential generation in the middle of Figure 3. The component collects the requested healthcare data from the institution's HIT system, e.g., a specific EMR at an institution. This data may reside in a data warehouse, the institution's EMR, or a separate staging server for shareable medical data. The generated certificates and data are transferred to the user. When the transfer has completed the connection may be terminated. The user then adds the certificates to the digital wallet and is able to read the medical data.

At any point during this process, the data collection and delivery component may be required to perform some ancillary action as required by the security policy. As healthcare data is protected by laws such as HIPAA and hospitals have significant financial investment in the generation of medical data (through MRIs, X-rays, or other analysis), the ability to fine-tune how an HIT system responds to a request for healthcare data is required. The security policy may require logging the transaction in a low risk audit log, in a high risk audit log, or dispatching a notification that a high risk transaction has occurred. As audit logs tend to be large and difficult to review (AHIMA, 2011), this ability working in tandem with a multi-level auditing system greatly assists in discovering and performing actions regarding mishandled data.

Healthcare Example

In this section, we present a comprehensive healthcare example in trust profiling and negotiation, shown in Figure 4. Suppose that Jane with a physician role is working at Family Medicine Center (FMC) and St. Francis Hospital (SFH). Jane has received one identity certificate from FMC where she works in ambulatory care, and one identity certificate from SFH where she works as a practicing physician. Jane also possesses attribute certificates for the physician role under each identity certificate. Note that these different certificates are shown in the upper portion of Figure 5 in Jane's current trust profile (smaller dashed box). Since Jane is known as a physician at both FMC and SFH, her access to the EMR of each is unrestricted, with improper access being determined through audits of the EMR's data access logs. Over the course of Jane's career at these healthcare organizations, she accesses PHI from the EMR of each organization utilizing a mHealth application provided for her appointments with patients, and each access is recorded in her trust profile, encoded in attribute certificates attached to the appropriate identity certificate, which travels with Jane to each organization.

Figure 4. The trust profile pre-negotiation process for Jane with Hartford Hospital

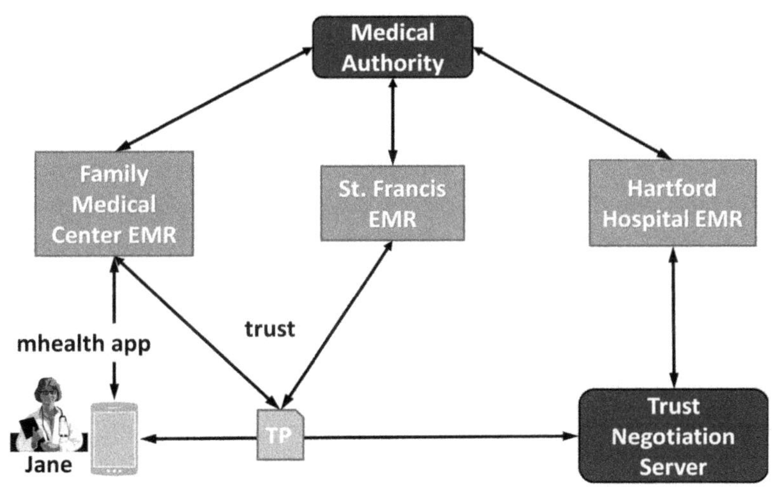

Figure 5. The trust profile post-negotiation process for Jane with Hartford Hospital

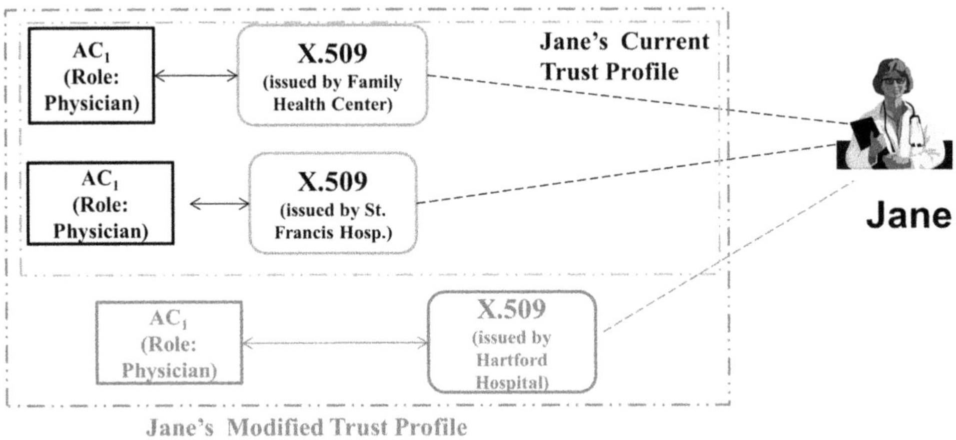

Jane is treating a patient that she has seen previously at FMC. The patient has recently received cardiology treatment from The Henry Low Heart Center at Hartford Hospital (HH). HH has never had previous contact with Jane, and Jane has never had previous contact with HH. Jane locates the patient's new PHI through a MPI and determines that she needs access to her patient's records in the EMR at HH to treat the patient. Through her mHealth application, a request for trust negotiation is initialized to HH's HIT system for data regarding the patient's cardiology treatment including medical notes by the treating physician and tests such as electrocardiograms or cardiac ultrasounds. The identity of the EMR is verified as belonging to HH through a TLS/SSL verification of its X.509 certificate and the medical authority's certificate. Jane selects portions of her trust profile indicating that she: has successfully accessed the patient's health data from FMC's EMR, is affiliated with SFH, and has a long history of accessing data in St. Francis' EMR utilizing a mHealth application,. The mHealth application selects the certificates within Jane's digital wallet that contain records of the selected history in the trust profile and sends them to HH's EMR.

HH's EMR receives the certificates and forwards them and the request to its trust negotiation server. At this point, the process completes the actions of the validation component (prior section and Figure 3) which requests proof of ownership to Jane, which is forwarded to Jane's trust agent. The validation component successfully reads the messages and completes the validation process by extracting the trust profile from the attribute certificates and passes the trust profile to the security component. In the next step, the security component reads the request and determines that to release the requested data, the trust profile must contain at least one record of the owner in a physician, specialist, or nursing role; Jane has physician role usage for FMC and SFC EMRs. Additionally, if the trust profile indicates that the owner has accessed the patient's medical data elsewhere, the transaction will be marked as low risk and recorded in a low risk audit log; Jane has accessed the patient's data in the FMC EMR. Conversely, if there is no indication of patient treatment by Jane, the transaction will be marked as high risk, recorded in a high risk audit log, and an e-mail will be sent to HH's auditor. A subsequent check will be initiated to see if there is any other evidence that would warrant granting Jane access, e.g., the trust profile indicates that the Jane has accessed health records under her physician role, including health records for many other patients in two EMRs. In this case, the security component could decide that this is sufficient evidence to allow Jane to have access to the HH EMR for the patient whose data is being requested.

In the final step, the data collection and delivery component creates a record of access granted to the HH EMR during the trust negotiation process. Since this is the first request for access by Jane to the HH's EMR, the data collection and delivery component requests a public key from Jane. Jane's mobile device sends a request to the trust agent to generate a new public/private key pair and sends the public key to the data collection and delivery component, which creates a certificate signing request for a new X.509 identity certificate for HH for Jane and sends it to the local CA for signing. The new certificate is shown in the bottom portion of Figure 5 with the larger dashed box that includes the current and new certificates. The private key is added to Jane's digital wallet. The records of access for the data requested by Jane are encoded in attribute certificates signed by the AA, attached to the newly generated identity certificate, and are sent to Jane at which point she adds the certificates describing this new entry to her trust profile and digital wallet. In parallel, the data collection and delivery component contacts the HIT system (HH EMR), gathers the requested data, and sends the requested data and generated certificates back to Jane. Jane now possesses the requested data and the certificates detailing the access she has gained to HH's EMR. The identity and attribute certificates are added to Jane's trust profile, which now contains proof of access to EMRs owned by FMC, SFH, and the newly approved HH. In future requests for data through trust negotiation, Jane is able to present this new certificate as a credential.

DESIGN AND PROTOTYPING OF TRUST PROFILES

The trust profile functionality as presented in this chapter has been integrated into the Connecticut Concussion Tracker (CT^2) mHealth Android app as a proof of concept prototype. CT^2 tracks concussions for grades kindergarten through high school and is a collaboration between the Departments of Physiology and Neurobiology, and Computer Science & Engineering at the University of Connecticut and Schools of Nursing and Medicine. The CT^2 mHealth app shown in Figure 6 contains the login screen (first screenshot) and additional screens: find students (Home tab), all students are assigned to a user (List tab), add a new concussion incident (Student tab), enter information on the concussion (Cause tab), enter student symptoms within 48 hour (Symptom tab), record the status of the student over time (Follow-up tab), and indicate when student can return to various activities at school (Return tab).

The data collected by the CT2 mHealth app is stored in a remote server running a custom MySQL database that contains tables for: student records, records of a student's concussion incident, the school, symptoms of the concussion incident, follow-up information, and records for when the user is allowed to participate in activities. The server provides outside access to the database through a REST API written in PHP using Slim. To demonstrate the trust process, the installation has been augmented with a trust negotiation agent that accepts trust negotiation requests and performs the required certificate validation checks. Once the user's credentials sent by the CT2 mHealth app are accepted, the trust negotiation agent passes new test certificates back to the user, which the user has the option of adding to his/her certificate store, expanding his/her trust profile. The initial screen of CT2 mHealth (Figure 6) has been upgraded to support the adaptive trust process in the 1st screen in Figure 7.

Figure 6. Select screens Connecticut concussion tracker (CT^2) app

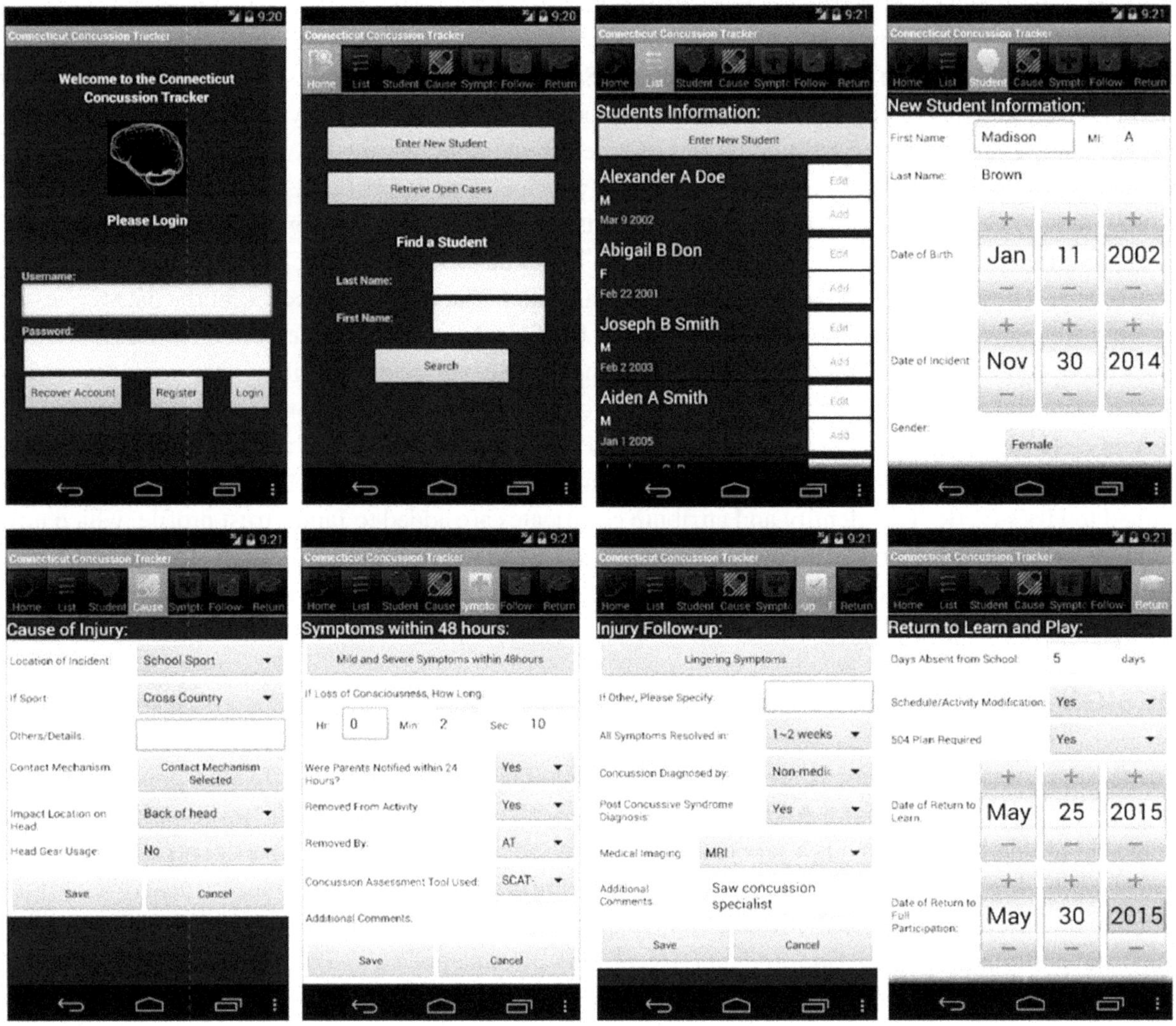

The prototype trust profile has been developed as a series of attribute certificates attached to an identity certificate. The trust profile contains verifiable information (provided by the certificate signers) related to the specific concussion records the user has accessed in the past and the circumstances regarding access. The access records contain information on:

- The user's role (for CT^2 roles include Nurse, Athletic Trainer, Coach, and Parent).
- The user's action in regards to the data (for CT^2 read a concussion record, create a concussion record, edit a concussion record, etc.).
- The id of the specific record that was requested by/sent to the user (for CT^2, the concussion id of the record).

Figure 7. Trust negotiation screens in the modified CT^2 app

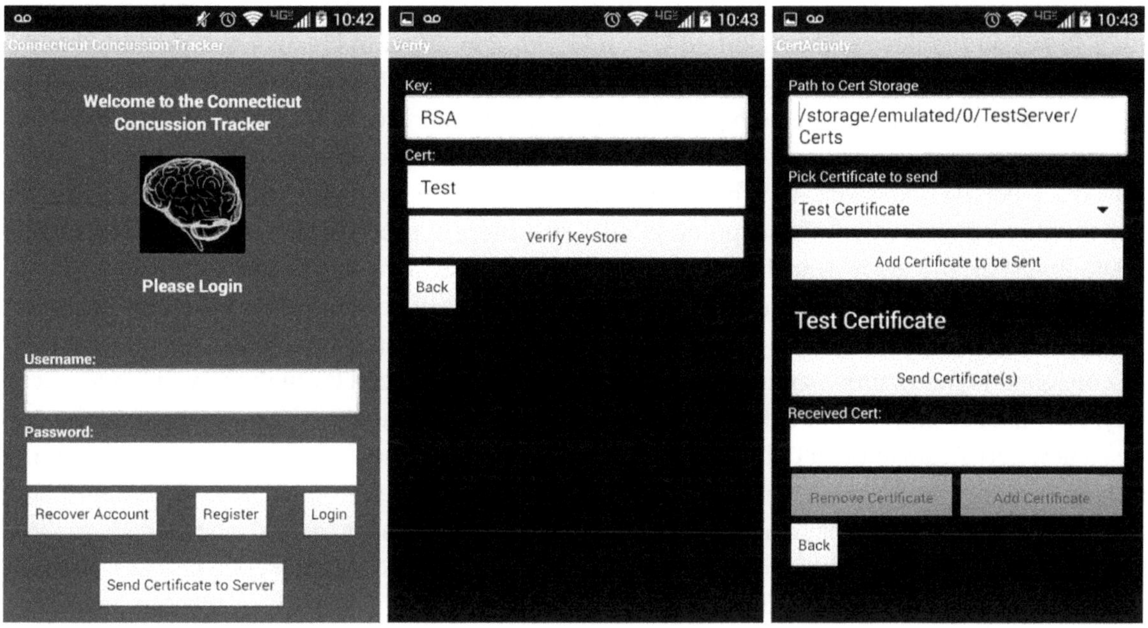

- The specific individual that that record is referring and the user's reason for access (for CT^2, the student id associated with the concussion id along with an action such as add a new symptom that showed up 48 hours or later).
- A timestamp detailing the time of access.

When the trust negotiation portion of the app is first run, the app generates necessary folder structures to hold the user's data (for CT2, a user with a role of Nurse, Coach, etc.). Within the prototype, the user's certificates reside on the mobile device in a KeyStore folder that the app reads during the credential selection phase. The trust negotiation agent is verified using standard SSL over an HTTPS connection. A successful connection indicates that the user has connected to the proper trust negotiation server. Certificates that a mobile device utilizes for verification of the trust negotiation agent are placed in a Trusted folder on the device rather than as a regular certificate on the device through the settings menu. When the trust negotiation component of the CT^2 mHealth app is first executed, the user has the option of creating a test public/private key pair and initial identity certificate for the trust negotiation process. These certificates are stored within the app folder in the KeyStore folder. Any new certificates created during the trust negotiation process and sent back to the user are processed by the mobile device and placed in a Certificates folder. The concussion data is received by the mobile app, processed, and displayed on the screens presented in Figure 6.

The CT^2 mHealth home screen shown in Figure 6 (1st screen, 1st row) provides options for a username/password combination, an account recovery option, and a login button. A user has been assigned a specific role (e.g., Nurse, Coach, etc.) that adjusts which screens in Figure 6 are available and whether or not a screen can be read or read/edited. In support of adaptive trust negotiation, the modified CT^2 mHealth app login screen in Figure 7 (1st screen) has a Send Certificate to Server button which sends the user with a given role to the first trust negotiation screen that verifies the user's trust store. The validation process (see Figure 3 again) begins by testing for the existence of a public/private key pair owned by the user and the presence of a properly formatted public key certificate. If these two elements are not found on the device, the modified CT^2 mHealth app offers to create a test public/private key pair and a test certificate for the user, shown in screen 2 of Figure 7. If the public/private key pair and associated identity certificate are both present, the verification process checks for proper formatting of the certificate and the encryption keys to ensure that they are the proper format.

Once the personal keystore is verified, the CT^2 mHealth app continues on to screen 3 in Figure 7 where the user is able to select his/her preinstalled certificates (if no certificates were installed, then only the default test certificate is available). The user selects one or more certificates to be sent to the trust negotiation server from a dropdown box labeled "Pick Certificate to send". Once the user has selected the certificates, he/she presses the Send Certificate(s) button and the certificates are uploaded. This send command is transmitted to our trust negotiation server to proceed through the validation, security policy, and, data collection and delivery components shown in Figure 3. Upon completion, the server issues a new certificate to the mobile device. The app receives the new certificate and displays it in the Received Cert: box in Figure 7. The user can then decide to remove the certificate, or add it to his/her digital wallet for future trust negotiation attempts. Note that the changes that have been made to the CT^2 mHealth app are at a programmatic level; we had the available Android app code and server/MySQL that allowed us to make these changes. We are currently exploring a way to encapsulate our adaptive trust negotiation via trust profiles into a device level app that can easily be referenced and used by others apps that require only minimal changes.

The addition of a trust negotiation feature to this mHealth app greatly simplifies the process of obtaining or adding patient data to the CT^2 database. This allows users to access concussion information or insert relevant concussion data without the need for a lengthy pre-registration process. Since the app is intended to be used by many stakeholders across the state, including school teachers and coaches, the reduction in the amount of necessary account registrations will result in decreased work for system administrators and increased access to the app's features. The manual selection of certificates works well when the user access history is small, but as the user adds to the trust profile, it quickly becomes difficult to manage and choose the best set of credentials. A search filter could improve the user's ability to find credentials that match the server's policy. Additionally, support for credential access policies (Winsborough, Seamons, & Jones, 2000) could be extended to the trust profile and incorporated into the app. Credential access policies allow the user to specify the conditions under which a credential can be released. For instance, the user can specify that the five latest records in the trust profile that match the intended request as closely as possible (student, role under which data was accessed, etc.) are to be released. This would automate the process of credential selection, enabling the creation of an ever-growing trust profile without requiring the user to manage it directly during trust negotiation.

FUTURE TRENDS

In this section, we explore three future trends that have the potential to augment trust negotiation: *spatio-temporal access control* where a user's permissions are restricted based on his/her geographic location and time; *biometrics* that utilizes a user's unique biological data to determine identity; and *single sign-on (SSO)* to manage multiple virtual identities.

Spatio-temporal access control is an access control model where user permissions change as the user moves to different geographic locations at different times. For example, if a user moves from the Family Medical Center to St. Francis Hospital (see Figure 1 again), the permissions of the user would change from patient data in the EMR at the center to the EMR at St. Francis. Similarly, if the device were to leave the premises entirely, the device would lack the permissions to access the EMR. Location-based access control as an extension to RBAC in (Bertino, Catania, & Damiani, 2005) is combined with the user's login information to determine the time that a user is working. The user is only able to log in successfully if he/she is scheduled to work at the time of the request for data access. Both of these types of access control could be integrated into the trust profile to enhance the owner's credentials and automate the credential selection process. When the physician moves to St. Francis, the owner will be able to present those portions of the trust profile that demonstrate the physician's history of accessing patient data at Family Medical Center.

Biometrics are being integrated with mobile devices for unique identification via: fingerprint scans, retina scans, gait recognition (Mantyjarvi, Lindholm, Vildjiounaite, Makela, & Ailisto, 2005), touch patterns on a smartphone display (Xu, Zhou, & Lyu, 2014), knuckle patterns, accelerometer data, keystrokes (Hwang, Cho, & Park, 2009), and voice recognition (Baloul, Cherrier, & Rosenberger, 2012). A user must register his/her biometrics in advance before he/she can be authenticated. Biometric authentication introduces new issues should the user's biometrics become unavailable in the event of extensive injury or are stolen (Nexus, n.d.; CNN Money, 2015)· Cancelable biometrics (Ratha, Connell, & Bolle, 2001) secures biometric data servers against attacks for users' biometric data by allowing users to revoke old biometric data and create new biometric data in the event that the server is compromised. This work has been extended to fingerprints (Ratha N., Connell, Bolle, & Chikkerur, 2006) and irises (Zuo, Ratha, & Connell, 2008) (Pillai, Patel, Chellappa, & Ratha, 2010). Biometrics may be used to provide additional assurance of identity during the trust negotiation process by acting as a passphrase to unlock the user's private keys.

*Single sign-on*enables a user to log in to multiple services with one log in without the difficulty of needing to remember multiple complex passwords from multiple services. During an SSO log in attempt, the user must authenticate with an SSO service, usually with a username/password combination. The SSO validates the user's identity and automatically logs him/her in to services where the user has authorized the SSO to manage account credentials on his/her behalf. Kerberos (Neuman & Ts'o, 1994) and Shibboleth are popular SSO systems utilized for log in to multiple servers in distributed systems. True SSO (Pashalidis & Mitchell, 2003) allows a many-to-many association between owned user identities and authenticated services. The multiple identities granted by a True SSO are useful for situations such as online shopping, where online stores may track user purchases that the user has purchased as a gift for someone else and adjust targeted ads accordingly. The True SSO multiple identity approach is akin to our digital wallet approach where a user obtains multiple identities from various healthcare organizations for use in authentication. There is current research in adapting SSO to healthcare such as (Heckle, Lutters, & Gurzick, 2008) that presented findings on the reaction of staff to the introduction of an SSO system in a hospital and (Mauro, Sunyaev, Leimeister, Schweiger, & Krcmar, 2008) that described a system to manage doctors' smart cards. SSO can simplify trust negotiation by performing the trust negotiation process as a trusted third party on behalf of the entities the user is attempting access resulting in a token the user can then present as credentials to the HIT systems he/she is attempting to access.

CONCLUSION

This chapter has presented *adaptive trust negotiation* in a mobile context and incorporated the concept of a *trust profile*. The *trust profile* provides detailed credentials that allows IT staff to create fine grained, adaptive security that can react dynamically via adaptive trust negotiation to protect sensitive data while still allowing data to be safely disseminated to legitimate users. Rather than basing the required credentials on what a user *is* (e.g., physician, nurse, psychiatrist), by knowing the actions that the user has been authorized to make, new organizations can better ascertain the user's level of trustworthiness and adjust its security policies accordingly in a dynamic fashion. The *Background* section briefly reviewed role-based access control, (the type of access a user may be allowed), identity certificates (a set of *trust profile* credentials), attribute certificates (to encode *trust profile* credentials in an endorsable format), and trust agents (to offload intensive cryptographic calculations to a more powerful server). The *Trust Profiling for Adaptive Trust Negotiation Section* described the trust profile in four parts: a general overview of the trust profile and its use in creating trust, the physical structure of the trust profile, a method for processing the trust profile in order to validate and extract the credentials contained within, and a healthcare example that describes the trust negotiation process and trust profile utilization. In *Design and Prototyping of Trust Profiles*, a prototype of our approach to trust negotiation in healthcare incorporated into the CT^2 concussion tracking mHealth app was presented. To complete the chapter, the *Future Trends* section discussed emerging trends in healthcare authentication and information exchange including: spatio-temporal access control, biometrics, and single sign-on.

REFERENCES

AHIMA. (2011, March). Security Audits of Electronic Health Information (Updated). *Journal of American Health Information Management Association*, *82*(3), 45–50.

Aitken, M. (n.d.). *Patient Apps for Improved Healthcare: From Novelty to Mainstream*. Retrieved from http://www.imshealth.com/portal/site/imshealth/menuitem.762a961826aad98f53c753c71ad8c22a/?vgnextoid=e0f913850c8b1410VgnVCM10000076192ca2RCRD

Artz, D., & Gil, Y. (2007, June). A Survey of Trust in Computer Science and the Semantic Web. *Journal of Web Semantics*, *5*(2), 58–71. doi:10.1016/j.websem.2007.03.002

Baloul, M., Cherrier, E., & Rosenberger, C. (2012). Challenge-based speaker recognition for mobile authentication. *2012 BIOSIG - Proceedings of the International Conference of the Biometrics Special Interest Group (BIOSIG)* (pp. 1-7). Darmstadt: IEEE.

Bertino, E., Catania, B., & Damiani, M. (2005). GEO-RBAC: A spatially aware RBAC. In *SACMAT '05 Proceedings of the tenth ACM symposium on Access control models and technologies* (pp. 29-37). Stockholm, Sweden: ACM.

Biometrics. (n.d.). *Biometrics*. Retrieved from http://dictionary.reference.com/browse/biometrics

CNN Money. (2015). *OPM hack's unprecedented haul: 1.1 million fingerprints*. Retrieved from http://money.cnn.com/2015/07/10/technology/opm-hack-fingerprints/

Conn, J. (2014). EHR makers' mobile medical apps grow in popularity. *Modern Healthcare*, *29*(November). Retrieved from http://www.modernhealthcare.com/article/20141129/MAGAZINE/311299981 PMID:25671868

Elkhodr, M., Shahrestani, S., & Cheung, H. (2011). *Enhancing the security of mobile health monitoring systems through trust negotiations. In Local Computer Networks (LCN), 2011 IEEE 36th Converence on* (pp. 754–757). Bonn: IEEE.

Farrell, S., & Housley, R. (2002, April). *An Internet Attribute Certificate Profile for Authorization*. Retrieved from The Internet Engineering Task Force (IETF®): https://www.ietf.org/rfc/rfc3281.txt

Fernández-Alemán, J., Señor, I., Lozoya, P., & Toval, A. (2013). Security and privacy in electronic health records: A systematic literature review. *Journal of Biomedical Informatics*, *46*(3), 541–562. doi:10.1016/j.jbi.2012.12.003 PMID:23305810

Ferraiolo, D. F., Sandhu, R., Gavrila, S., Kuhn, D. R., & Chandramou, R. (2001). Proposed NIST standard for role-based access control. *ACM Transactions on Information and System Security*, *4*(3), 224–274. doi:10.1145/501978.501980

GAA-API. (n.d.). *Generic Authorization and Access-control API (GAA-API)*. Retrieved from http://gost.isi.edu/info/gaaapi/

Gartner. (2015). *Gartner Says Global Devices Shipments to Grow 2.8 Percent in 2015*. Retrieved from http://www.gartner.com/newsroom/id/3010017

Heckle, R., Lutters, W., & Gurzick, D. (2008). Network Authentication Using Single Sign-on: The Challenge of Aligning Mental Models.*Proceedings of the 2nd ACM Symposium on Computer Human Interaction for Management of Information Technology* (pp. 6:1-6:10). San Diego, CA: ACM. doi:10.1145/1477973.1477982

Herzberg, A. (2003, May). Payments and Banking with Mobile Personal Devices. *Communications of the ACM*, *46*(5), 53–58. doi:10.1145/769800.769801

Himiss. (2014). *How mHealth is Changing Health and Healthcare*. Retrieved from http://www.himss.org/ResourceLibrary/mHimssRoadmapLanding.aspx?ItemNumber=30562

Housley, R., Polk, W., Ford, W., & Solo, D. (2002, April). *Internet X.509 Public Key Infrastructure Certificate and CRL Profile*. Retrieved from The Internet Engineering Task Force: http://www.ietf.org/rfc/rfc3280.txt

Hwang, S., Cho, S., & Park, S. (2009). Keystroke dynamics-based authentication for mobile devices. *Computers & Security*, *28*(1-2), 85–93. doi:10.1016/j.cose.2008.10.002

IHS. (n.d.). *Health Information Exchange and Master Patient Index*. Retrieved from https://www.ihs.gov/hie/index.cfm?module=dsp_hie_mpi

iTunes. (n.d.). *iTunes App Store Medical Apps*. Retrieved from https://itunes.apple.com/us/genre/ios-medical/id6020?mt=8

Lewis, N. (2011). 80% Of Doctors Use Mobile Devices At Work. *Information Week, 21*(October). Retrieved from http://www.informationweek.com/mobile/80--of-doctors-use-mobile-devices-at-work/d/d-id/1100880

Mantyjarvi, J., Lindholm, M., Vildjiounaite, E., Makela, S.-M., & Ailisto, H. (2005). Identifying users of portable devices from gait pattern with accelerometers. *Acoustics, Speech, and Signal Processing, 2005. Proceedings. (ICASSP '05). IEEE International Conference on.* IEEE. doi:10.1109/ICASSP.2005.1415569

Mauro, C., Sunyaev, A., Leimeister, J., Schweiger, A., & Krcmar, H. (2008). A Proposed Solution for Managing Doctor's Smart Cards in Hospitals Using a Single Sign-On Central Architecture.*Hawaii International Conference on System Sciences, Proceedings of the 41st Annual* (pp. 2565-266). Waikoloa, HI: IEEE. doi:10.1109/HICSS.2008.33

Mavridis, I., Georgiadis, C., Pangalos, G., & Khair, M. (2001, January-March). Access Control based on Attribute Certificates for Medical Intranet Applications. *Journal of Medical Internet Research, 3*(1), e9. doi:10.2196/jmir.3.1.e9 PMID:11720951

Montopoli, B. (2013). *For criminals, smartphones becoming prime targets*. Retrieved from http://www.cbsnews.com/news/for-criminals-smartphones-becoming-prime-targets/

Na, S., & Cheon, S. (2000). Role Delegation in Role-based Access Control.*Proceedings of the Fifth ACM Workshop on Role-based Access Control* (pp. 39-44). Berlin, Germany: ACM. doi:10.1145/344287.344300

Neuman, B., & Ts'o, T. (1994, September). Kerberos: An Authentication. *IEEE Communications Magazine, 32*(9), 33–38. doi:10.1109/35.312841

Nexus. (n.d.). *Fingerprint security on Nexus devices*. Retrieved from https://support.google.com/nexus/answer/6300638?hl=en

Pashalidis, A., & Mitchell, C. (2003). A Taxonomy of Single Sign-On Systems. *8th Australasian Conference, ACISP. 2727*. Wollongong, Australia: Springer-Verlag Berlin Heidelberg.

Pillai, J., Patel, V., Chellappa, R., & Ratha, N. (2010). Sectored Random Projections for Cancelable Iris Biometrics. *Acoustics Speech and Signal Processing (ICASSP), 2010 IEEE International Conference on* (pp. 1838-1841). Dallas, TX: IEEE. doi:10.1109/ICASSP.2010.5495383

Ratha, N., Connell, J., & Bolle, R. (2001). Enhancing security and privacy in biometrics-based authentication systems. *IBM Systems Journal, 40*(3), 614–634. doi:10.1147/sj.403.0614

Ratha, N., Connell, J., Bolle, R., & Chikkerur, S. (2006). Cancelable Biometrics: A Case Study in Fingerprints. *Pattern Recognition, 2006. ICPR 2006. 18th International Conference on, 4*. doi:10.1109/ICPR.2006.353

Ray, I., & Toahchoodee, M. (2007). A spatio-temporal role-based access control model.*Proceedings of the 21st annual IFIP WG 11.3 working conference on Data and applications security* (pp. 211-226). Redondo Beach, CA: Springer-Verlag.

Ryutov, T., Zhou, L., Neuman, C., Leithead, T., & Seamons, K. (2005). Adaptive Trust Negotiation and Access Control. *SACMAT '05 Proceedings of the tenth ACM symposium on Access control models and technologies* (pp. 139-146). New York: ACM.

Sabater, J., & Sierra, C. (2005, September). Review on computational trust and reputation models. *Artificial Intelligence Review*, *24*(1), 33–60. doi:10.1007/s10462-004-0041-5

Seabrook, H., Stromer, J. N., Shevkenek, C., Bharwani, A., Grood, J., & Ghali, W. A. (2014). Medical applications: A database and characterization of apps in Applie iOS and Android Platforms. *BMC Research Notes*, *7*(1), 573. doi:10.1186/1756-0500-7-573 PMID:25167765

SlimFramework. (n.d.). *Slim framework*. Retrieved from http://www.slimframework.com/

State of Connecticut. (n.d.). *An Act Concerning Young Athletics and Concussions*. Retrieved from http://www.cga.ct.gov/2014/act/pa/pdf/2014PA-00066-R00HB-05113-PA.pdf

Sundelin, T. L. (2003). *Surrogate Trust Negotiation: Solving Authentication and Authorization Issues in Dynamic Mobile Networks*. Brigham Young University.

van der Horst, T. W., Sundelin, T., Seamons, K. E., & Knutson, C. D. (2004). Mobile Trust Negotiation: Authentication and Authorization in Dynamic Mobile Networks. *Proc. of the Eighth IFIP Conference on Communications and Multimedia Security*.

van der Horst, T. W., Sundelin, T., Seamons, K. E., & Knutson, C. D. (2005). Mobile Trust Negotiation. In D. a. Chadwick (Ed.), *Communications and Multimedia Security* (Vol. 175, pp. 97–109). Springer. doi:10.1007/0-387-24486-7_7

Vawdrey, D. K., Sundelin, T. L., Seamons, K. E., & Knutson, C. D. (2003). Trust Negotiation for Authentication and Authorization in Healthcare Information Systems. *Engineering in Medicine and Biology Society, 2003.Proceedings of the 25th Annual International Conference of the IEEE* (pp. 1406-1409). IEEE.

Ventola, C. L. (2014, May). Mobile Devices and Apps for Health Care Professionals: Uses and Benefits. *Pharmacy and Therapeutics, 39*(5), 356-364. Retrieved from http://www.ncbi.nlm.nih.gov/pmc/articles/PMC4029126

West, D. (2012) How Mobile Devices are Transforming Healthcare. *Issues in Technology Innovation, 19*. Retrieved from http://www.brookings.edu/~/media/research/files/papers/2012/5/22-mobile-health-west/22-mobile-health-west.pdf

Winsborough, W. H., Seamons, K. E., & Jones, V. E. (2000). *Automated trust negotiation. In DARPA Information Survivability Conference and Exposition, 2000. DISCEX '00. Proceedings* (pp. 88–102). Hilton Head, SC: IEEE; doi:10.1109/DISCEX.2000.824965

X.509. (n.d.). *Standard*. Retrieved from https://tools.ietf.org/html/rfc5280

Xu, H., Zhou, Y., & Lyu, M. R. (2014). Towards Continuous and Passive Authentication via Touch Biometrics: An Experimental Study on Smartphones.*Symposium On Usable Privacy and Security (SOUPS 2014)* (pp. 187-198). Menlo Park, CA: USENIX Association.

Yu, J., Wang, G., & Mu, Y. (2012). Provably Secure Single Sign-on Scheme in Distributed Systems and Networks. *TRUSTCOM '12 Proceedings of the 2012 IEEE 11th International Conference on Trust, Security and Privacy in Computing and Communications* (pp. 271-278). Washington, DC: IEEE.

Zuo, J., Ratha, N., & Connell, J. (2008). Cancelable Iris Biometric. *Pattern Recognition, 2008. ICPR 2008. 19th International Conference on* (pp. 1-4). Tampa, FL: IEEE. doi:10.1109/ICPR.2008.4761886

KEY TERMS AND DEFINITIONS

Adaptive Trust Negotiation: Trust negotiation in which the request receiver adjusts its security policies based on the user's credentials.

Attribute Certificate: A structured tamper-resistant file that is associated with an identity through an identity certificate and that lists data in a key-value pairs.

Certificate Authority (CA): An entity endorsed by another authority that vets user identities and signs identity certificates.

Digital Wallet: A collection of credentials a user earns through being granted access to secure systems.

Electronic Medical Record (EMR): A collection of credentials a user earns through being granted access to secure systems.

Health Information Exchange (HIE): The sharing of health data between stakeholders over a secure medical network, or the computer system that facilitates data sharing.

Identity Certificate: A structured tamper-resistant file that is used to identify an individual and provide assurance for secure connections.

Root Authority: An entity that signs user certificates with a self-signed certificate, users must add the certificate to their certificate store to establish trust.

Trust Negotiation: The process two entities without prior contact undertake to establish trust based on credentials other than identity.

Chapter 6
Role-Based Access Control for Mobile Computing and Applications

Yaira K. Rivera Sánchez
University of Connecticut, USA

Steven A. Demurjian
University of Connecticut, USA

Joanne Conover
University of Connecticut, USA

Thomas P. Agresta
University of Connecticut Healthcare Center, USA

Xian Shao
University of Connecticut, USA

Michael Diamond
Pomona College, USA

ABSTRACT

The proliferation of mobile devices has changed the way that individuals access digital information with desktop applications now performed seamlessly in mobile applications. Mobile applications related to healthcare, finance/banking, etc., have highly sensitive data where unsecure access could have serious consequences. This chapter demonstrates an approach to Role-Based Access Control (RBAC) for mobile applications that allows an information owner to define who can do what by role, which is then enforced within a mobile application's infrastructure (UI, API, server/database). Towards this objective, the chapter: motivates the usage of RBAC for mobile applications; generalizes the structure and components of a mobile application so that it can be customized by role; defines a configurable framework of locations where RBAC can be realized in a mobile application's infrastructure; and, proposes an approach that realizes RBAC for mobile security. To demonstrate, the proposed RBAC approach is incorporated into the Connecticut Concussion Tracker mobile application.

INTRODUCTION

The proliferation of mobile devices in all aspects of daily living has fundamentally altered the way that individuals interact with mobile applications. Evidence includes: the worldwide shipments of 1.9 billion

DOI: 10.4018/978-1-5225-0945-5.ch006

phones and 230 million tablets outpacing PC/laptop sales (300 million estimate) (Gartner, 2015; Cisco, 2014); a report of smartphone usage in the U.S. where 64% of adults own a Smartphone, 42% own a Tablet, and 32% own an e-reader (Pew Research Center, 2012; Smith, 2015); and, predictive statistics that tablet users will surpass 1 billion worldwide in 2015 (eMarketer, 2015) and total devices will exceed 12.1 billion by 2018 (Radicati, 2014). Mobile applications now span a broad spectrum of complexity, including games, social networking, email, web browsing, financial management, health and fitness, pharmaceutical, etc. For both personal and business usage, there is a need to protect secure information ranging from protected health information (PHI) and personally identifiable information (PII) to confidential work product that is displayed, accessed, modified, and stored. Mobile health (mHealth) applications in healthcare and fitness are numerous and diverse: tracking medications (myCVS (CVS Pharmacy, 2015), MedWatcher (2012), etc.); personal health records (PHR) (CAPZULE PHR (Capzule, 2012), MTBC PHR (2011), etc.); fitness applications that work with phones and wearables (Cohen, 2015); Apple's HealthKit app (iOS 9, 2014) and the Google Fit fitness tracker (Google Play, 2013), to track activity, heart rate, blood pressure, etc. (Kelly, 2014); and, Apple's ResearchKit (Apple, 2015), an open source framework for mobile applications to support medical research. Patients also seek to have access via their mobile devices to the electronic medical records (EMRs) utilized by medical providers and health information technology (HIT) systems that contain medical testing results (Care360, 2014) or results from imaging testing (My Imaging Records App, 2013). All of these systems must adhere to the Health Insurance Portability and Accountability Act (HIPAA) (HHS.gov, 2013) for the security, availability, transmission, and release of a patient's medical information.

To augment the usability of these mHealth applications, there is also a desire by patients to attain privacy control to different individuals at varying levels of granularity over their electronic health and fitness information in various locations (Caine & Hanania, 2013). For a given patient, this effort highlights the potential recipients of the information (e.g., primary physicians, spouse, family, emergency medical providers, etc.) and the type of information to be controlled (e.g., contact info, current conditions, medications, recent test results, genetic information, etc.). In such a setting, patients are also interested in actually defining specific fine-grained access control by role (Sujansky, Faus, Stone, & Brennan, 2010), for example: a family member may view my medication list (but not all of them), a medical provider may view my medication list and history of hospital visits (but not modify), my personal physician may both view and modify my health care and fitness data, etc. These efforts highlight a strong need to achieve fine grained role-based level of security to allow patients to define who can see and/or modify what portions of their health/fitness data, where the mobile application itself can be customized based on role to meet the permission definition provided by the patient (Peleg, Beimel, Dori, & Denekamp, 2008). Securing information of mHealth applications for a diverse set of stakeholders may benefit from the usage of role-based access control (RBAC) (Ferraiolo & Kuhn, 1992). Such an inclusion allows permissions established by the information owner to be defined for other authorized users by role and use this as a basis to have the mobile application deliver only authorized information, and permitted view and/or modify capabilities. Note also that RBAC has been heavily adopted in healthcare, where a recent published literature review (Fernández-Alemán, Señor, Lozoya, & Toval, 2013) had 35 efforts utilizing access control methods and 27 of these specifically utilized RBAC.

One challenge in such an approach is to identify locations where RBAC can be incorporated within the mobile application's infrastructure (UI, API, Server/Database), which can be conceptualized as a configurable framework for RBAC permission definition and enforcement for mobile applications, as shown in Figure 1. In the figure, there are three options for including RBAC. The first option, *direct UI modifications,* shown in the left side of Figure 1, would be to modify the existing mobile application itself with RBAC permissions on screens, UI widgets, etc., which would involve code-level changes so that the look-and-feel of the UI would change based on the user and his/her role. The second option, *intercepting API calls,* shown in the middle of Figure 1, would be to define RBAC permissions on the API (REST, web, cloud) and/or database calls of the mobile application and intercept them in order to include RBAC permission checks that determine the filtered information returned to the mobile application or control information that can be stored in the mobile application's server. This may require minimal changes on the way that the mobile application calls the backend or the way that the backend calls are intercepted by the access control code. Finally, the third option, *direct server modifications,* shown in the right side of Figure 1, involves making changes to the backend of the existing mobile application (e.g., server for database, cloud, web, etc.) that would retain the view of the mobile application's API to the mobile application and embed RBAC on the server side. There are different ways to realize the three options with the mobile application infrastructure. One approach is via *application containers*, which separates a group of mobile applications that share/contain highly-sensitive data by placing them in an encrypted virtual container at the device level. Another approach is *application wrappers*, which wrap security policies on individual mobile applications at the application level (Symantec, 2014). We believe that the usage of application wrappers is the preferred approach for the options for three reasons: wrappers can be used to realize finer-granularity security policies than could be attained using application containers; wrappers can be utilized on third-party mobile applications; and, wrappers have a lower potential of damaging the functionality of a mobile application. This paper primarily focuses on option 1, *direct UI modifications*; the other two options are discussed in part as they represent ongoing work.

The remainder of the chapter has five sections. In the *Background and Motivation* section we review the NIST RBAC standard, RBAC in the healthcare domain, and RBAC in mobile computing, including work by both researchers and practitioners in applying RBAC to healthcare. In the *RBAC Approach for Mobile Applications* section, we present a generalizable approach to achieve RBAC security for mobile applications in support of the direct UI modifications option of Figure 1 that is realized with a corresponding RBAC model in SQL, which can be utilized to both define and enforce security on a mobile application. In the *RBAC in Connecticut Concussion Tracker Mobile Application* section, we demonstrate the attainment of the RBAC model of the prior section in the Connecticut Concussion Tracker (CT²) application, providing detailed screen shots and a scenario of usage that fully demonstrates the ability to have the mobile application appear different to users based on their role. This includes the detailing

Figure 1. A configurable framework for RBAC and its interactions with the mobile application infrastructure

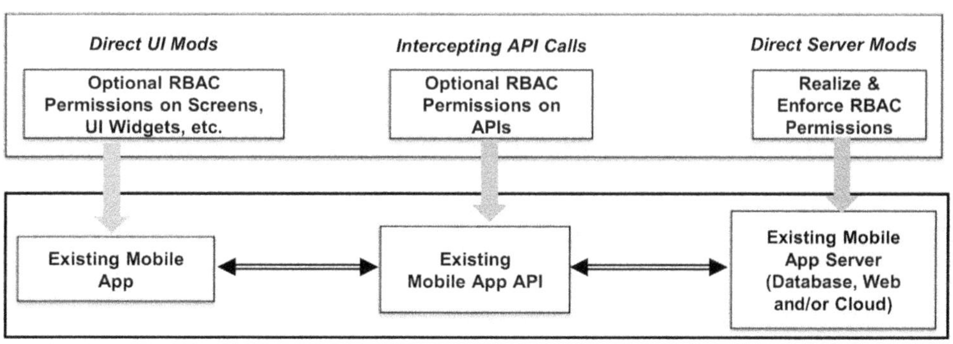

of the direct UI modifications option as achieved by an application wrapper as well as a brief discussion of the intercepting API calls and direct server modifications options. Next, the *Future Trends* section presents emerging efforts that may impact RBAC for mobile applications, including: other types of access control such as Attribute-Based Access Control (NIST Computer Security Division, 2013), and BiLayer Access Control (Alshehri & Raj, 2013); as well as emerging approaches for mobile application security (MobileIron (2015)). Finally, the *Conclusion* section summarizes the contributions of the chapter.

BACKGROUND AND MOTIVATION

This section provides background information on: role-based access control (RBAC); security issues in healthcare; the usage of RBAC in healthcare; and, general security issues in mobile computing. To begin, *access control* is a security process and mechanism that allows user permissions to be granted or denied against the resources (objects) of a system or application. *Authorization,* a part of access control, is the process that is specifically associated with allowing or denying access to a resource (object). One of the dominant approaches, role-based access control (RBAC), was proposed by David Ferraiolo and Richard Kuhn (Ferraiolo & Kuhn, 1992) and transitioned to the National Institute for Standards and Technology (NIST) (Ferraiolo, Sandhu, Gavrila, Kuhn, & Chandramouli, 2001) that was adopted in 2004. The main concepts of the NIST RBAC standard (Sandhu, Ferraiolo, & Kuhn, 2000) are conceptualized in Figure 2 (SlideShare, 2012) with four reference models. $RBAC_0$ in the middle portion of Figure 2 is comprised of: *users* that perform a specific function within an organization, *roles* that are assigned to users based on their responsibilities, and *permissions* that define which operations/objects within a system/application a role can have access to. Users can have one or more roles, and roles can contain one or more permissions to objects. $RBAC_1$, shown in the upper middle portion of Figure 2, supports the ability of roles to be organized in a hierarchy. $RBAC_2$, shown in the upper left of Figure 2, provides the definition of constraints, such as separation of duty (SoD), mutual exclusion (ME), and cardinality. Lastly, $RBAC_3$, shown at the bottom part of Figure 2, captures the concept of sessions that represent the lifetime of a particular user, role, and permissions in a dynamic runtime application.

Figure 2. General structure of the RBAC model
SlideShare, 2012.

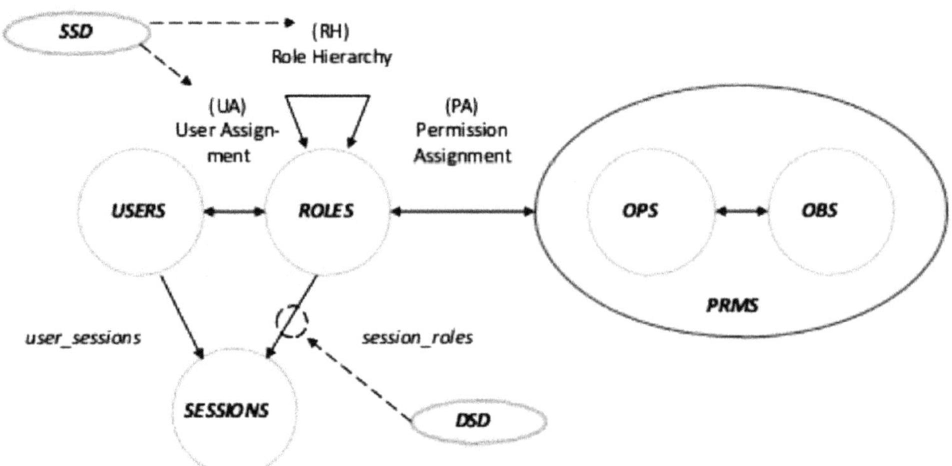

Next, we highlight two key issues involving security for healthcare. The first issue of incorporating health technologies such as Electronic Medical Records (EMRs) in healthcare organizations concerns privacy and security (Boonstra & Broekhuis, 2010). Improper disclosure of data in healthcare systems can have serious effects on patients, which can include personal embarrassment, prejudice, ostracization from family and community groups, as well as issues with insurability (Rindfleisch, 1997). In addition, unauthorized individuals that obtain access to patient information could use it for their own profit. For instance, in 2013, a billing technician at a hospital spent several months looking for people that had recently been in car accidents and then sold that information to an attorney. The attorney would then contact the individuals who were involved in the car accident and offered them legal representation (Wiech, 2013). This issue highlights the need for access control in systems that contain highly sensitive information, such as hospitals and health insurance companies. A second issue involves ensuring that healthcare systems comply with several security standards. One of the main requirements can be found in §164.312(a)(1) of the standard (Scholl et al., 2008) that indicates that an approach for access control should be identified in order to allow access to authorized users and no one else. The HIPAA Privacy Rule also states that health care organizations should only disclose the minimum required information in order for the individual to realize an action (Health Information Privacy, 2002). In fact, as part of the Health Level Seven (HL7) International Standard (Health Level Seven International, 2010), a scenario-based role engineering process has been proposed and adopted (HL7 Security Technical Committee, 2007).

The third background area, RBAC in healthcare, has numerous efforts to extend RBAC to address limitations in regards to supporting healthcare. One effort extended RBAC in order to create an authorization model for Healthcare Information Systems (Hsu & Pan, 2013). The proposed model has the ability to assign different access permissions and to define new ones, with the consideration of roles having the ability to access certain permissions in emergent situations, and using RBAC to create an authorization mechanism for this case. Another approach proposes the creation of an access control model that contains elements from the mandatory, discretionary, and role-based access control models (Gajanayake, Iannella, & Samaha, 2014) for providing security and privacy to EMRs. On the practitioner side, one effort (Science Application International Corporation, 2004) in the United States involves the implementation of a large RBAC system for health care for Armed Service veterans and Native Americans: Kaiser Permanente, Department of Veterans Affairs, Department of Defense, and the Indian Health Service. Another example has added RBAC functionality to an existing medical database (Slevin & Macfie, 2007). This was done in order to observe the feasibility of the implementation, and to include RBAC as a security mechanism within the National Program for Information Technology (NPfIT) of the UK National Health Service (NHS). A third effort example implemented an Electronic Patient Record (EPR) in the Hospital S. João in Porto, Portugal using an RBAC model in order to provide a secure authorization mechanism (Ferreira, Chadwick, & Antunes, 2007).

The final background area involves security issues for mobile computing and applications. Mobile devices and computing are being improved on a seemingly daily basis in terms of hardware and software, increasing capabilities, features, and capacity. This in turn has resulted in the rise of new security risks. In the worst-case, a mobile device may be lost or stolen, necessitating the availability of techniques to control and securely access highly sensitive data. For example, healthcare data stored in a mobile device is being created, retrieved, and manipulated from multiple sources and by varied applications and this sensitive information must be protected from disclosure. This security requirement is juxtaposed against a recent survey (West & Miller, 2009) where the majority of people wanted email access with providers (74%), diagnostic test results electronically (67%), and access to their EMR (64%). These tasks will require a great amount of security as the information to be shared is highly sensitive and pertains to specific people from multiple sources and ultimately residing on a patient's mobile device.

From a security perspective, many high-end mobile devices contain biometric controls (e.g., fingerprint, voice recognition, facial recognition) and input controls (e.g., pattern, PIN, password) for authentication and authorization. All work under the assumption that one person is the owner of the device. To protect sensitive information many corporations are relying on the bring-your-own-device (BYOD) concept (e.g., corporate email, corporate applications) through their personal mobile devices or through a mobile device issued by the corporation itself. While this enhances productivity and increases employee satisfaction and engagement after hours (Wainwright, 2012), a study of mobile devices utilized in a corporate setting (Paganini, 2015) showed that approximately 14,000 applications installed on these devices were unsafe and compromised sensitive device data such as phone location, phone contacts, and SMS message logs. One of the methods that companies who have utilized BYOD policies have relied on to secure employees' mobile devices is to have defined roles for employees that limit access to mobile applications. This insures that employees only have access to what they need, and lowers the risk of downloading malicious applications. Finally, as part of BYOD, corporations utilize the concept of Mobile Device Management (MDM) (Gartner, 2012), which consists of software that manages software distribution, policies, inventory, security, and services.

RBAC APPROACH FOR MOBILE APPLICATIONS

In the literature, there are numerous access control models and approaches that utilize and extend RBAC (Motta & Furuie, 2003; Peleg, Beimel, Dori, & Denekamp, 2008; Russello, Dong, & Dulay, 2008). In addition, several approaches (Schefer-Wenzl & Strembeck, 2013; Santos-Pereira, Augusto, Correia, Ferreira, & Cruz-Correia, 2012; Abdunabi, Sun, & Ray, 2014) have emerged that propose RBAC for a mobile setting. However, while these approaches add capabilities such as spatio-temporal or context-aware techniques, they do not specifically address the way that the mobile application itself (UI, API, Server/Database) is impacted depending on the role that a user assumes for a particular mobile application session. To address this issue, this section provides an approach and detailed examination for achieving RBAC for mobile computing and applications by leveraging the NIST RBAC model, as presented in Figure 2, to the user interface (UI) of a mobile application; this is supporting the direct UI modifications option as shown in Figure 1. *User interface (UI)* can be defined as the means users utilize to interact with computer systems. The specific objective is to allow a mobile application to dynamically customize the capabilities of the mobile application's UI that are available based on a user's role, to both permit a user to perform needed tasks using the mobile application while simultaneously limiting and/or disabling and/or removing capabilities and features that are not allowed at certain times or in certain situations. In our approach, the "objects" of the NIST $RBAC_0$ will correspond to the various components of a mobile application's UI.

The remainder of this section is organized into a five-part discussion. In part one, we motivate the proposed RBAC approach by providing a sample mobile application in healthcare that illustrates the intended capabilities of our RBAC approach in terms of customizing the look-and-feel of the UI of a mobile application on a role-by-role basis. Part two generalizes the features and characteristics of the UI of a mobile application in order to be able to control what is viewable and/or modifiable on a role-by-role basis. Part three defines the different types of permissions that are available against the screens and components of a mobile application in order to allow a role to be defined with a set of permissions that limits the look-and-feel of the UI by role. Part four defines a relational database structure to store the content of a generalized mobile application that is comprised of a UI with screens and their components, to allow permissions to be defined that customize the mobile application look-and-feel based on role. Part five explains the programmatic changes that must be made to the mobile application itself to allow for the screens and their components to be customized.

In the first part of this section, we present a mobile application utilized by personnel at a pharmacy to fill and process prescriptions for customers with a UI that has five screens to: look up the status of a prescription (Screen 1), enter a new prescription to be filled (Screen 2), fill and dispense the prescription with the appropriate medication (Screen 3), look up to see if a medication is in inventory (Screen 4), and order medications for inventory (Screen 5). The five screens could be linked by next and back buttons or could be five different tabs on one screen. There are two types of users: *pharmacy technicians* that interact with the customer to receive and enter the prescription; and, *pharmacists* that have the legal authority to fill and dispense the prescription. A pharmacy technician would be limited to Screens 1, 2, and 4, while a pharmacist would have access to all five screens. To achieve this in practice, the mobile application's UI would have to be adaptable based on role in order to determine which components of the application the user with a role is allowed to see in order to avoid improper disclosure of information. Our solution in the proposed RBAC approach is to place these access control policies in a database where they will be easier to manage and change as permissions are created, modified, deleted, etc. To support such an approach, there is a need to control not only which screens are available to which users but to also control the actual capabilities on each screen which may limit some users to be able to view data on a screen while others could view/modify it.

This leads to the second part of this section that enumerates the general structure of a mobile application in terms of its features and characteristics so that the way that the mobile application looks and feels can be dynamically customized on a role-by-role basis. Specifically, a mobile application's UI will be comprised of a series of inter-connected *screens* where each screen contains a portion of the functionality. Each screen of the UI will have a set of different *components* consisting of information that is displayed (cannot be changed) and information that can be entered by a user including: text field (TF), button (BN), drop down (DD), checkboxes (CB), radio buttons (RB), spinner (SP), date picker (DP), etc. A mobile application can have one or more screens and screens can have one or more of the aforementioned components. Definitions 1 to 4 formalize these concepts:

Defn. 1: A mobile application, *MA*, consists of a user interface (UI) that contains a set of *n* screens, *S*, that are organized as either tabs (users can click among tabs) or a sequence of inter-connected screens which are linked with next and previous buttons.

Defn. 2: Each screen, S^i, has a set of *k* screen components, *SC*, denoted, SC_k^i, that allow a user to select, enter, and manipulate data in a MA.

Defn. 3: A component, *C*, is a portion of a screen that can be displayed and/or entered by users and includes but is not limited to: a text field (TF) to enter information; a button (BN) to effect the state of the application (save, cancel, next, previous, etc.); a drop down (DD) where one value is chosen; a set of checkboxes (CB) where multiple values can be chosen; a set of radio buttons (RB) to select only one of a number of options; a spinner (SP) to select values; a date picker (DP) to enter calendar dates; etc.

Defn. 4: Each MA has a screen set, *SS*, that is classified as either:

- A collection of tabs where each tab is a screen, where there is an order among the tabs in the way that they are displayed left to right within the MA.
- A collection of screens where each screen has an appropriate list of buttons to navigate among screens that is augmented with the *screen interactions, SI,* necessary to switch among the various screens.

Utilizing these definitions as a basis, users will be authorized to access a subset of the screens with defined permissions for each screen in order to limit and control the access to each screen's components, where our RBAC approach can enable/disable the components based on a user's role. As result, a user as owner by his/her role could have full access to the application (e.g., pharmacist in the prior can utilize all five screens) while another user would be restricted to a dynamically customized version of the mobile application (e.g., pharmacy technician limited to Screens 1, 2, and 4). The end result is the ability to control which components of the application's UI users can have access (view/edit) to depending on their role. The advantage of this is that user permissions can be configured based on his/her role; therefore, the application does not need to be configured for each individual user, but will operate by role against the user instance that has been authorized.

This third part of the proposed RBAC approach provides the ability to define permissions against a generalized structure of a mobile application's UI screens and their components that is capable of customizing which screens and their respective components are available in the mobile application, depending on the role a user assumes. Towards this end, the left side of Figure 3 contains a sample screen from the UI where a set of components (text fields, spinners, date pickers, drop down boxes, and buttons) is shown. The screens and the components are the objects in $RBAC_0$ that will be authorized as screen, component, and screen interaction permissions to a particular role. This essentially defines what a role can and cannot do in terms of screen, component, and screen interaction permissions and determines whether the user with such role can access and/or view a certain component. As an example, the screen with all capabilities on the left side of Figure 3 is customized as shown on the right side of Figure 3 where a text field (the search bar), the drop down, the checkboxes, and button number 2 have been disabled. In this case, the user with the role would be able to view information but would not be able to make any changes to the aforementioned disabled components. The permissions that are defined on the components of a screen are placed in two main categories: *on/off permissions* that are for components that can be 'on' (enabled) or 'off' (disabled); and, *data permissions* that are for components that can be 'view', 'edit', or 'edit once'. On/Off permissions are defined for the different components: button (BN), radio button (RB), drop down (DD), checkbox (CB), date picker (DP), spinner (SP), and text fields (TF), while data permissions are defined for text fields (TF). A text field has to be On in order for view/edit/edit once to be defined. A mobile application can have one or more screens and screens can have one or more of the aforementioned components and permissions on a user/role basis. Definitions 5 to 9 formalize these concepts:

Defn. 5: A screen permission, $sp = < s, p >$, where $s \in SS$ is a screen and *p* is permission, is utilized to define whether a screen s in SS as given in Defns. 2 and 4, that is part of a mobile application MA is allowable (*p=true*) or not (*p=false*).

Figure 3. A screen with components (left) that are customized (right)

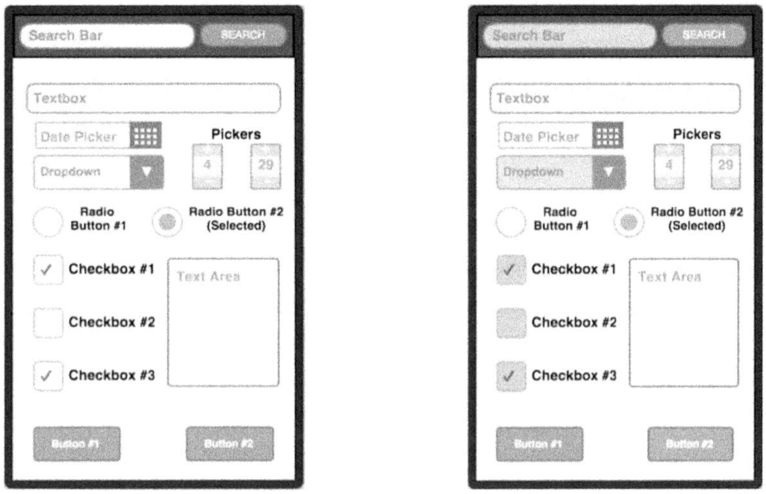

Defn. 6: A component permission, $cp =< sc, p >$, where $sc \in SC_k^i$ is a screen component (Defn. 2), is utilized to define permissions on various components of each screen S (Defn. 3). There are two types of component permissions:

- On/off permissions for button (BN), radio button (RB), drop down (DD), checkbox (CB), date picker (DP), spinner (SP), or text field (TF). The permission values for each component are: *p=enabled* or *disabled.*
- Data permissions for text fields (TF). The permission values for a text field are: *p=view*, *edit*, or *edit once*.

Defn. 7: A screen interaction permission set, $SIP =< si_1, \cdots, si_e >$, defines all permitted screen interactions, where each $si = [s_x, s_y]$ is a pair of screens that means that screen s_x interacts with screen s_y.

Defn. 8: A role, *R*, is assigned a set of role permissions, R_p, for a subset $m \leq n$ screens of *SS* in MA as: $R_p =< \sigma, \chi, \delta >$ where: $\sigma =< sp_1, \cdots, sp_m >$ are the *m* screens (Defn. 5) assigned to the role *R*; $\chi =< cp_1, \cdots, cp_j >$ are *j* component permissions (Defn. 6) for all *m* screens; and $\delta = SIP$ are the screen interactions (Defn. 7) for non-tabbed UIs.

Defn. 9: A user, *U*, can be assigned a role *R* that customizes the UI of the MA based on the role permissions (Defn. 8).

The fourth part of the proposed RBAC approach is the design of an entity-relationship diagram representation of the database model in Figure 4 that is utilized to capture screen, component, and screen interaction permissions (see Defns. 5, 6, and 7), for each role (see Defn. 8), by representing the generalized structure of the mobile application (via the entities screens, screen_sequence, and screen_components) and the permissions (via the entities roles, screen_access, and component_access). In Figure 4, the *roles* entity contains the name of the roles that are available as well as a unique role_id assigned to each one of these. This role_id is utilized to determine if a role has access to a specific screen of a UI, to a component in that UI screen, and the sequence of screens it is allowed to view. This role_id would leverage the prior pharmacy example, where the pharmacy technician role would only be allowed to access UI screens 1, 2, and 4. If a role by role_id has access to a specific screen, then the second step of the permission process would be to define components of the screen that such role can access; otherwise, if role_id is not assigned to a screen, the screen and its components are hidden. In this model, users can only have one role, and roles can have one or more permissions. In order to capture permissions, the entities screen_access and component_access are utilized, where: the *screen_access* entity supports the definition of the permission for a role with respect to the entire screen and the *component_access* entity supports the definition of the permissions of the role with respect to the components of a screen. To bring the concepts together, Figure 5 illustrates the authorization process that assigns a user one or more roles (while limiting a user to one identified role per session) and then defines screens, their components, and their interactions on a role-by-role basis against all mobile application screens/components. Note that both the on/off and data permissions are for the components that are captured in the component_access entity.

Figure 4. Proposed entity-relationship diagram for capturing permissions on screens and components

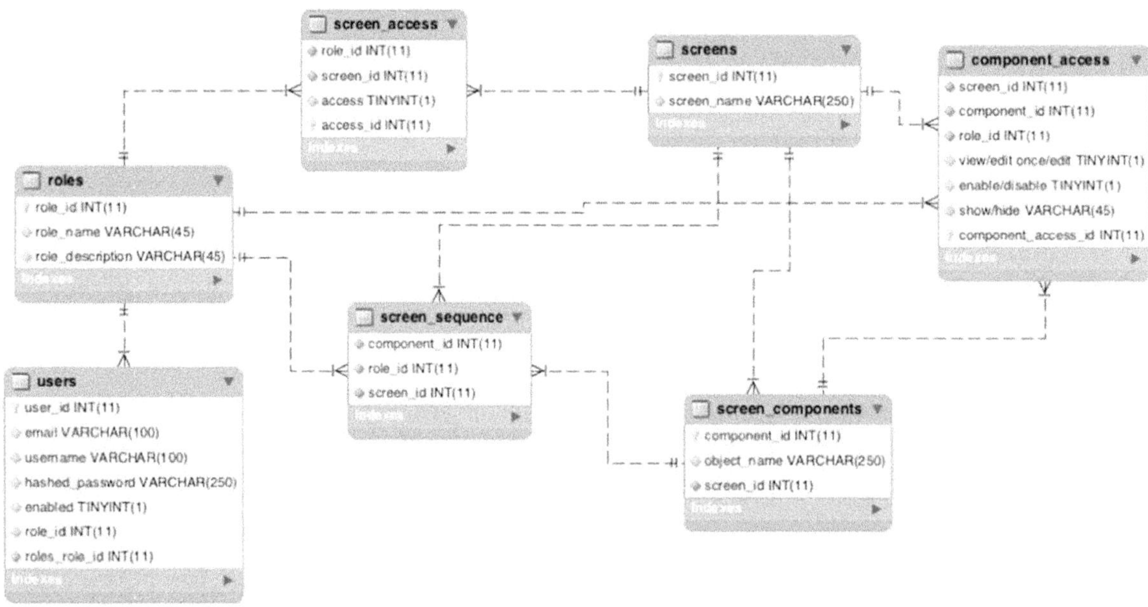

Figure 5. Authorization process for roles with respect to screens and components

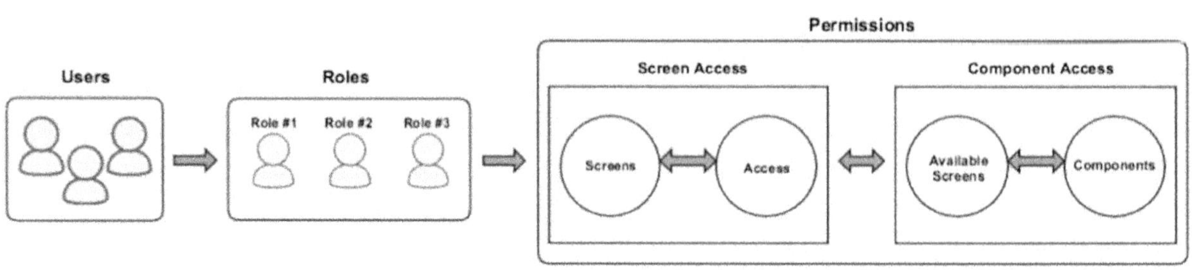

The authorization policies for a generalized mobile application as defined via the process in Figure 5 are then enforced as shown in Figure 6. The enforcement process begins with the mobile application authenticating the user, verifying the credentials, and then retrieving the role of the user to customize the mobile application (Defn. 9). The right hand side of Figure 6 (Proposed Approach box) utilizes the data in the screen access and component access entities (see Figure 4 again) to determine a custom version of the UI of the mobile application by role (Defn. 8). Notice that the screen access instances are shown for a role (like the pharmacy technician role of the prior example), with screens 1, 2, and 4 authorized. The *Accessible Screens* part of Figure 6 illustrates a basic idea of the screen permissions (Defn. 5). Also, the *Component Access* table on the right side of Figure 6 illustrates the components of the aforementioned screens that have been authorized to a specific role, thereby realizing the component permissions (Defn. 6). There are no screen interaction permissions in Figure 6, since we assume the mobile application consists of a set of tabs.

The fifth part of our approach involves programmatic changes that are made just one time; since the permissions are taken from the database of Figure 4, these can be changed and the mobile application's UI will adjust the look-and-feel based on roles/permissions without any additional code-level changes. The entity-relationship diagram in Figure 4 was realized as a relational database using MySQL. To support interactions from the mobile application to the MySQL database of permissions, an RBAC Application Programming Interface (API) was created that can be utilized to support the application wrapper for the direct UI modifications option as illustrated in Figure 1. The RBAC API calls are invoked within the mobile application's source code and return data from the queries in JSON format. There are three main API calls that will constitute the wrapper that have been defined against the entities in Figure 4 and realized in the MySQL database, which are briefly described along with the incorporation of their usage within the code of the mobile application. The API calls are as follows:

- **Is_Screen_Allowable(role_id, screen_id):** returns a Boolean value true (JSON format: [{"access":"1"}]) if the role represented by role_id has the permission to display a particular UI screen screen_id, and false otherwise (JSON format: [{"access":"0"}]). This first API call queries the screen_access entity of Figure 4 which was illustrated in Figure 6, and would return true for screens 1, 2, and 4, and false for screens 3 and 5. This requires a change to the mobile application

Figure 6. Enforcement process for a mobile application

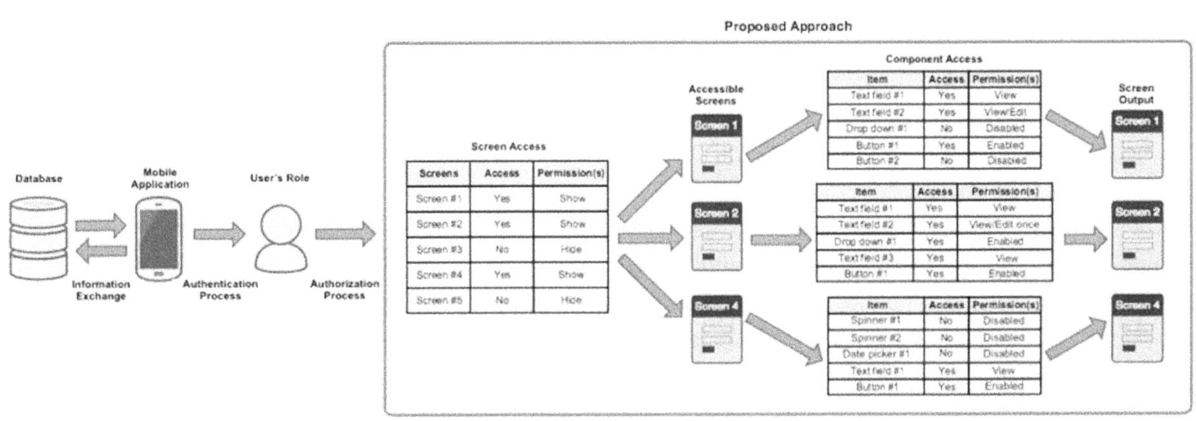

code to include a conditional statement that only displays a particular screen of the UI based on the screen_id and role_id if there is permission defined in the screen_access entity.

- **Return_Allowable_Components(role_id, screen_id):** Returns a table of component permissions for the role_id/screen_id combination that identifies the on/off and data permissions on buttons, spinners, date spinners, radio buttons, checkboxes, drop downs, and text fields. The API call begins by querying the component_access entity of Figure 4, which was illustrated in Figure 6 by the three permission tables for screens 1, 2, and 4 (*Component Access* table in the figure). Then, the API call retrieves the component permissions for a single screen of the UI that is authorized to that role. Finally, the API call is invoked for the allowable screens following the API call Is_Screen_Allowable(role_id, screen_id). As part of the process to display an allowable screen **s**, the mobile application's code is modified with conditional statements for the various components of each screen. Specifically, for the on/off permissions, the button (BN), radio button (RB), drop down (DD), checkbox (CB), date picker (DP), and/or spinner (SP) components are disabled for all non-allowed actions which are the 'no' entries as shown in the component_access table of screen 1 for Figure 6. For the data permissions, each text box is set accordingly based on View, Edit, or Edit once. Note that the APIs are not called in sequence rather are utilized in multiple locations throughout the mobile application's source code.
- **Return_Allowable_Screen_Interactions(role_id):** Utilizes a role_id to look up all of the allowable screens (via a database query to the screen_access entity for role_id) and using this information, returns the sequences of permissible movement/interactions among all allowable screens of the UI. First, the API call utilizes the screen_sequence entity in Figure 4 in order to find all of the allowed interactions among screens of the UI for role_id and only enable those interactions that occur among the allowable screens. The API call is then utilized to set behavior related information to buttons (BN) on a particular screen. For a button that is enabled, the information in screen_sequence for the given role and its allowable screens will allow the button to cause the screen to be reached; a button not enabled will not link to another screen.

To this point, we have shown the different components that our approach contains and the way that these are incorporated in a mobile application in support of the direct UI modifications option via an application wrapper. Next, we review the way that a mobile application maintains its functionality after adding these direct changes. First, we identify the screens and the objects of a screen of the mobile application to which RBAC is to be applied. Then, we assign a unique id to each of these components and store them as tables in a database as shown in Figure 4. The role of the user is retrieved and stored in a secured session variable (passed over https) through the means of an API call (part of the application wrapper). Using this as a basis, to enforce the policies established in the database with the mobile application, we create a set of API calls. These calls will return if a component can be shown/edited or if it needs to be disabled/hidden. Each call will check if the component that we are trying to apply the policy to exists in the database; if it doesn't, then the API call will return that the component is enabled for the role that is in session as no RBAC permissions were found in the created policy. By making these changes, the functionality of the mobile application will not be affected since the API calls that are being added to the source code of the mobile application always return a value regardless if the component does not have a policy stored in the database anymore. In addition, storing the role of users in a secure session variable will prevent a user from tampering its role and it does not require any changes in the source code.

Lastly, note that the approach presented in this section supports the direct UI modifications option that was part of the Figure 1 configurable framework for RBAC. The direct UI modifications option focuses on RBAC of the UI of a mobile application's screens and their components, and then customizes the look-and-feel by role that is defined with varied screen, component, and screen interaction permissions stored in a database. When permissions change, only the database needs to be changed, and the mobile application will adjust appropriately. However, one disadvantage of the direct UI modifications option is that programmatic changes are required in the mobile application itself through the addition of condition statements and calls to APIs that return allowable screens and permissions on components in order to adjust the look-and-feel of the mobile application (via the application wrapper). The other two options in Figure 1, intercepting API calls and direct server modifications, are both under investigation in our work. Consider that the intercepting API calls option would require minimal or no changes to the UI of the mobile application other than for the need to identify a given role for a session being initiated by a user. In this case, we incorporate the functionality of our API calls into REST or API services that are utilized to intercept the API calls. Here, we are not changing the look-and-feel of the mobile application, but essentially disabling the delivery of content to the user. Recalling the pharmacy example, the pharmacy technician could see all five screens, but information on screens 3 and 5 would be blocked in the display of data. For the case where the pharmacy technician attempts to utilize screens 3 and 5, if they do attempt to take a positive action to search or insert information, this would be intercepted at the server side to disallow the attempt. Basically, the RBAC checks on defined permissions that have been discussed for the approach in this chapter would be before the REST/API calls in the case of the intercepting API calls option and after the REST/API calls for the direct server modifications options. For both of these options, insert/update actions need to be before in order to prevent changes to be made and inserts to be stopped. For retrieval actions that are not allowed, we perform the RBAC checks after the information is retrieved from the database to filter out some or all of the information; this would be done by intercepting the calls in the intercepting API calls option and by including these checks in the server code in the third option. By embedding these checks into the REST/API calls themselves at the server side for the direct server modifications option, the mobile application remains primarily unchanged other than requiring the mobile application to identify itself by role.

RBAC IN CONNECTICUT CONCUSSION TRACKER MOBILE APPLICATION

In order to evaluate the proposed RBAC approach for mobile computing and applications, we apply the model and algorithm as presented in the prior section to the Connecticut Concussion Tracker (CT^2) mobile application. CT^2 is a collaboration between the Departments of Physiology and Neurobiology, and Computer Science & Engineering at the University of Connecticut and Schools of Nursing and Medicine in support of a new law passed in the state of Connecticut to track concussions of kids between ages 7 to age 19 in public schools (CT Law HB6722) (Connecticut General Assembly, 2015). As shown in Figure 7, the Android mobile application CT^2 (first screenshot) consists of a UI with 7 additional tabbed screens: 'Home', 'List', and, 'Student', respectively, in the top row of the figure; and 'Cause', 'Symptoms', 'Follow-up', and 'Return', respectively, in the bottom row. Briefly, we explain each screen. The 'Home' tab allows the user to enter a concussion, to retrieve an open case, or to find a student by typing the student's last name and first name. If a user has access to 'Retrieve Open Cases', this returns the 'List' tab in the top row, which contains the list of students the user has permission to view and, for each student gives him/her the option to add a concussion or edit an existing one (if he/she has permission to access these components). The 'Student' tab, the last screenshot in the top row, allows the user to input the student's general information (e.g., name, birthdate, school) and the date that the incident occurred. The 'Cause' tab, the first screenshot of the bottom row, allows the user via drop down options to specify where the injury was caused, with what it was caused, etc. After the user saves the data he/she entered in the 'Cause' tab, he/she can proceed to the 'Symptoms' tab (second screenshot, bottom row), where the symptoms the student had within 48 hours and other pertinent data are entered. To finish, the 'Follow-up' and 'Return' (third and fourth screenshot, respectively, bottom row) tabs allow users to record the status of the student over time (Follow-up) and when the student can return to various activities at school (Return).

To illustrate the proposed approach of the chapter, four different roles for CT² are defined: the *Nurse* role represents the actions that a school nurse would take in order to manage a student's concussion

Figure 7. CT² mobile application

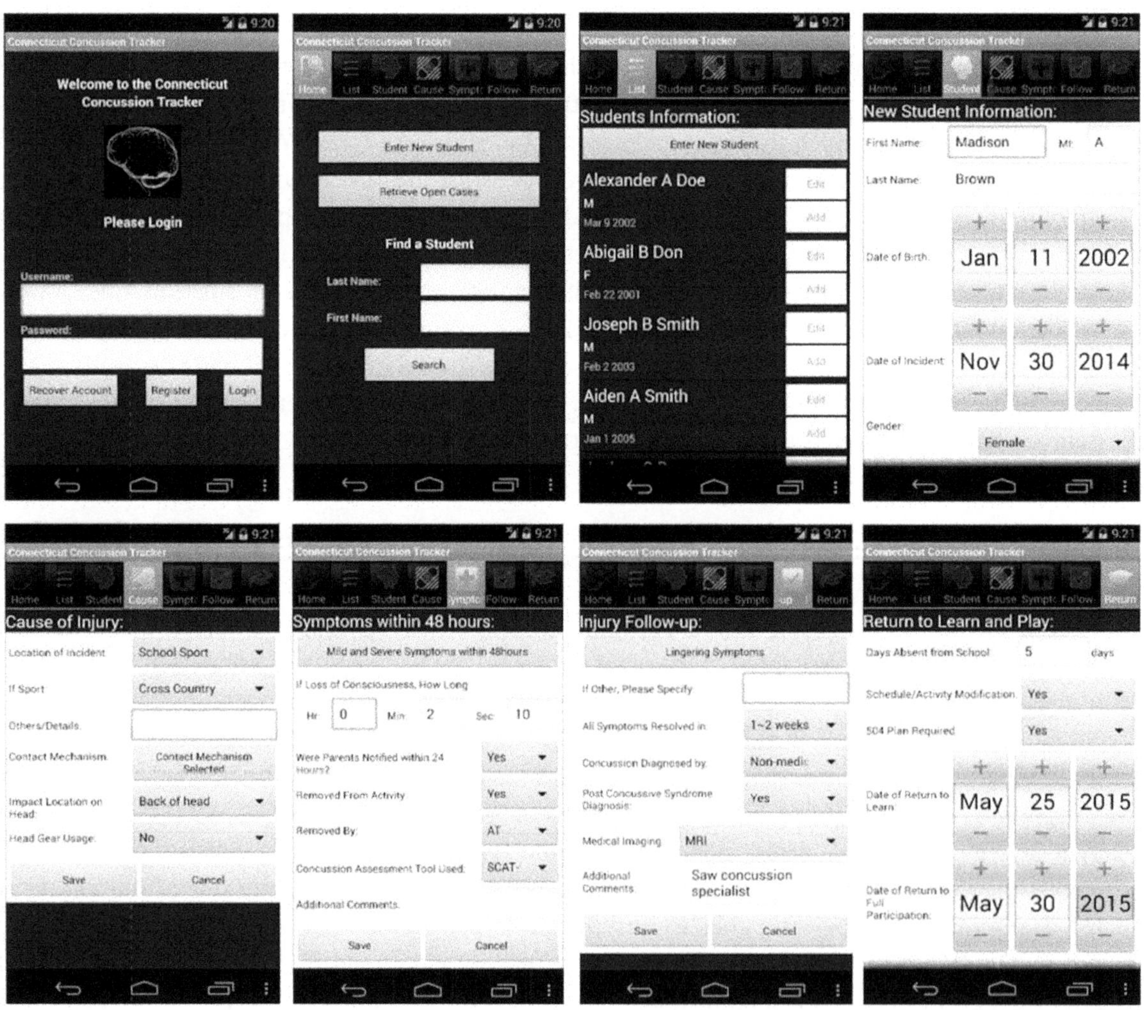

incident from its occurrence to its resolution; the *Athletic Trainer* (*AT*) role represents the actions that a trainer would take at an athletic event including a limited preliminary assessment if a concussion incident occurs at the event; the *Coach* role represents the actions a coach of a sport would take to report a concussion incident at an athletic event with very limited information on the student; and, the *Parent* role which represents the actions to both report a concussion incident on his/her child while attending the athletic event or to track the current status of his/her children that have ongoing concussions. While all of these roles are allowed to enter basic information to report the concussion (fourth screenshot, top row in Figure 7), there are limitations in terms of the screens and the components on the screens that each role is allowed to utilize. For instance, a user with the role *AT* can add students (fourth screenshot, top row in Figure 7), input the cause of the concussion incident (first screenshot, bottom row) that are related to a sport, and add symptoms (second screenshot, bottom row). However, the *AT* role cannot edit information after it is saved (edit once permission), has access to view the *Follow Up* screen but can't edit it (third screenshot, bottom row), and does not have access to the *Return* screen (fourth screenshot, bottom row). On the other hand, a user with role *Nurse* has permission to utilize all of the tabs shown in Figure 7. *Coach* and *Parent* roles will have also limited access to view and/or edit information.

The four roles can be defined in terms of their ability to access the UI of the CT^2 application screens and the components (text fields, spinners, date pickers, drop down boxes, and buttons) on a screen-by-screen basis to establish both the on/off permissions and the data permissions as discussed in the prior section. This information represents the privileges or permissions that are authorized to each role, which when assigned to a given user, results in the CT^2 application being customized in terms of the screens that are displayed and the components that are enabled. Figure 8 contains a table that summarizes all of the screen, on/off, and data permissions for all screens (tabs) and components for the CT^2 application for the four roles: Nurse, Athletic Trainer, Coach, and Parent with respect to their permissions. In terms of permissions, the entire screen can either be shown or hidden as a first level of control. For screens that are shown, the different components can be enabled/disabled (button (BN), radio button (RB), drop down (DD), checkbox (CB), date picker (DP), and/or spinner (SP)) or can be view, edit, or edit once (text field (TF)) via the on/off (values of enable and disable) and data (values of view, edit, and edit once) permissions. The edit once data permission means that the user can input data in the text field one time and, after he/she saves such data, he/she cannot modify it. The edit once option also applies to buttons, drop downs, and spinners, since there are cases where the user selects an option from one of these and it can't be modified later on by him/her. If a screen is hidden from the role, then all of the components of the screen are hidden by default. One of the benefits of our approach is that we can adjust the level of granularity we want to provide to the mobile application since we can add the RBAC proposed approach to specific screens, to specific components, or to all of these (as we do in this case).

The component permissions (Defn. 6) as defined in Figure 8 are inserted as tuples into the CT^2 database that manages all of the concussion incidents, students, and users. To accomplish this, the entity-relationship model as presented in Figure 4 was integrated into the CT^2 database that tracks all of the information on the tabs shown in Figure 7. This includes database tables for: a student, a student's concussion incident, the school and its location, symptoms, information for following up on the concussion, and tracking when the student can return to various activities; note that we have omitted these tables from this discussion since the issue is to authorize and define permissions with respect to information on the various screens and their components. For completeness, we review the entities in Figure 4 in greater detail from a database table perspective in order to support the discussion on the permission process. The roles table assigns a unique id to each of the roles and, stores the name of the role as well as

Figure 8. Tabular summary of role permissions in CT²

Screens/Components	Roles and their Permissions			
	Nurse	Athletic Trainer	Coach	Parent
Home Tab	**Show**	**Show**	**Show**	**Show**
'Enter New Student' BN	Enabled	Enabled	Enabled	Enabled
'Retrieve Open Cases' BN	Enabled	Enabled	Enabled	Enabled
'Last Name' TF	View/Edit	View/Edit once	View/Edit once	View/Edit once
'First Name' TF	View/Edit	View/Edit once	View/Edit once	View/Edit once
'Search' BN	Enabled	Enabled	Enabled	Enabled
List Tab	**Show**	**Show**	**Show**	**Show**
'Enter New Student' BN	Enabled	Enabled	Enabled	Enabled
'View Student Info' BN	Enabled	Enabled	Enabled	Enabled
'Edit' BN	Enabled	Disabled	Disabled	Disabled
'Add' BN	Enabled	Enabled	Enabled	Enabled
Student Tab	**Show**	**Show**	**Show**	**Show**
'First Name' TF	View/Edit	View/Edit once	View/Edit once	View/Edit once
'Middle Initial' TF	View/Edit	View/Edit once	View/Edit once	View/Edit once
'Last Name' TF	View/Edit	View/Edit once	View/Edit once	View/Edit once
'Gender' DD	View/Edit	View/Edit once	View/Edit once	View/Edit once
'Date of Birth' DP	View/Edit	View/Edit once	View/Edit once	View/Edit once
'Date of Past Concussions' DP	View/Edit	View/Edit once	View/Edit once	View/Edit once
'State' DD	View/Edit	View/Edit once	View/Edit once	View/Edit once
'City/Town/Region' DD	View/Edit	View/Edit once	View/Edit once	View/Edit once
'District' DD	View/Edit	View/Edit once	View/Edit once	View/Edit once
'School' DD	View/Edit	View/Edit once	View/Edit once	View/Edit once
'Save' BN	Enabled	Enabled	Enabled	Enabled
'Cancel' BN	Enabled	Enabled	Enabled	Enabled
Cause Tab	**Show**	**Show**	**Show**	**Show**
'Location of Incident' DD	Enabled	View/Edit once	View/Edit once	View/Edit once
'If Sport' DD	Enabled	View/Edit once	View/Edit once	View/Edit once
'Others/Details' TF	View/Edit	View/Edit once	View/Edit once	View/Edit once
'Contact Mechanism' DD	Enabled	View/Edit once	View/Edit once	View/Edit once
'Impact Location of Head' DD	Enabled	View/Edit once	View/Edit once	View/Edit once
'Head Gear Usage' DD	Enabled	View/Edit once	View/Edit once	View/Edit once
'Save' BN	Enabled	Enabled	Enabled	Enabled
'Cancel' BN	Enabled	Enabled	Enabled	Enabled
Symptom Tab	**Show**	**Show**	**Hide**	**Show**
'Mild and Severe Symptoms' BN	Enabled	View/Edit once	-	View
'Hour' TF	View/Edit	View/Edit once	-	View
'Minute' TF	View/Edit	View/Edit once	-	View
'Second' TF	View/Edit	View/Edit once	-	View
'Were Parents Notified?' DD	Enabled	View/Edit once	-	View
'Removed From Activity' DD	Enabled	View/Edit once	-	View
'Removed by' DD	Enabled	View/Edit once	-	View
'Concussion Assessment Tool' DD	Enabled	View/Edit once	-	View
'Additional Comments' TF	View/Edit	View/Edit once	-	View
'Save' BN	Enabled	Enabled	-	Disabled
'Cancel' BN	Enabled	Enabled	-	Disabled
Follow Up Tab	**Show**	**Show**	**Hide**	**Show**
'Lingering Symptoms' BN	Enabled	View	-	View
'If Other, Please Specify' TF	View/Edit	View	-	View
'All Symptoms Resolved in' DD	Enabled	View	-	View
'Concussion Diagnosed by' DD	Enabled	View	-	View
'Post Concussive Syndrome Diagnosis' DD	Enabled	View	-	View
'Medical Imaging' DD	Enabled	View	-	View
'Additional Comments' TF	View/Edit	View	-	View
'Save' BN	Enabled	Disabled	-	Disabled
'Cancel' BN	Enabled	Disabled	-	Disabled
Return Tab	**Show**	**Hide**	**Hide**	**Hide**
'Days Absent From School' TF	View/Edit	-	-	-
'Schedule/Activity Modification' DD	Enabled	-	-	-
'504 Plan Required' DD	Enabled	-	-	-
'Date of Return to Learn' DP	Enabled	-	-	-
'Date of Return to Full Participation' DP	Enabled	-	-	-
'Save' BN	Enabled	-	-	-
'Cancel' BN	Enabled	-	-	-

an optional description. The users table in the CT^2 database models information on each user and it was modified to include the role_id foreign key in order to determine the role of the user once he/she logs in to the application. After a user logs in successfully, the role_id is obtained to determine which students and which screens of the UI the user has access to. The screens table assigns a unique screen_id to each of the screens of the UI as well as a name so that it would be easier to identify and manage them while the screen_components table stores the information that is used to view which components belong to which screen. The screen_access table contains the screen_id, the role_id, and a Boolean value named access, and determines if the specified role has access to the specified screen (access = 1) or not (access = 0). The screen_sequence table is utilized to determine the screen(s) that a role has access to once it modifies a component. Finally, the component_access table stores the information that specifies the screen in which a component is located, which role id can access the component, and the permissions the role has over this component in terms of on/off and data permissions that are utilized to enable/disable, view, edit, or edit once. The permissions as captured in the upgraded CT^2 database in Figure 4 support the three API calls of the prior section and are executed in the CT^2 mobile application's code to enforce the permissions of each role to determine the screens and their components for each user by role. CT^2 utilizes a MySQL database to store its data and relies on API commands to retrieve data for display on the UI and to store new concussion incidents (or changes) into the database. Figure 9 illustrates the result of the screen and component permissions for the *Nurse*, *Parent*, and *Coach* roles. The *Nurse* role has all of the tabs active (first screen of figure 9). The *Parent* role has limited access to the tabs and also, although users with the *Parent* role can view the 'Symptom' and 'Follow Up' tabs, they are not allowed to update that information (second screen of Figure 9). Finally, the *Coach* role has the first four tabs active and has limited access to the information it can view/modify (third screen of Figure 9).

Figure 9. Result of RBAC in CT^2 (nurse, parents, and coach view)

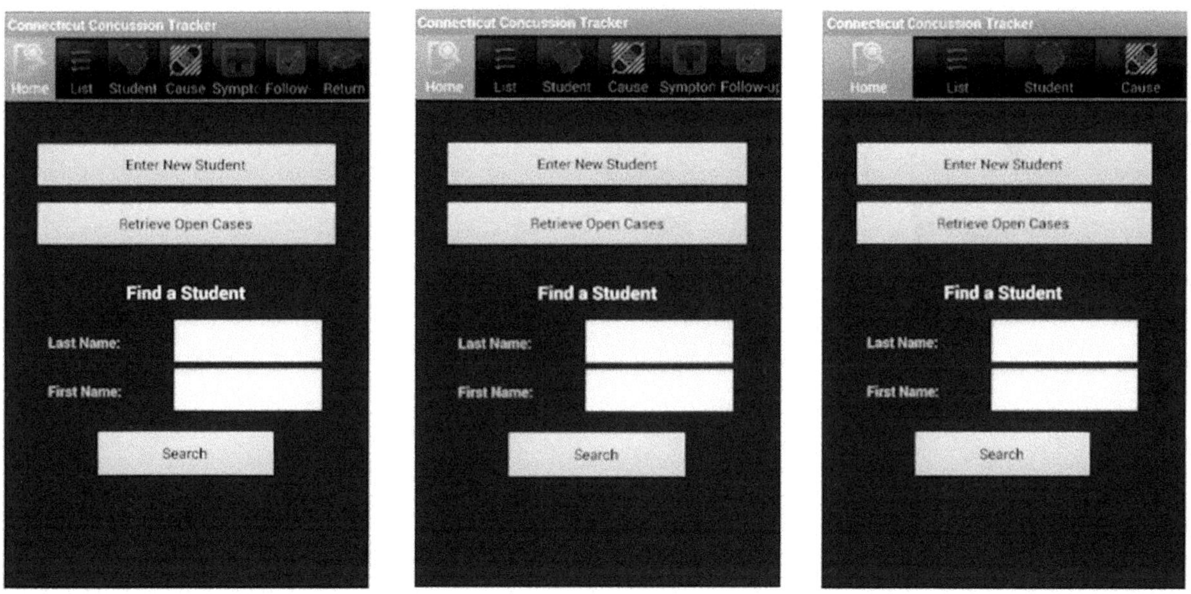

FUTURE TRENDS

This section explores future trends with a two-fold focus: novel access control methods that have the potential for future usage in a mobile setting; and, device level security methods related to authorization and authentication.

The Attribute-Based Access Control (ABAC) method (Coyne & Weil, 2013) is a context-aware approach that considers the user's responsibilities and can place users in groups/roles based on their attributes. For example, a physician has a different context when he/she is seeing patients in the medical practice, attending patients at a hospital or clinic, or answering phone calls after hours and on weekends; this context differs based on what he/she does. One of the main advantages of ABAC over RBAC is the ability of granting dynamic permissions by checking the user's attributes to determine if the user has access to that component at that particular moment. Figure 10 illustrates the overall architecture of the ABAC model where: a user tries to access a component and the Access Control Mechanism processes the user's attributes by analyzing these against an access control policy, subject attributes, component attributes, and environmental conditions (The Johns Hopkins University Applied Physics Laboratory, 2014).

Figure 10. Overall architecture of the ABAC model

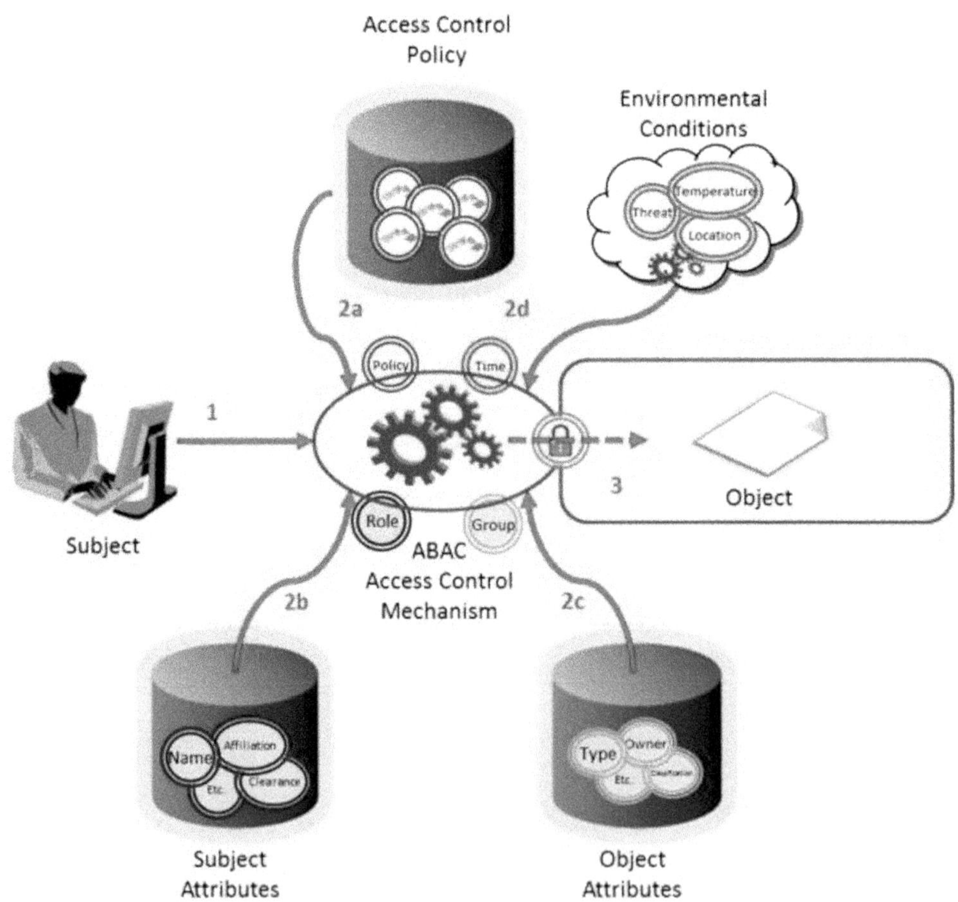

The BiLayer Access Control (BLAC) method (Alshehri & Raj, 2013) provides secure authorization as a fusion between RBAC and ABAC that combines attributes with roles. BLAC has additional advantages to RBAC of supporting attributes and policies where is verified utilizing two layers. The first layer checks the individual against pseudo roles (i.e., the list of individual's attributes) while the second layer checks the requested component against rules within the policies that are related. Figure 11 illustrates the high-level components of the BLAC model that: assigns a pseudo role to the subject; evaluates this against the subject's attributes, the environmental attributes, and the component's attributes; verifies all attributes against the policy; and determines whether the subject has access to the requested component or not. The use of pseudo roles could be leveraged in the physician example of ABAC to define pseudo roles based on where the physician treats patients.

To complete this discussion, there are emerging device level security approaches for mobile computing related to authorization and authentication. The Oauth2 authorization framework (Hardt, 2012) has the purpose of granting access to a third-party application to obtain limited access to an HTTP service. One of the advantages of Oauth2 is that there is no need to store the user's credentials (e.g., username and password) to *reauthorize* in the future since an access token is employed. Such a process is performed through four different roles: *resource owner* manages the access to a protected source; *resource server* utilizes access tokens to determine which requested resources a user can view; *client* is the application that makes resource requests on behalf of the resource owner; and, *authorization server* issues access tokens to the client once it has properly authenticated the resource owner as well as obtained authorization. A good amount of websites that require users to authenticate (i.e., Facebook, Gmail, Twitter, etc.) utilize the Oauth2 protocol's API to authorize them after the authentication process takes place.

Another security solution for mobile devices is Stormpath (2011) (API and open source SDKs) that utilizes the OAuth2 password grant type scheme to authenticate users to manage part of the authentication process and to be able to issue an access token for the user that is requesting access to the mobile

Figure 11. High-level components in BLAC model
Alshehri & Raj, 2013.

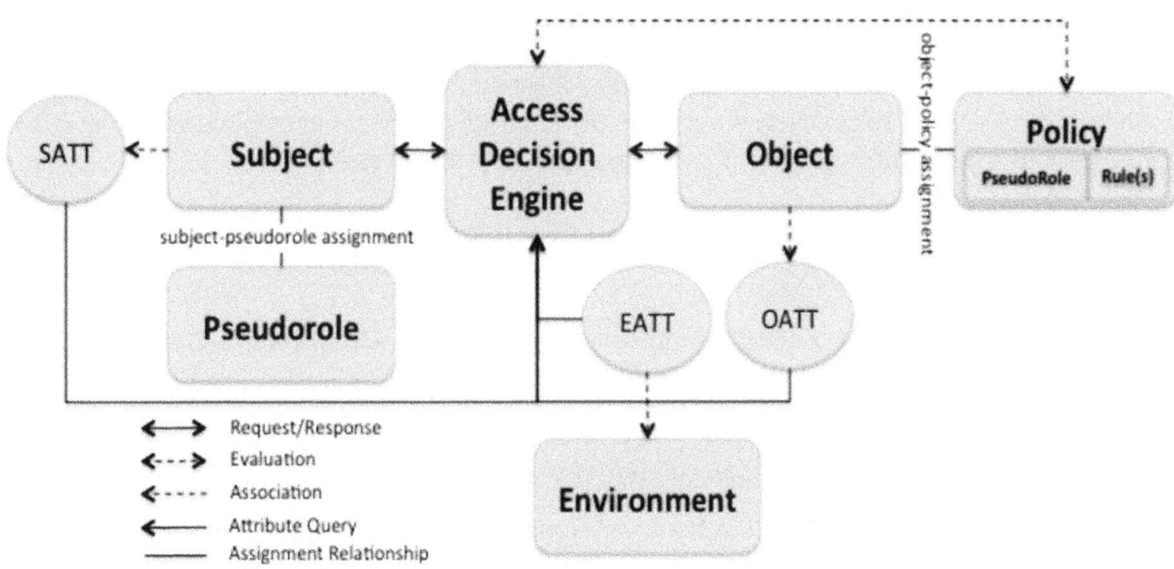

application. One of Stormpath's features is a user authorization management mechanism for mobile applications and support for single sign-on and multi-factor authentication (Stormpath, 2015). Another approach, MobileIron (2015), is a solution that seeks to secure a group of mobile applications that have highly secure information for a particular enterprise. MobileIron focuses on providing security on mobile devices, mobile applications, and mobile content of enterprises through the means of a solution called Enterprise Mobility Management (EMM) (MobileIron Solutions EMM, 2015). EMM is composed of three approaches: Mobile Device Management (MobileIron Solutions MDM, 2015), Mobile Application Management (MobileIron Solutions MAM, 2015), and Mobile Content Management (MobileIron Solutions MCM, 2015). These three approaches manage enterprise settings, which users can utilize to secure mobile apps and, to obtain enterprise access to corporate data via VPNs.

CONCLUSION

This chapter has presented a generalizable approach to RBAC for mobile applications that proposed a configurable framework of three options in Figure 1, which can be utilized to both define and enforce security on a mobile application in differing locations of the infrastructure. The direct UI modifications option was the focus of this chapter to the UI to be dynamically customized by role via the usage of an application wrapper. To organize the presentation, the *Background and Motivation* section reviewed NIST RBAC and the usage of RBAC in mobile computing and healthcare. With this as a basis, the *RBAC Approach for Mobile Applications* section presented an RBAC approach capable of defining roles with screen and component permissions of the mobile application's UI, realized with a corresponding RBAC model in SQL, to dynamically customize the UI of a mobile application in support of the direct UI modifications option. This included an API as an application wrapper which was employed to make programmatic changes to the mobile application with conditional statements that would enable/disable by role the screens and components of a UI. This also briefly discussed the intercepting API calls and direct server modifications options of the configurable framework of Figure 1, illustrating the other ways to include RBAC that has less of an impact on the mobile application itself. To demonstrate the feasibility and utility of our work, the *RBAC in Connecticut Concussion Tracker Mobile Application* section, implements our proposed approach in the CT^2 mobile application, making it appear different (in terms of UI) to users based on their role, with permissions illustrated for four different roles (Nurse, Parent, Coach, Athletic Trainer). To complete the chapter, the *Future Trends* section reviewed emerging RBAC approaches and security infrastructures for authorization and authentication.

REFERENCES

Abdunabi, R., Sun, W., & Ray, I. (2014). Enforcing spatio-temporal access control in mobile applications. *Computing*, *96*(4), 313–353. doi:10.1007/s00607-013-0340-2

Alshehri, S., & Raj, R. (2013). Secure Access Control for Health Information Sharing Systems.*2013 IEEE International Conference on Healthcare Informatics, ICHI 2013*. doi:10.1109/ICHI.2013.40

Apple. (2015). *ResearchKit and CareKit*. Retrieved from http://www.apple.com/researchkit/

Boonstra, A., & Broekhuis, M. (2010). Barriers to the acceptance of electronic medical records by physicians from systematic review to taxonomy and interventions. *BMC Health Services Research BMC Health Serv Res*, *10*(1), 231. doi:10.1186/1472-6963-10-231 PMID:20691097

Caine, K., & Hanania, R. (2013). Patients want granular privacy control over health information in electronic medical records. *Journal of the American Medical Informatics Association*, *20*(1), 7–15. doi:10.1136/amiajnl-2012-001023 PMID:23184192

Capzule. (2012). *Capzule PHR*. Retrieved from http://www.capzule.com/

Care360. (2014). *MyQuest*. Retrieved from https://myquest.questdiagnostics.com/web/home

Cisco. (2014). *Cisco Visual Networking Index: Global Mobile Data Traffic Forecast Update, 2014-2019*. Retrieved from http://www.cisco.com/c/en/us/solutions/collateral/service-provider/visual-networking-index-vni/mobile-white-paper-c11-520862.html

Cohen, J. (2015, January 7). *11 Health And Fitness Apps That Achieve Top Results*. Retrieved from http://www.forbes.com/sites/jennifercohen/2015/01/07/the-11-top-health-fitness-apps-that-achieve-the-best-results/#11f5c21a1aca

Connecticut General Assembly. (2015). *Substitute for Raised H.B. No. 6722*. Retrieved from https://www.cga.ct.gov/asp/CGABillStatus/CGAbillstatus.asp?which_year=2015&selBillType=Bill&bill_num=HB6722

Coyne, E., & Weil, T. (2013). ABAC and RBAC: Scalable, Flexible, and Auditable Access Management. *IT Professional*, *15*(3), 14–16. doi:10.1109/MITP.2013.37

eMarketer. (2015, January 9). *Tablet Users to Surpass 1 Billion Worldwide in2015*. Retrieved from http://www.emarketer.com/Article/Tablet-Users-Surpass-1-Billion-Worldwide-2015/1011806

Fernández-Alemán, J., Señor, I., Lozoya, P., & Toval, A. (2013). Security and privacy in electronic health records: A systematic literature review. *Journal of Biomedical Informatics*, *46*(3), 541–562. doi:10.1016/j.jbi.2012.12.003 PMID:23305810

Ferraiolo, D., & Kuhn, R. (1992). Role-Based Access Control. In *Proceedings of the NIST-NSA National (USA) Computer Security Conference*.

Ferraiolo, D., Sandhu, R., Gavrila, S., Kuhn, D., & Chandramouli, R. (2001). Proposed NIST standard for role-based access control. *ACM Transactions on Information and System Security*, *4*(3), 224–274. doi:10.1145/501978.501980

Ferreira, A., Chadwick, D., & Antunes, L. (2007). Modelling access control for healthcare information systems. In *Proceedings of the Doctoral Consortium at the 9th International Conference on Enterprise Information Systems (ICEIS)*.

Gajanayake, R., Iannella, R., & Sahama, T. (2014). Privacy oriented access control for electronic health records. *Electronic Journal of Health Informatics*, *8*(2), e15.

Gartner. (2012). *Mobile Device Management (MDM)*. Retrieved from http://www.gartner.com/it-glossary/mobile-device-management-mdm

Gartner. (2015, March 19). *Gartner Says Global Devices Shipments to Grow 2.8 Percent in2015*. Retrieved from http://www.gartner.com/newsroom/id/3010017

Google Play. (2013). *Fitness Tracker*. Retrieved from https://play.google.com/store/apps/details?id=com.realitinc.fitnesstracker

HL7 Security Technical Committee. (2007, September). *HL7 Role-Based Access Control (RBAC) Role Engineering Process*. Retrieved from http://csrc.nist.gov/groups/SNS/rbac/documents/hl7_role-based_access_control_(rbac).pdf

Hardt, D. (Ed.). (2012). *The OAuth 2.0 Authorization Framework*. Internet Engineering Task Force (IETF). Retrieved from https://tools.ietf.org/html/rfc6749

Health Information Privacy. (2002). *Minimum Necessary Requirement*. Retrieved from http://www.hhs.gov/hipaa/for-professionals/privacy/guidance/minimum-necessary-requirement/index.html

Health Level Seven International. (2010). *Introduction to HL7 Standards*. Retrieved from http://www.hl7.org/implement/standards/

HHS.gov. (2013). *Health Information Privacy*. Retrieved from http://www.hhs.gov/hipaa/index.html

Hsu, W., & Pan, J. (2013). The Secure Authorization Model for Healthcare Information System. *Journal of Medical Systems*, *37*(5), 1–5. doi:10.1007/s10916-013-9974-z PMID:24061706

iOS 9. (2014). *Health: An innovative new way to use your health and fitness information*. Retrieved from https://www.apple.com/ios/health/

Kelly, S. M. (2014, June 27). *In Google Fit vs. Apple HealthKit, Fitness Apps Stay Neutral*. Retrieved from http://mashable.com/2014/06/27/healthkit-google-fit-apps/#nf_r0U7flmqC

MedWatcher. (2012). *MedWatcher*. Retrieved from https://medwatcher.org/

MobileIron. (2015). *MobileIron*. Retrieved from https://www.mobileiron.com/en

MobileIron Solutions EMM. (2015). *Enable business transformation MobileIron's Enterprise Mobility Management (EMM) Platform*. Retrieved from https://www.mobileiron.com/en/solutions/enterprise-mobile-management-emm

MobileIron Solutions MAM. (2015). *Do more with secure mobile application management (MAM)*. Retrieved from https://www.mobileiron.com/en/solutions/mobile-application-management-mam

MobileIron Solutions MCM. (2015). *Secure mobile content management (MCM) keeps data safe and business moving*. Retrieved from https://www.mobileiron.com/en/solutions/mobile-content-management-mcm

MobileIron Solutions MDM. (2015). *Mobile Device Management (MDM): The foundation for a secure mobile enterprise*. Retrieved from https://www.mobileiron.com/en/solutions/mobile-device-management-mdm

Motta, G., & Furuie, S. (2003). A contextual role-based access control authorization model for electronic patient record. *Proceedings of the IEEE Transactions on Information Technology in Biomedicine*, *7*(3), 202–207. doi:10.1109/TITB.2003.816562 PMID:14518734

MTBC PHR. (2011). *MTBC PHR*. Retrieved from https://phr.mtbc.com/

My Imaging Records App. (2013). *My Imaging Records App*. Retrieved from http://myimagingrecords.com/index.html

NIST Computer Security Division. (2013). *Attribute-Based Access Control*. Retrieved from http://csrc.nist.gov/projects/abac/

Paganini, P. (2015, June 17). *Mobile App Security: Threats and Best Practices*. Retrieved from https://www.veracode.com/blog/2015/05/mobile-app-security-threats-and-best-practices-sw

Peleg, M., Beimel, D., Dori, D., & Denekamp, Y. (2008). Situation-Based Access Control: Privacy management via modeling of patient data access scenarios. *Journal of Biomedical Informatics*, *41*(6), 1028–1040. doi:10.1016/j.jbi.2008.03.014 PMID:18511349

Pew Research Center. (2012, February 23). *Mobile Technology Fact Sheet*. Retrieved from http://www.pewinternet.org/fact-sheets/mobile-technology-fact-sheet/

Pharmacy, C. V. S. (2015). *myCVS On the Go*. Retrieved from http://www.cvs.com/mobile-cvs

Radicati, S. (2014). *Mobile Statistics Report, 2014-2018*. Retrieved from http://www.radicati.com/wp/wp-content/uploads/2014/01/Mobile-Statistics-Report-2014-2018-Executive-Summary.pdf

Rindfleisch, T. C. (1997). Privacy, information technology, and health care. *Communications of the ACM*, *40*(8), 92–100. doi:10.1145/257874.257896

Russello, G., Dong, C., & Dulay, N. (2008). A Workflow-Based Access Control Framework for e-Health Applications.*22nd International Conference on Advanced Information Networking and Applications - Workshops (AINAW 2008)*. doi:10.1109/WAINA.2008.131

Sandhu, R., Ferraiolo, D. F., & Kuhn, R. (2000). The NIST Model for Role Based Access Control: Toward a Unified Standard. In *Proceedings of the Fifth ACM Workshop on Role-based Access Control (RBAC '00)*. doi:10.1145/344287.344301

Santos-Pereira, C., Augusto, A. B., Correia, M. E., Ferreira, A., & Cruz-Correia, R. (2012). A Mobile Based Authorization Mechanism for Patient Managed Role Based Access Control. LNCS, 7451, 54-68.

Schefer-Wenzl, S., & Strembeck, M. (2013). Modelling Context-Aware RBAC Models for Mobile Business Processes. *International Journal of Wireless and Mobile Computing*, *6*(5), 448. doi:10.1504/IJWMC.2013.057387

Scholl, M. A., Stine, K. M., Hash, J., Bowen, P., Johnson, L. A., Smith, C. D., & Steinberg, D. I. (2008). *An Introductory Resource Guide for Implementing the Health Insurance Portability and Accountability Act (HIPAA) Security Rule*. Gaithersburg, MD: National Institute of Standards & Technology. doi:10.6028/NIST.SP.800-66r1

Science Application International Corporation (SAIC). (2004, May 11). *Role-Based Access Control (RBAC) Role Engineering Process*. Retrieved from http://csrc.nist.gov/groups/SNS/rbac/documents/HealthcareRBACTFRoleEngineeringProcessv3.0.pdf

Slevin, L. A., & Macfie, A. (2007). Role Based Access Control for a Medical Database. In *Proceedings of Software Engineering and Applications* (pp. 195–199). SEA.

SlideShare. (2012). *Constrained RBAC diagram*. Retrieved from http://image.slidesharecdn.com/rbac6576-121205031439-phpapp01/95/rbac-18-638.jpg?cb=1354677352

Smith, A. (2015, April 1). *U.S. Smartphone Use in2015*. Retrieved from http://www.pewinternet.org/2015/04/01/us-smartphone-use-in-2015/

Stormpath. (2011). *Stormpath*. Retrieved from https://stormpath.com/

Stormpath. (2015). *User Authorization Management*. Retrieved from https://stormpath.com/product/authorization

Sujansky, W. V., Faus, S. A., Stone, E., & Brennan, P. F. (2010). A method to implement fine-grained access control for personal health records through standard relational database queries. *Journal of Biomedical Informatics*, *43*(5), 46–50. doi:10.1016/j.jbi.2010.08.001 PMID:20696276

Symantec. (2014, October 7). *Securing Mobile App Data - Comparing Containers and App Wrappers*. Retrieved from https://www.symantec.com/content/en/us/enterprise/white_papers/b-securing-mobile-app-data-comparing-containers-wp-21333969.pdf

The Johns Hopkins University Applied Physics Laboratory. (2014, December 19). *Digital Policy Management Framework for Attribute-Based Access Control*. Retrieved from https://www.ise.gov/sites/default/files/DigitalPolicyFramework-ABAC.pdf

Wainwright, A. (2012, June 21). *7 Benefits of BYOD on Enterprise Wireless Networks*. Retrieved from http://www.securedgenetworks.com/blog/7-Benefits-of-BYOD-on-Enterprise-Wireless-Networks

West, D., & Miller, E. A. (2009). Digital Medicine: Health Care in the Internet Era. Brooking Institution Press.

Wiech, D. (2013, April 17). *Role-Based Access Control for Healthcare Data Security*. Retrieved from http://healthcare-executive-insight.advanceweb.com/Features/Articles/Role-based-Access-Control-for-Healthcare-Data-Security.aspx

KEY TERMS AND DEFINITIONS

Access Control: Method that determines who or what has access to resources in an environment.

Application Wrapper: Management layer applied to a mobile application in order to incorporate additional functionality that can be called by the application with a limited overhead and impact.

Authorization: Procedure that allows an individual to access/obtain something.

Electronic Medical Record (EMR): Digital version of a patient's paper health record.

Health Information Technology (HIT) systems: Electronic systems that are utilized with the expectation that they will lower costs, improve efficiency, and reduce errors, in addition to providing better care and services for consumers in the healthcare domain.

Health Insurance Portability and Accountability Act (HIPAA): Law which main purpose is to protect the confidentiality and security of healthcare information, as well as assist the healthcare industry to control administrative costs.

Health Level Seven International (HL7): Organization that delivers a framework and several standards in order to manage electronic health information.

Highly Sensitive Data: Information that if exposed or modified without proper authorization could lead in a critical effect on the performance, assets, or reputation of its owner (e.g., medical data, credit card data, social security number, etc.).

Mobile Computing: Devices that allow people to access software/data regardless of their location.

Role-Based Access Control (RBAC): Approach utilized to control access to a computer system through the means of roles assigned to the users of such system.

Chapter 7
A Spatio-Situation-Based Access Control Model for Dynamic Permission on Mobile Applications

Xian Shao
University of Connecticut, USA

Steven A Demurjian
University of Connecticut, USA

Thomas P Agresta
University of Connecticut Health Center, USA

ABSTRACT

As users are now able to take their mobile devices from location to location, there has been a transition from a static program running on a PC/laptop to a dynamic application that can adapt based on a variety of conditions and criteria. This highlights an emerging need to support dynamic permissions of mobile applications as a user moves from location to location based and perform different actions in particular situation. This chapter presents a Spatio-Situation-Based Access Control model that extends role-based access control to secure sensitive data for mobile applications with the ability to make dynamic authorization decisions according to the time/location and the particular situation being encountered by a user. To demonstrate the feasibility of the work, a realistic healthcare scenario examines the complex workflow of treating a patient by a physician utilizing a mobile health (mHealth) app to access patient data, as she/he moves among multiple locations at different times throughout the day/week requiring access to different patient data repositories at different times.

INTRODUCTION

Mobile devices and applications have dramatically altered the way that users interact electronically, moving away from a PC/laptop-based world. The idea of sitting in one place to do computing has evolved to one where movement is the norm and not the exception. In acknowledgement of this ever-changing

DOI: 10.4018/978-1-5225-0945-5.ch007

environment, one effort has identified the need for different programming and usability models to develop mobile applications (Harrison, Flood, & Duce, 2013), particularly in the context where individuals are moving from location to location. The usage of location has dramatically changed the way that applications are written – allowing there to be an engagement of interactions that allows users to effortlessly find businesses, restaurants, hotels, shopping venues, etc (Ramey, 2013). This is primarily supported by the ubiquity of GPS chips in mobile devices (Venturebeat Staff, 2014). Mobile devices are also impacting business to business (B2B), business to employee (B2E), and business to consumer (B2C) interactions (Cryderman, 2011) by: providing disposable applications that can target B2C for very short durations which eliminates a need to maintain application versions over time; and, increasing application communications including bi-directional communications for B2C where the user's feedback can impact the application behavior and the application itself. In all of these activities, the location of the app, and the movement of a user from location to location has changed the view from a static program running on a PC/laptop to a dynamic application that can adapt based on a variety of conditions and criteria.

One area where location-based behavior would be particularly useful and beneficial is in mobile applications for healthcare, coined "mHealth", in 2009 (Torgan, 2009) with a recognition of its potential impact (Himss, 2014). Mobile applications in healthcare are touted for their potential to improve access to care, patient engagement, and safety, while also requiring new models for physicians to use in their medical practices (Savitz, 2012). For patients, those applications may serve as a diagnostic tool, monitoring glucose level for diabetes or weight for obesity, health/fitness tracking, health information, diseases monitoring, etc.; all of these capabilities can provide new and relevant information for physicians (Haberle, 2014). To give an idea of the scope, a report from the IMS Institute for Healthcare Informatics (Aitken, 2013) found 43,700+ medical applications in the Apple application store, with approximately 69% targeting consumers/patients and 31% for use by medical providers (e.g., physicians, nurses, therapists, specialists, etc.); this was further summarized (Posada, 2014). Medical providers are also increasingly utilizing mobile applications in their medical practices for information management, social networking for consulting physicians, drug and medical information, medical education and training, and patient management and monitoring (Lee Ventola, 2014). All of these various mHealth applications for patients/consumers and medical providers will require Health Insurance Portability and Accountability Act (HIPAA) (HHS, 1996) compliant storage of data collected by patients and the potential interaction of data stored in numerous health information technology (HIT) systems that include an Electronic Health Record (EHR) (PrognoCIS, 2010), an Electronic Medical Record (EMR) (OpenEMR, 2012), and/or a Personal Health Record (PHR) (Microsoft, 2007). All of these mHealth applications contain sensitive information that is collected, recorded, stored, processed, and transferred which require a high degree of privacy safety per both HIPAA and new Food and Drug Administration (FDA) guidelines for mobile medication applications (FDA, 2015).

For mHealth applications for patients/consumer and medical providers, the location of an individual could dictate the type of information that is available from a particular health IT system. For example, consider a patient with a mobile application to track his/her medical data (e.g., demographics, contact info, current conditions, medications, recent test results, mental health information, genetic information, etc.). The information that he/she is able to see and/or modify may vary and depend on his/her location. A patient visiting a family medicine physician would see the demographics, contact info, current conditions, medications, recent test results, etc., but not see the mental health information unless the patient was visiting his/her psychiatrist. Likewise, a medical provider utilizing a tablet may see different data when he/she is at his medical practice (basic patient data of his/her patients in the practice's EMR) vs.

when at a hospital (specific medical test results only available in that setting in the hospital's health IT systems) vs. when at home during the night or weekend when covering emergency calls a group of physicians (basic patient data from patients that may be in different EMRs). A medical provider in his/her day-to-day activities may move among multiple physical locations, seeing the patient initially at his/her medical practice, admitting the patient to a hospital for testing and treating, and following the patient at the hospital in different departments (locations) until the patient is discharged to either home or an after-care facility. In all of these different locations, the information that the medical provider accesses will be specific to a department in a particular location at a given time and he/she may not be allowed to access data from a prior specific place at a later time and in a different location. Further, within a large hospital medical complex spreading across multiple city blocks and comprised of multiple separate and connected buildings, the medical provider may be limited to access specific data at certain locations within the medical complex. In all of these situations, there will be a critical need to deliver customized data to mHealth applications that is targeted towards what each stakeholder (patient, consumer, medical provider, etc.) is allowed to view and/or edit.

In this chapter, our focus is on assessing variants of Role-Based Access Control (RBAC) (Ferraiolo et al, 2001) utilized in traditional computing settings for their suitability in a mobile context. This includes approaches that have been proposed for healthcare such as: Spatio-temporal Access Control (Ray & Toahchoodee, 2007), Situation-Based Access Control (Peleg et al., 2008), Workflow-Based Access Control (Russello, Dong & Dulay, 2008), and Collaboration and Workflow-Based Access Control (Le et al., 2012). Using this as a basis, the chapter proposes and discusses the Spatio-Situation-Based Access Control (SSBAC) model that combines features from existing access control models with new capabilities for the dynamic enforcement of security as a user moves among various locations with his/her mobile device and associated applications over time and distance. While GPS capabilities in mobile devices can obtain a very good estimate of location, this GPS/spatio information may not be accurate enough to differentiate between two locations that are very close to one another within a large medical complex. This proposed SSBAC demonstrates the way that this concept can be applied to an mHealth application, to allow, for example, a medical provider to only access/edit patient data when he/she is in the hospital at specific time, thereby providing a location-based access control. The result is to constrain access to different health IT systems as a medical provider moves in both space/time and by the situation so that the mHealth application can dynamically adapt to the environments and allow or deny the access to specific data.

This chapter presents our work on a Spatio-Situation-Based Access Control (SSBAC) model for mobile computing built on top of RBAC and leverages, combines, and extends two access control approaches: Spatio–temporal Access Control and Situation-Based Access Control (Ray & Toahchoodee, 2007; Peleg, M., Beimel, Dori & Denekamp, 2008). SSBAC utilizes a combination of the location and time to check privileges (spatio) coupled with the type of mobile application and/or its scenario of usage (situation), extended by introducing the *Zone* concept that represents a specific geographic area such as a building or a complex that contains standalone and inter-connected buildings. The decision on granting access to roles takes into account the location of users while simultaneously considering the situation in which the data-requester is operating his/her mobile application in a particular Zone. Incorporating contextual factors in mobile environments is a challenging problem, as these environments are inherently dynamic (Kirkpatrick, Damiani & Bertino, 2011). To accomplish this, there needs to be techniques that are able to monitor and react to changes in a user's location. This challenge can be described as enforcing continuity of usage constraints (Kirkpatrick, Damiani & Bertino, 2011). SSBAC adopts the capabilities of the Spatio-temporal RBAC Access Control model (STRBAC), (Ray & Toahchoodee, 2007) which has been designed and targeted for pervasive computing applications and extends RBAC with time and location. SSBAC also employs concepts from the Situation-Based Access Control model (SitBAC) (Peleg, M., Beimel, Dori & Denekamp, 2008), also based on RBAC, which defines scenarios where a patient's data access is permitted or denied. The main concept of SitBAC that SSBAC utilizes is the Situation Schema, which is a design pattern consisting of the entities that include: the data-requester, the patient, the data source (EMR), the legal-authorization to the protected patient data, etc.; this supports the situation aspect of SSBAC. To illustrate both the spatio and situation aspects of SSBAC, suppose the physician is utilizing an EMR at his/her office. If the physician is moving from his/her office to a hospital in a distinguishable geographic location (far enough away for differentiation via GPS and/or cell tower pinging), then the spatio aspect of SSBAC can be utilized to determine the correct appropriate privileges and allowed/prohibited access. If the office and hospital that are located in buildings across the street from one another (geographically indistinguishable), the physician may be prohibited from access to the office EMR while in the hospital. In this case, the proposed SSBAC needs to utilize the situation (what the user/physician is doing with the mobile/mHealth application) to dynamically determine the correct appropriate privileges and allowed/prohibited access. An orthogonal concern for SSBAC is the work on access control models with "break-the-glass" capabilities in the medical domain, when in emergent situations, a physician could obtain access to protected data (Brucker & Petritsch, 2009). This important issue, while not the subject of the paper, is a crucial expected capability in the medical domain.

The remaining of the Chapter has five sections. In the *Background,* work on RBAC is reviewed along with Spatio-temporal Access Control (Ray & Toahchoodee 2007), and Situation-Based Access Control model (Peleg et al., 2008). In the *Spatio-Situation-Based Access Control (SSBAC) Model* section, we introduce and formally define the model for SSBAC that combines STRBAC and SitBAC and adds a new capability to define Zones which are geographic locations; specify constraints that focus on when a user playing a role can access a permissions based on location; and, present a platform-independent software architecture for SSBAC. To demonstrate the capabilities of SSBAC, the *Scenario of Usage for SSBAC in Healthcare* section provides a detailed example using an mHealth application via a platform-independent implementation architecture that maps the high-level proposed access control model for SSBAC to its enforcement mechanism in mobile applications. Next, the *Future Literature Review*section presents emerging efforts related to dynamic access control for mobile devices, including: combining the location of a user with his/her social connections for Geo-Social RBAC (Baracaldo, Palanisamy & Joshi, 2014); reviewing a technique that consider the tasks that a user performs in conjunction with the role (Mallare & Pancho-Festin, 2013); a context aware approach that considers the behavior and associated workflow of a user (Yarmand, Sartipi & Down, 2008); and, a review of the extension of access control using break-the-glass (Brucker & Petritsch, 2009). Note that the last three topics are targeting the healthcare domain. Finally, *Conclusion* section completes the chapter.

BACKGROUND

This section provides background on NIST RBAC (2015) and Spatio-temporal RBAC (Ray & Toahchoodee, 2007) and an approach for modeling of situations in healthcare Situation-based Access Control (SitBAC) (Peleg et al., 2008). RBAC (Ferraiolo & Kuhn, 2009) associates individual users with roles which are then assigned permissions. A user in RBAC model is a human being, a role is a job function or job title within the organization, and a permission is an approval of a particular operation to be performed on one or more objects. The relationship between user, roles and permission is shown in Figure 1. There are five basic data elements in RBAC: user, role, object, operation and permission (Ferraiolo et al, 2001). The type of operations and the objects that RBAC controls depend on the type information/system; within a database system, operations include read/write/insert/delete. There are four levels of RBAC models (Sandhu, Ferraiolo & Kuhn, 2000). The *Flat RBAC* level has the core concepts of RBAC is that users are assigned to roles, permissions (operations on objects) are assigned to roles and users acquire permission by being members of roles. The *Hierarchical RBAC* level expands the ability of roles support the design of role hierarchies all roles.

The *Constrained RBAC* level adds separation of duties (SoD) which requires that for particular sets of transactions, no single individual is allowed to execute all transactions within the set (Ferraiolo, Barkley & Kuhn, 1999). There are two types of SoD: *static SoD* and *dynamic SoD* (Simon & Zurko, 1997). Static SoD places a restriction on the users that a role can play, meaning that a user authorized to role A would be prohibited from role B. Dynamic SoD weakens static SoD by allowing a user to be authorized to both roles A and B but to prohibit use of both roles at the same time in a session. As an example, consider a physician role (write a prescription) and a pharmacist role (dispense a prescription for a patient). Static SoD would require that a user couldn't both write a prescription for a patient and dispense the medication, while dynamic SoD allows the case for user to have both roles in an emergent situation to both write and dispense a prescription. *Symmetric RBAC,* extends roles/role hierarchies to include an interface for permission-role review with respect to a defined user or role returning.

The *Spatio-temporal Role-Based Access Control (STRBAC)* model (Ray & Toahchoodee, 2007) in Figure 2 extends NIST RBAC with features that can support pervasive computing with two constraints: *location* to constrain where an action can occur; and, *time* to constrain when an action can occur. In STRBAC, time and location (bottom portion of Figure 2) are applied to: user, role/subject, object, operation, and permission. In STRBAC, a user assigned to a role is permitted to access the permissions of the role if the user is in a designated location at a specified time. If a user moves or the time of access expires, then the access is not allowed. As shown in Figure 2 (redrawn from Abdunabi, Sun & Ray, 2014), the operation is dependent on: role of the user, permission of the role, pre-designed location, and

Figure 1. The NIST RBAC model concepts and interactions

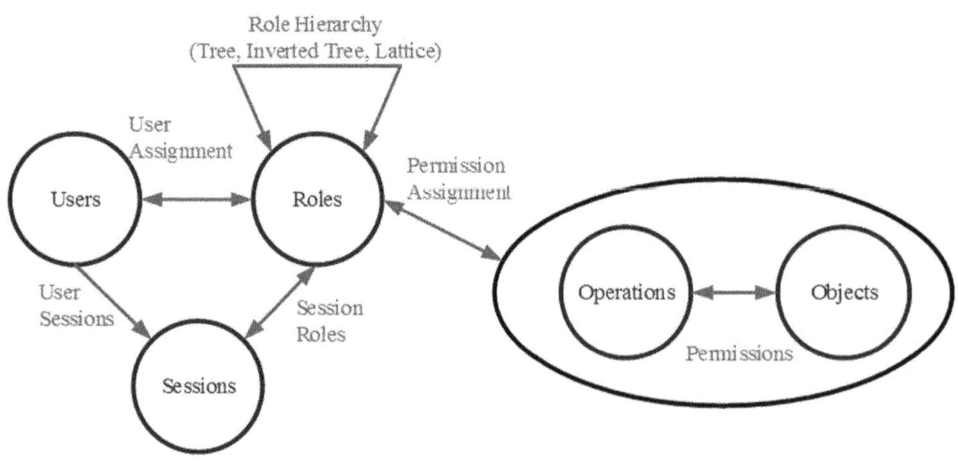

specified time. STRBAC can be utilized for applications where spatial and temporal information of a subject and an object must be taken into account before granting or denying access. For example, Tom is a physician assistant who works the 3:00pm to 11:00pm shift during the week and 7:00am to 7:00pm on weekends. To ensure that the EMR is secure, he can only see authorized patients by role when he is in the hospital and during his working hours.

Situation-Based Access Control (SitBAC) (Peleg et al., 2008) is targeted for formulating securer data-access policies, where health organizations can specify their regulations involving access to a patient's data according to the context of the request. SitBAC was created using input from a study (Peleg et al., 2008) that elicited various access-request scenarios from patients, senior physicians, nurse, secretaries, etc. SitBAC's central modeling concept is the *Situation*, a formal representation of a patient's data-access scenario (Beimel & Peleg, 2011) consisting of six entities (Peleg et al., 2008) supplemented with *Refineables* and *Relations*. Briefly:

- **Data-Requestor:** A (human) entity requesting access to the patient's data.
- **Patient:** A (human) entity who is the subject of the requested data.
- **EHR:** The Electronic Health Record where the patient's data is maintained.
- **Task:** The operation on the data that the data-requestor wishes to carry out (e.g., view medications and update prescriptions).
- **Legal-Authorization:** A legal document authorizing the requested task (e.g., a social worker needs to view a patient's diagnosis in order to approve payment).
- **Response:** The data access decision (approved/denied) made with respect to a situation.

A Situation can have multiple sub-situations. In the case of healthcare, different departments in a hospital (e.g., admitting, emergency room, inpatient services, operating room, etc.) all share patient data.

Figure 2. The spatio-temporal role-based access control (STRBAC) model

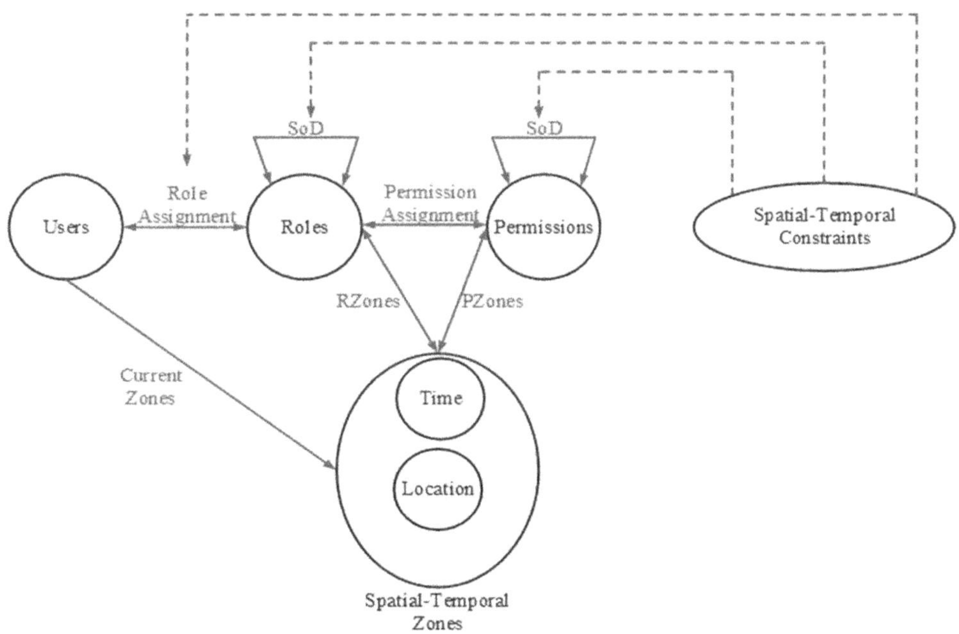

The legal-authorization entity, which represents all of the required factors of the application is needed for both ordinary and complex scenarios.

Refineables and *Relations* are utilized to augment the six aforementioned entities and are defined properties of the conceptual SitBAC model that belong to each entity and are structured hierarchically. A *refineable* can be assigned with an allowed value (Peleg et al., 2008). In healthcare, a refineable property for a patient would be demographic information such as name, age, address, etc.; the role of a physician is also a refineable property. A *Relation* is utilized to define a dependency between two entities or two refineables (Abdunabi, Sun & Ray, 2014). An Entity-to-Entity relation (E2E) relates two of the six aforementioned entities. For example, two possible E2E entity pairs are: EHR-Patient entity pair is a relation type that is assigned with record-of that allows the relationship between an EHR record and a specific patient; and, a Data-requestor-Patient entity pair that associates a stakeholder (role) with a patient such as ER-Doctor-of-Patient, Surgeon-of-Patient, etc. A Refineable-to-Refineable relation (R2R) defines a relationship (e.g., equal-to, different-from, etc.) between two refineables, e.g., the patient's treatment location refineable could be equal-to a physician's workplace location refineable.

Figure 3 presents a sample situation diagram of a patient data access that has: a data-requestor sends a request to access a patient record in the EHR to perform some operation; the authorization of the request operation is checked; if denied, a response is sent to the requestor; if accepted, the requestor receives the patient record form the EHR. Figure 4 presents an example of the unfolding of the data-requestor

Figure 3. A sample SitBAC diagram

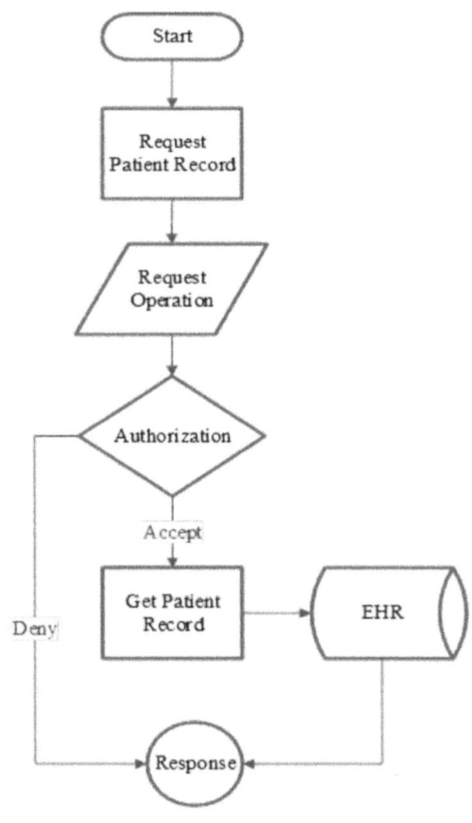

Figure 4. Unfolding data-requestor entities

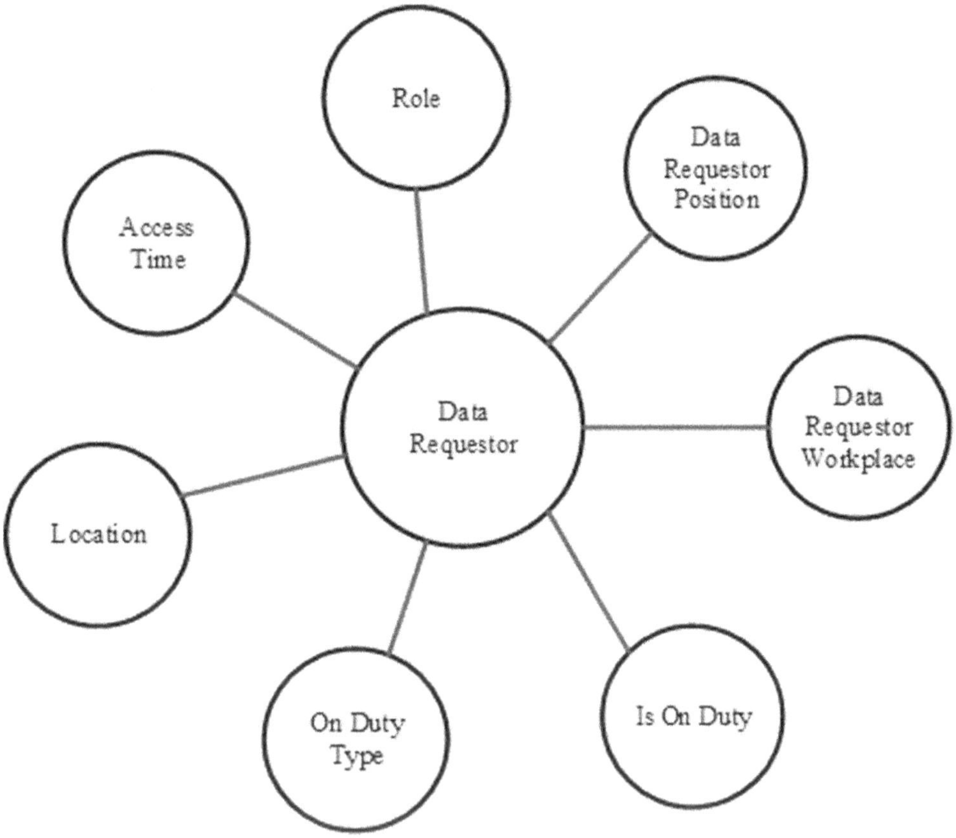

entities, that has: the data-requestor's role; his/her position and the workplace he/she works; whether he/she is on duty; the on duty type; the location; and, the time he/she has the permission to access.

A SPATIO-SITUATION-BASED ACCESS CONTROL (SSBAC) MODEL

In this section, we present our proposed Spatio-Situation-Based Access Control SSBAC model that: extends Spatio-temporal Role-Based Access Control (STRBAC) (Ray & Toahchoodee, 2007) by augmenting their time and location with an adaptable location (Spatio); integrates the Situation-Based Access Control (SitBAC) (Peleg et al., 2008) ability to define a situation schema (SS); and, introduces a *Zone* concept that represents a wide geographic area defined for the Situation Schema that then enforces privileges by role for users that are in the Zone. The result allows a user assigned a role to access its permissions only when the user is in a designated location at a specified time (see Figure 2 again) with the permissions defined using a situation (see Figure 3 again) where the user is functioning as a requestor that involves the respective entities (see Figure 4 again). The proposed SSBAC model defines the new Zone concept and also realizes the Spatio and Situation Schema extensions as constraints akin to SoD constraints. SSBAC as presented in this section has a restrictive approach in terms of the information that is available as a user moves from location to location. As a result, it may be the case that SSBAC may be too restrictive for use by physicians that want as much data available as possible on a patient in a given situation. This would be true in emergent situations when a physician is trying to collect as much as possible information on a patient. While not currently considered in our proposed model, there is a clear need for an approach that would support break-the-class control (Brucker & Petritsch, 2009) in SSBAC. We also acknowledge that the proposed SSBAC as presented does not at this time consider a myriad of issues that may be of relevance including: a malicious user that is intending to utilize the mobile app for nefarious purposes; a user that employs spoofing to mask the location of the phone which needs to be handled in order to insure that the true location is being utilized with SSBAC; and, other typical attacks that are emerging for mobile platforms. The focus of this paper is on spatial and situation access control as a model that extends RBAC; all of these other issues will need to be addressed in the future.

The remainder of this section discusses our proposed SSBAC in three different parts. Part one formally defines the SSBAC model which leverages and adapts STRBAC and SitBAC to support the determination of access based on role, location, and situation. Part one also introduces the *Zone* concept as a geographic location where permissions by role area defined and enforced. Part two formally defines: *RBAC Entity Spatio-Situation constraints* for users, roles, and permissions (objects, operations); and, *RBAC Assignment Spatio-Situation constraints* for user-role, role-permission, and separation of duties. Part three defines the overall SSBAC software architecture for enforcement including the use of a cryptographic protocol and algorithms.

The SSBAC Model

The Spatio-Situation-Based Access Control (SSBAC) model given in Figure 5, is adapted from STRBAC (Ray & Toahchoodee, 2007) and SitBAC (Peleg et al., 2008) and augmented with a *Zone* concept that is defined as a geographic location for defining and enforcing security. The essence of SSBAC is to obtain the access permission only when the user is in the proper location and in the defined situation schema. When the user moves or the user isn't involved in the situation schema (SS), the access is denied. To support this process, we assume that the mobile device utilized by the user has a GPS function and WIFI or data connectivity, and the device has a sufficient antivirus software to insure that the location information cannot be misrepresented by malicious intrusions. In the bottom portion of Figure 5, the Spatio-Situation zones (SSZones), encapsulates the time and location (spatio) from STRBAC, the situation schema from SitBAC, and a new Zone concept to represent the geographic location of the situation schema. Note that Zones of situation schemas must be non-overlapping. Definitions 1 to 3 are for Zone, Situation Schema, and Application, respectively:

Defn. 1: A Zone $Z = < Z_{ID}, Z_{NAME}, Z_{LOCATION} >$ is identified by Z_{ID} is a unique identifier, Z_{NAME} the name, and $Z_{LOCATION}$ the geographic location on a map.

Defn. 2: A situation schema $SS = < SS_{ID}, SS_{NAME}, SS_{LIFETIME}, SS_{ZONE} >$ is identified by SS_{ID} is a unique identifier, SS_{NAME} the name, $SS_{LIFETIME}$ is the time when the schema is active, and the SS_{ZONE} as the geographic location.

Defn. 3: An application A is comprised of set of k SSs pairs $A = \{SS_1, SS_1, SS_2, ..., SS_k\}$.

Note for the purposes of this paper, we do not concretely define the values of a location. In practice, a $Z_{LOCATION}$ could be defined by using GPS coordinates for the boundaries of the Zone. In addition,

Figure 5. The proposed spatio-situation-based access control (SSBAC) model

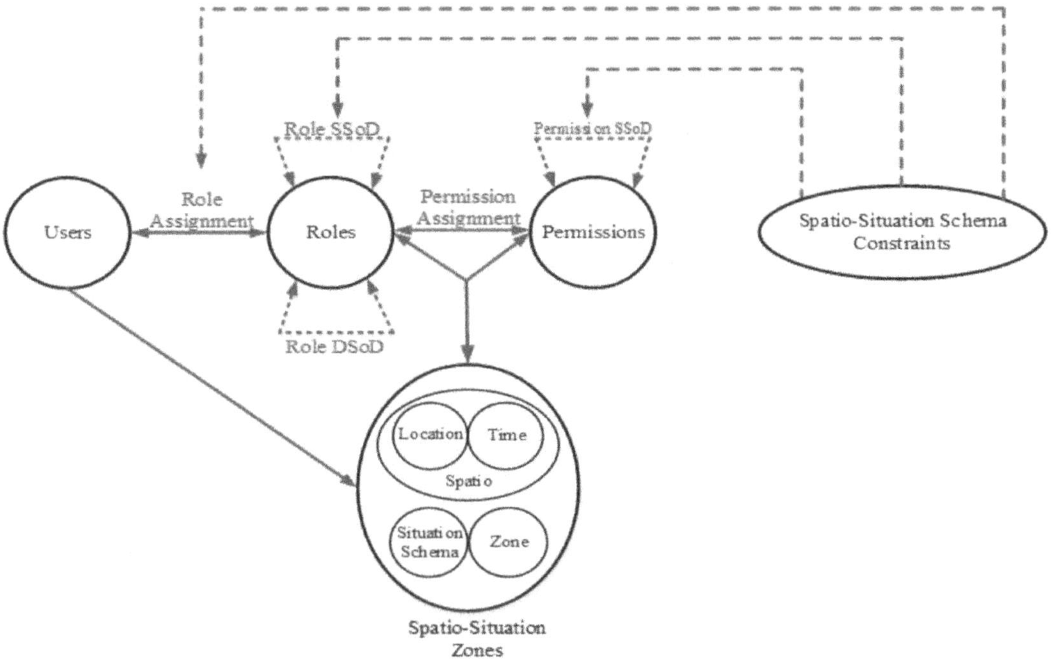

$Z_{LOCATION} = \infty$ means that the Zone has no defined location (e.g., available anywhere). Note also that $SS_{LIFETIME}$ will be the combination of an hour range (9am-5pm, 3pm-3am, etc.) with days of the week (Monday-Friday, Saturday/Sunday, etc.). $SS_{LIFETIME} = 24\ hrs$ means that the situation schema is always active.

The SSBAC model entities for users, roles, and permissions are associated with an SS_{ZONE} in order to define when and where these entities are accessible. Definitions 4 to 11 define roles, users, user-role assignment, objects, permissions, and role-permission assignments:

Defn. 4: A role *r* is defined as a two-pair $r = < r_{ID}, r_{NAME} >$ representation of the responsibilities for a job task against one or more of the SSs.

Defn. 5: Let $R = \{r_1, r_2, ..., r_j\}$ be defined as the set of j roles for a given application A, where $r_j \in R$ and $r_j = < r_{ID_j}, r_{NAME_j} >$.

Defn. 6: A user u is defined as a tuple $< u_{ID}, u_{NAME}, u_{r_{ID}} >$, where $u_{r_{ID}}$ is the role as defined in Defn. 4 that will have the ability to access portions of an application and uniquely identifies each user by u_{ID}. The association of $u_{r_{ID}}$ with user *u* is called the user-role assignment. Note that a user can be authorized to more than one role, but is limited to playing one role in any application session.

Defn. 7: Let $U = \{u_1, u_2, ..., u_j\}$ be defined as the set of j users for a given application A, where $u_j \in U$ and $u_j = < u_{ID}, u_{NAME}, u_{r_{ID}} >$.

Note that *User-role assignment* means that a user has to be assigned to role to activate a role. For example, an individual can active the physician role only when he/she is assigned to the role of the physician. Next, *Role-Permission assignment* means each role has been assigned one or more permissions (operations on objects). For example, the physician role will be given permissions to read, write, delete, and edit a patient's medical record in the hospital's EHR. For permissions we can define:

Defn. 8: An object *o* is defined as a tuple $< o_{ID}, o_{NAME} >$ where o_{ID} is a unique identifier for the element and o_{NAME} is the name of the element in the data repository under the control of a SS.

Defn. 9: Let $O = \{read, aggregate, insert, update, delete\}$ be the set of operations that can be performed against a data element that is under the control of a SS. For permissions, each $op \in O$ will be assigned to individual roles.

Defn. 10: A permission *p* is represented by the four tuple $p =< p_{ID}, ss_{ID}, o_{ID}, op, p_{ZONE} >$, where p_{ID} is the unique identifier of the permission (akin to u_{ID} and r_{ID}), ss_{ID} is the identifier of a SS, o_{ID} is the identifier of an element on which operation *op* can be performed, and p_{ZONE} is the Zone that the permission is defined within ss_{ID}.

Note that since permissions are in Zones, they are also restricted to being accessible only during the lifetime $SS_{LIFETIME}$.

Defn. 11: Each role, *r*, has a set of *n* permissions called the role-permission assignments $r_{RPA} = \{< r, p_{ID_1} >, \ldots, < r, p_{ID_n} >\}$, where role *r* as defined in Defn. 5, and p_{ID} is the identifier of a permission *p* as defined in Defn. 10.

Notice that permissions are defined for each situation schema to represent the allowed actions for each data requestor against the involved repository (see Figures 3 and 4 again). Roles are then assigned one or more permissions which could involve one or more situation schemas. A user is assigned multiple roles but only allowed to play a single role during any session, unless this is overridden by dynamic separation of duties. Since permissions are defined against situation schemas which have Zones, each permission corresponds to a situation schema and its Zone. A role that contains multiple permissions will therefore likely access multiple situation schemas and their respective zones. At any point in time, a user has a location, and this location will occur within some Zone in which a situation schema has been defined. Within that situation schema/Zone combination, the user is restricted to those permissions that are defined for the role being played by the user. For a given situation schema/Zone combination, there will be a set of roles defined, where each role maps to a set of permissions that are relevant for the data-requestor (user) for all of the different situation schema. This leads to the following three definitions:

Defn. 12: A given role *r* has a set of *m* Zones that it utilizes, defined as $r_{ZONES} = \left\{z_{p_{ID_1}}, \ldots, z_{p_{IDm}} >\right\}$ where each $z_{p_{ID_j}}$ is the zone for a permission in r_{RPA}.

Defn. 13: A given user $u = < u_{ID}, u_{NAME}, u_{r_{ID}} >$ playing a particular role $u_{r_{ID}}$ has a set of $u_{ZONES} = r_{ID\ ZONES}$ that is the same as the Zones for role.

Defn. 14: The *Allowed User Access (AUA)* in SSBAC for a user *u* with respect to a role and a permission is determined with a six-tuple: $u_{AUA} =< u, r, p, ctime, clocation, SS >$ where the user *u* playing a role *r* in is allowed to access a particular permission *p* if the *clocation* of *u* is in the Zone SS_{ZONE} and the *ctime* of *u* is in the lifetime $SS_{LIFETIME}$.

As a result of Definition 14, SSBAC provides the ability to define policies that allow the user to perform an action based where the user is located at a particular time against a situation schema that resides in a particular zone and has a defined duration.

In addition, SSBAC as shown in Figure 5 contains three types of separation of duties (SoD) constraints that extend their NIST/STRBAC capabilities by considering location: *Role SSoD, Role DSoD,* and *Premission SSoD. Role SSoD* is static SoD and prevents the same user from activating conflicting roles in a certain spatio-situation zone. For example, the same person cannot activate both the physician and nurse roles in one department that is in a specific location within the hospital. *Role DSoD* is dynamic SoD and allows the same user to assign conflicting roles in a certain spatio-situation Zone, but the conflicting roles will not be activated at the same time. This means that within a Zone a user may move between two different roles that in other situations would be prohibited. *Permission SSoD* prevents the same role from being assigned to conflicting permissions in a certain spatio-situation zone. For example, a physician named Tom cannot be given to permission to allow Tom to conduct surgery on a patient in the operating room at a hospital and also be able to meet with a patient in his medical office at a different location (Zone), at the same time. The surgery on a patient would be under the control of situation schema of the hospital while the meeting with a patient would be under the control of situation schema of the medical office. Each of these two situation schemas have their own respective Zones.

SSBAC Constraints

The second part of the SSBAC model is to define Spatio-Situation constraints for the zone concept in the proposed SSBAC model. There are three types of *Spatio-Situation* constraints: *RBAC Entity Spatio-Situation constraints, RBAC Assignment Spatio-Situation constraints;* and, *Separation of duties (SoD) Assignment Spatio-Situation constraints. RBAC Entity Spatio-Situation constraints* involve RBAC entities users, roles and permissions (objects, operations) that are associated with spatio-situation zones, namely, CurrentZones, RZones, and PZones, respectively. To assist in the discussion, recall that we are assuming that a mobile device is being utilized by an individual assigned a role to access a mHealth application that is accessing data for his/her job responsibilities in a particular organization, e.g., Tom playing a physician role to access patient data. Thus, in SSBAC, we assume that a user utilizes the mobile device which is capable of tracking his/her location and the mobile device in running a mHealth application that realizes a particular situation schema (SS) (see Figure 3 again). Conceptually, RBAC Entity Spatio-Situation constraints are utilized to allow different behavior based on both user and role. In terms of user behavior, the *user* to be associated with a spatio-situation zone that is utilized to obtain the user's location and insure that the user is utilizing the proper situation schema at a particular time in that Zone via

$$SS =< SS_{ID}, SS_{NAME}, SS_{LIFETIME}, SS_{ZONE} >.$$

In terms of role behavior, a *role* that has been assigned can only be activated in a specific location (Zone) and the proper situation schema as dictated by a *permission* that determines where and in which situation schema a permission can be activated via

$$p = < p_{ID}, ss_{ID}, o_{ID}, op, p_{ZONE} > .$$

Note that since a role has multiple permissions, the permissions can be partitioned (without overlap) into a collection of sets where each set of permissions goes to the same PZone. So for a given role, permissions of that role map to multiple PZones. A user has the same set of Zones as its role, and at any point in time, a user is in a CurrentZone which is the Zone of the SS that the user is physically located within. To realize RBAC Entity Spatio-Situation constraints, three functions are defined:

Defn. 15 - The Permission Zone Function PZone: permission(objects, operations) → SSZone is defined to represent that each permission can be invoked in a specific spatio-situation schema Zone.

Defn. 16 - The Role Zone Function RZones: role → PZones is defined to represent that a role consists of multiple permissions and each permission maps to its own Zone, then a role, maps to multiple PZones which determine the set of spatio-situation schema zones (and their coordinates) where the given roles can be assigned.

Defn. 17 - The Current Zone Function Currentzone: user →SSZone is defined as the Zone that the user is located in at a particular time when attempting to access to a permission assigned to the user's role.

RBAC Assignment Spatio-Situation constraints are defined between the user, role, and permission entities in SSBAC, namely: user-role assignment, role-permission assignment, and separation of duties (SoD) assignment. The RBAC Assignment Spatio-Situation constraint is both location and situation schema dependent, which means that a user can be assigned to a role when the individual is in a specific location and under a proper situation schema (see Figure 3) – see Definitions 7 and 12/13 that define the user and the zones of the user and role. The RBAC Assignment Spatio-Situation constraint means that the permission can only be assigned to a role when the individual is in a specific location and the proper situation schema – see Definition 10 which defined

$$p = < p_{ID}, ss_{ID}, o_{ID}, op, p_{ZONE} > .$$

For example, Tom who is a physician, can utilize the ER-physician role when he is in the ER of a hospital and in the proper situation schema (e.g., treating a patient that just had an automobile accident):

Defn. 18: The process of role-permission assignment is defined as: RolePermissionAssignment ⊆ roles × permissions × SSZones that constrains the set of all permissions for a given role within the available PZones (see Defn. 15) for the roles $r_{ZONES} = \left\{ z_{p_{ID_1}}, \ldots, z_{p_{IDm}} > \right\}$.

Defn. 19: The process of user-role assignment is defined as: UserRoleAssignment ⊆ users × roles × SSZones that constrains the set of all users, along with their respective roles, to a Zone which is the Allowable User Access per Defn. 14, namely,

$u_{AUA} =< u, r, p, ctime, clocation, SS >$.

Lastly, Separation of duties (SoD) Assignment Spatio-Situation constraints involve both roles and permissions, and for roles supports both static and dynamic separation of duties (SoD), as we have defined previously. The Separation of duties (SoD) Assignment Spatio-Situation constraints are:

Defn. 20: Let role *r* have permissions p_1 and p_2. Permission SSoD is defined as: $[p_1 \neq p_2; \Rightarrow \neg (r \times p_1 \times PZone \wedge r \times p_2 \times PZone)]$ means that the conflicting permissions p_1 and p_2 cannot be assigned to the same role in specific location and SS.

Defn. 21: Let user *u* have two assigned roles r_1 and r_2. Role SSoD is defined as: $[r_1 \neq r_2; \Rightarrow \neg (u \times r_1 \times RZone \wedge u \times r_2 \times RZone)]$ means that at most one conflicting role can be assigned to a user in specific location and SS.

Defn. 22: Let user *u* have two assigned roles r_1 and r_2. Role DSoD is defined as: $[r_1 \neq r_2; \Rightarrow \neg (u \times r_1 (active) \times RZone \wedge u \times r_2 (active) \times RZone)]$ means that two conflicting roles cannot be activated for the same user although they can be assigned to the same user in specific location and SS.

At runtime, the Spatio-Situation constraints as given in Definitions 15 through 22 utilize the current time and current location (Definitions 14 and 17) of the user to dynamically check if the user is allowed to access the permissions (Definition 15) for a given role (Definition 16) that verifies both role-permission (Definition 18) and user-role (Definition 19) assignment. If SoD has been defined, there are additional checks for Permission SSoD (Definition 20), Role SSoD (Definition 21), and Role DSoD (Definition 22). Note that while none of the Defns. 15 through 22 include time specifically, time is included implicitly, since each SS has a $SS_{LIFTIME}$ (see Definition 2) that is associated with the Zone of the schema. As a result, any user, role, or permission that is associated with a particular situation schema which means that they all acquire $SS_{LIFTIME}$.

SSBAC Software Architecture

The third part of the discussion presents the software architecture for the SSBAC model that is utilized in conjunction with the mobile device and associated application(s) in order to achieve the dynamic access control based on a user's time and location. In keeping with the principle of the $UCON_{abc}$ model (Park & Sandhu, 2004), we do the authorization process both before and during the access. The SSBAC software architecture shown in Figure 6 contains three major modules: *mHealth Application, Authorization Server,* and *Database Source*. The *Database Source* (bottom middle of Figure 6) module is where the various health IT systems reside on which the situation schema and permissions are defined. Note that the SSBAC software architecture as presented in Figure 6 will have multiple instances for each of the three main components when the architecture is deployed to various organizations (e.g., different medical offices, hospitals, etc.).

The *mHealth Application* (upper left of Figure 6) module has the components that are required to perform the Allowed User Access (Definition 14) as well as the various constraint checks with respect to location and time (Defns. 15 to 19) and separation of duties (Defns. 20-22). The *mHealth application* module has the primary responsibility to create an access request. The module creates an access request package for the Authorized User Access Definition 14) that includes: the user identification (name,

Figure 6. SSBAC software architecture

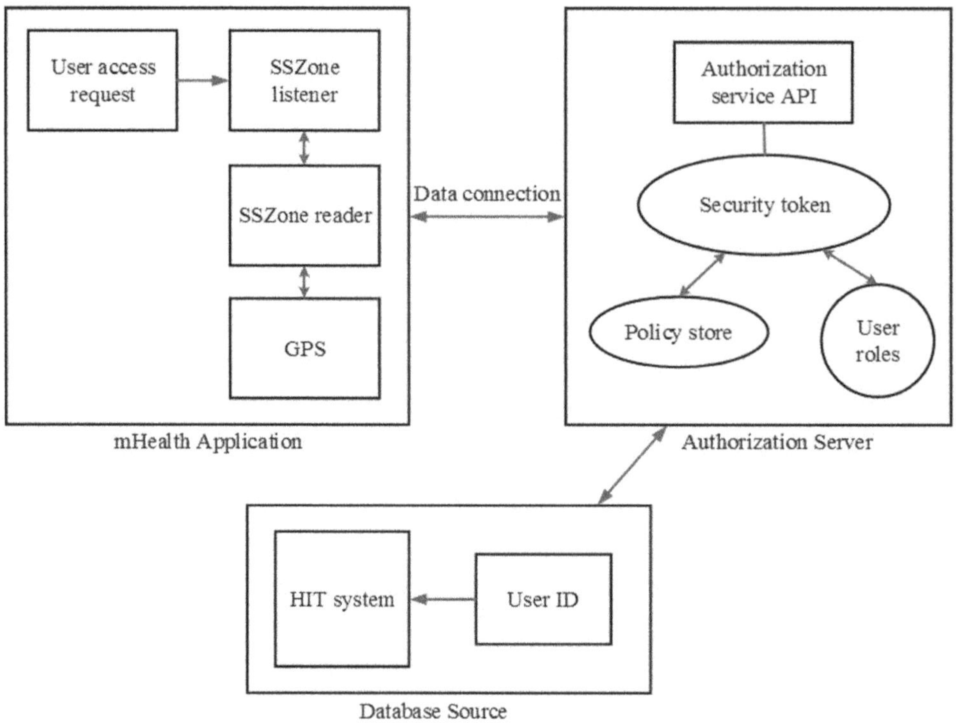

password), the current role, the location (CurrentZone), current time, and the access requests which are the underlying permissions; all of this information is sent to the Authorization Server module. Note that the requested permissions map to particular PZones and for the access to be successful, the user must be in the particular PZone of the permission during the allowable time ($SS_{LIFTIME}$). The user access request is captured by the SSZone listener who queues the request and sends in the order received to the SSZone reader. The SSZone reader works with the GPS component to check whether the user is in the proper situation schema, the correct location, and at the allowable time; this requires SSZone reader to interact with the Authorization Server module by passing in the Authorized User Access six-tuple:

$$u_{AUA} = < u, r, p, ctime, clocation, SS > .$$

Lastly, the *Authorization Server* (upper right of Figure 6) module has components that realize the Application's (Definition 2) security schemas (Definition 3), permissions, roles, and users (Defns. 3 to 13) as well as the necessary security software infrastructure (APIs, security tokens, etc.). The Authorization Server module receives:

$$u_{AUA} = < u, r, p, ctime, clocation, SS > .$$

The first step of the Authorization Server module is to check the security token (username/password) via the authorization service API. The next step is to check that the user's supplied role in the Authorized

User Access has been authorized. Then, the Authorization server performs the more complex checks that take the Authorized User Access's information (user, role, requested permission, current time, current location, and SS – Defns, 3 to 13) to determine if the Authorized User Access (Definition 14): satisfies functions PZone, RZone, and CurrentZone (Defns. 15 to 17); has been authorized to the role and permission in the current location (Defns. 18 and 19); and, satisfies additional checks for Permission SSoD (Definition 20), Role SSoD (Definition 21), and Role DSoD (Definition 22). If the checks are all successful, the Authorization Server module passes the success to the Database Source module which, based on the situation schema in the Authorized User Access, performs the access allowed by the permission. For example, when Tom requests access to the medical record of a patient he is treating in the Emergency Room (ER), the mHealth application checks whether Tom is in the ER at the hospital (Zone), and whether he has been authorized to the proper situation schema. If so, Tom gets the patient from the EMR.

SCENARIO OF USAGE FOR SSBAC IN HEALTHCARE

In this section, we provide a realistic scenario to that demonstrates the usage of the SSBAC, that is based on one co-author of this paper (T. Agresta) who is an MD in Connecticut. Specifically, over the course of a week's time, Tom works in a number of locations which have different health IT systems as shown in Figure 7:

Figure 7. Organization, locations, and situation schemas (SS)

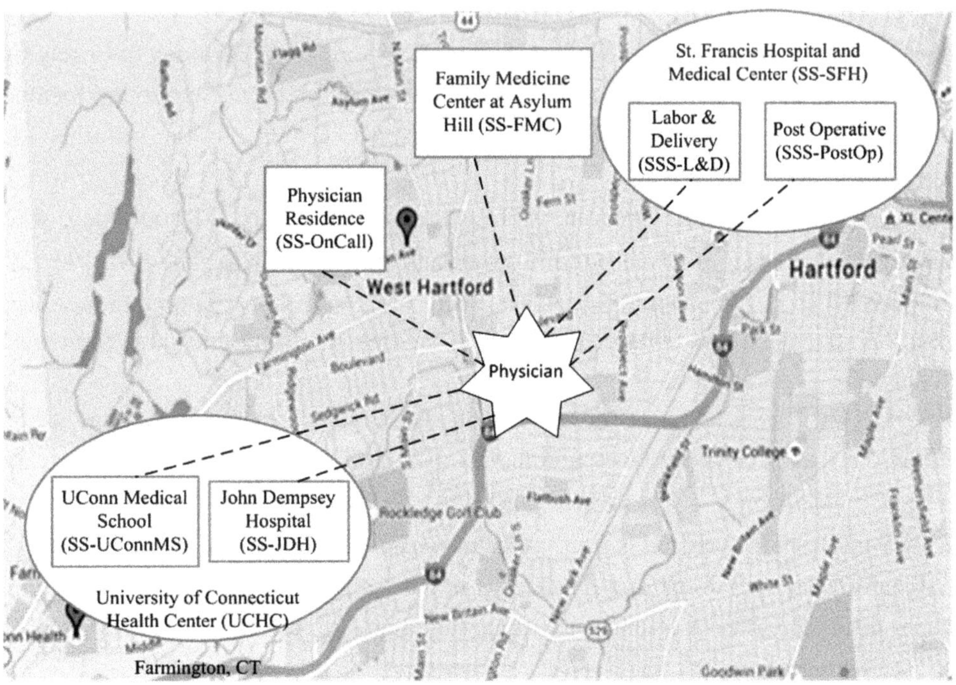

- **Family Medicine Center (FMC):** This is a medical practice in Hartford, CT; some of the physicians are associated with the UConn Medical School. Treats from children to elderly and includes services such as obstetrics (delivering children). Family Medicine Center has the GE Centricity EMR (2002).
- **Saint Francis Hospital and Medical Center (SFH):** Located across the street from Family Medicine Center, this is a full service hospital with emergency room, obstetrics department, cardiac center, cancer center, etc., spread across multiple separate and interconnected buildings (a medical complex). St. Francis Hospital has the EPIC HER (1979).
- **University of Connecticut Health Center (UCHC):** Located approximately 12 miles southwest of Harford, CT, an integrated academic medical center composed of standalone and connected buildings including: the John Dempsey Hospital (JDH), the UConn Medical (UConnMS) and Dental Schools, and clinical research facilities. John Dempsey Hospital has the NextGen EMR (2009).
- The Physician residence in West Hartford is also displayed in Figure 7.

Over the course of time, Tom with a Physician role moves among the four different locations and even within Saint Francis Hospital and Medical Center could be in different buildings for Labor and Delivery to deliver a baby and for Post-Operative Unit to visit a patient who just had surgery.

Given these different organizations, the next step in the example is to define a regional healthcare application ($A = \{SS_1, SS_1, SS_2, ..., SS_k\}$ in Definition 2) that consists of a set of situation schemas (see Figure 7) defined via

$$SS =< SS_{ID}, SS_{NAME}, SS_{LIFETIME}, SS_{ZONE} >:$$

- **SS-FMC:** This schema for Family Medical Center has $SS^{FMC}_{LIFETIME}$ Monday to Friday from 9:00am to 5:00pm to treat patients in an office setting at the Family Medical Center and located in its own building and has the GE Centricity EMR. The SS-FMC SS^{FMC}_{ZONE} is the building located at 99 Woodland Street, Hartford, CT.
- **SS-SFH:** This schema for the St. Francis Hospital has $SS^{SFH}_{LIFETIME}$ 24 hours/day, seven days per week, to treat patients in multiple buildings and has the EPIC EHR. The SS-SFH SS^{SFH}_{ZONE} is the building located at 114 Woodland Street, Hartford. CT. Within St. Francis, there exists two other sub-situation schemas (SSS) that are located in separate building and have their own health IT systems namely:
    - **SSS-L&D:** This schema is for the Labor and Delivery Department, has $SSS^{L\&D}_{LIFETIME}$ 24 hours/day, seven days per week and has its own L&D EMR that links mothers and children. The SSS-L&D $SSS^{L\&D}_{ZONE}$ is located within the SS-SFH SS^{FMC}_{ZONE}.
    - **SSS-PostOp:** This schema is for the Post-Operative Department, has $SSS^{PostOp}_{LIFETIME}$ 24 hours/day, seven days per week and has access to the Surgical EMR that has separate scheduling capabilities and information for the surgery and post-surgery care. The SSS-PostOp SSS^{PostOp}_{ZONE} is located within the SS-SFH SS^{FMC}_{ZONE}.

- **SS-UConnMS:** This schema for the UConn Medical School has $SS^{UConnMS}_{LIFETIME}$ Tuesday's and Thursday's from 9:00am-12:00 noon for teaching medical students and would have access to a Training EMR that contains fake patient data and medical cases for education purposes. The SS-UConnMS $SS^{UConnMS}_{ZONE}$ is the building located at 263 Farmington Avenue, Farmington, CT.
- **SS-JDH:** This schema for the John Dempsey Hospital has $SS^{JDH}_{LIFETIME}$ 24 hours/day, seven days per week, to treat patients and has the NextGen EMR. The SS-JDH SS^{JDH}_{ZONE} is the building located at 263 Farmington Avenue, Farmington, CT.
- **SS-OnCall:** This schema is for an handling phone calls from patients when the Family Medical Center is closed on night and weekends and has $SS^{JOnCall}_{LIFETIME}$ 5:01pm to 8:59am Monday-Friday and 24 hours/day on weekends. Here, the MD could be either home, at the hospital (St. Francis or John Dempsey), or taking calls in his office (Family Medical Center) before and after office hours. The unique characteristic of SS-OnCall is a physician may actually need to access to multiple health IT systems at different times, requiring access to multiple SSs.

Note that UConnMS and John Dempsey Hospital are two separate parts of the UCHC medical complex in separate building/wings. To define the Zone in this case it would be required to use more fine-grained mapping level coordinates.

Using this as a basis, assume that Tom is utilizing a mHealth application that is capable of accessing patient data in an EMR/her and as he moves among the different locations at a particular time, his role must be validated against each of the respective schemas based on his time and location which would access multiple EMRs/EHRs. Within St. Francis, Tom may be further constrained by the SSS-L&D and SSS-PostOp sub-situation schemas which have their own smaller non-overlapping sub-zones within the larger St. Francis Hospital Zone.

To understand the intricacies of the SSBAC model and approach as related to the aforementioned seven SSs, we postulate a typical schedule for Tom over a two-week period:

Week 1: Week at Family Medical Center

- Tom with Physician role works at Family Medical Center on a daily basis and sees patients from 9:00am-12noon and 1:00pm to 5:00pm and Monday, Wednesday, and Friday, which is within the SS^{FHC}_{ZONE} (building) and within the $SS^{FHC}_{LIFETIME}$ 9:00am to 5:00pm. Tom would operate under SS-FHC with the mHealth application accessing patients in the GE Centricity EMR.
- As needed, Tom can leave his office at Family Medical Center during lunch (12:00noon-1:00pm) or visit St. Francis Hospital after hours or on weekend, moving from one zone (SS^{FHC}_{ZONE}) to another (SS^{SFH}_{ZONE}) zone. While at St. Francis Hospital, Tom would operate under SS-FHC with the mHealth application accessing patients in the Epic EHR and is prohibited from SS-FHC's GE Centricity EMR. This is Role static separation from Definition 21.
- On Tuesday/Thursday Tom may drop in to see patients at St. Francis Hospital, while not at his teaching duties from 9:00am-12noon at the medial school UConnMS in $SS^{UConnMS}_{ZONE}$ and during $SS^{UConnMS}_{LIFETIME}$. At the medical school, Tom has a Faculty role and would operate under SS-UConnMS with the mHealth application accessing patients in the Training EMR.

- Also on Tuesday/Thursday, after those duties, Tom in the Physician role may do medical rounds with students to see patients at John Dempsey Hospital in SS_{ZONE}^{JDH} and during $SS_{LIFETIME}^{JDH}$, and would operate under SS-JDH with the mHealth application accessing patients in the NextGen EMR and is prohibited from SS-UConnMS's Training EMR.
- During this first week, Tom has the OnCallMD role to answer questions from patients from 5:01pm to 8:59am Monday-Friday and 24 hours/day on weekends which matches $SS_{LIFETIME}^{OnCall}$. Tom might be at his home, at FHC, checking on patients at St. Francis or John Dempsey hospital, or in any other location (no fixed Zone). The unique characteristic in this case is SS-OnCall must provide access to three different aforementioned health IT systems so that the mHealth accessing patients based on Tom's location.

Week 2: St. Francis Hospital Service

- Tom will be at St. Francis Hospital starting at 12:01am Monday morning and continuing through Sunday 12:00midnight. He may see patients throughout all of the departments of the hospital, in the Physician role, and is operating under SS-FHC with SS_{ZONE}^{SFH} and during $SS_{LIFETIME}^{SFH}$ with the mHealth application accessing patients in the Epic EHR.
- If one of Tom's women patients is admitted to have a child, Tom would switch to an Obstetrician role and proceed to the Labor & Delivery department in a separate building in the SSS-L&D sub-situation schema and transition to $SSS_{LIFETIME}^{L\&D}$ and $SSS_{ZONE}^{L\&D}$ with the mHealth application accessing the woman and the child whose records are in the L&D EMR. Note that within the Labor & Delivery department (building) he also has access to SS-SFH within the larger SS_{ZONE}^{SFH}, with the mHealth application accessing patients in the Epic EHR. Both roles active is role dynamic separation of duties from Definition 22.
- In a similar manner, if Tom visits patients in the Post Operative Department in a Physician role, Tom would be a separate building with the SSS-PostOp now in force ($SSS_{LIFETIME}^{PostOP}$ and SSS_{ZONE}^{PostOp}). The mHealth application would access the Surgical EMR, and, Tom would still access Epic since he remains within the larger SS_{ZONE}^{SFH}.

The scenario described above clearly outlines the need for the Zone concept and for the mHealth application to adapt and adjust as a user moves to different locations at different times. To keep the discussion at a high level, we have intentionally not provided details on the actual types of patient data and specific permissions. The reader is referred to (Caine and Hanania, 2013) for a detailed treatment of patient data information in the context of patients who desire want granular privacy control over health information in electronic medical records.

FUTURE LITERATURE REVIEW

The proposed SSBAC model as presented in this chapter combined information from the spatio-temporal access control model (STRBC) (Ray & Toahchoodee 2007) and situation-based access control model (SitBAC) (Peleg et al., 2008) and extended the approaches to add the concept of a Zone, a geographic location associated with a situation schema and its permissions and roles. In this section, we review other emerging efforts for dynamic access control for mobile devices, namely, two approaches that target the healthcare domain (Mallare & Pancho-Festin, 2013; Yarmand, Sartipi & Down, 2008), one that combines location with social connections (Baracaldo, Palanisamy & Joshi, 2014), and one that includes break-the-glass in access control (Bruker and Petritsch, 2009). *Workflow-Based Access Control* (WBAC) (Russello, Dong & Dulay, 2008) is a flexible access control mechanism that adapts the access rights of the subjects to the actual task that they have to fulfill and demonstrating the work in e-health. The requirements of entities' duties are expressed by means of workflow. WBAC ensures that entities can access the resources associated with a workflow task but only while such a task is active. *Collaboration and Workflow-Based Access Control* (Le et al., 2012) is an enhanced RBAC model which defines bridging entities and contributing attributes thereby extending access permissions to include workflow. The work also synthesizes a role-based access delegation model to target on specific objects and develops domain ontologies as instantiations of the general model to specific applications. This work was applied to a medical education setting for medical professionals and patients around disease education. Both of these efforts have the potential to augment the situation schema concept in a novel way. Specifically, by providing a more detailed workflow in a medical setting, it may be possible to augment the situation schema with the workflow information and to make the check not just based on Zone, but also based on what the user is doing in the Zone (e.g., the workflow step).

Another future work is *Geo-Social RBAC* (Baracaldo, Palanisamy & Joshi, 2014) that combines the location of a user with his/her social networking activities and location. The novelty of Geo-Social RBAC is that it includes geo-social constraints as part of the access control policy to capture both the locations of a user requesting access and his/her social connections. This is accomplished through the use of geo-social cardinality constraints that dictate how many people related by a particular social relation need to be present in a required location at the time of an access. The model also allows for the specification of geo-social and location trace constraints that may be used to dictate if an access needs to be granted or denied. By extending RBAC with social interactions that augment locations, privileges are based on the way that individuals interact and connect with one another. These connections could be utilized in a healthcare setting to determine which individuals have access to what information based on the degree that they are connected to a particular patient. Such an approach could be utilized to disseminate information on a patient's condition to interested parties. The end result of all three of these efforts is that they are providing additional dimensions of what a user does in terms of situations, workflow, social interactions, and location. These dimensions can then be leveraged to extend the capability of mobile devices to securely access the right information at the correct time as users move among locations and are involved in varied situations, a particular workflow, or participating in on-line social networking.

The other relevant future work involves the effort that extends an access control model with the break-the-glass capability (Bruker and Petritsch, 2009). *Break-the-glass* allows users to override the access control decision on demand and is an approach which extends user's access rights in exceptional cases. In the approach of (Bruker and Petritsch, 2009), SecureUML (Lodderstedt, Basin, & Doser, 2002) is extended to model break-the-glass and associated permissions and the generation of security policies achieved via the extensible access control model language (XACML) (OASIS, 2013). One of the advantages of this approach is that it supports the generation of break-the-glass UML policies for a concrete security architecture based on Java and XACML. The other advantage is the approach integrates a method for monitoring and logging the usage of emergency rights. The importance of the break-the-glass access has been highlighted in the HIPPA (HHS, 1996) standard. In the health care domain, there may be situations where it will be allowable for emergency room doctors and/or other physicians to "break the glass" to access the protected EHR in an emergent situation. In this case, a user would receive a warning and an audit trail would be generated to track the access. In these situations, integrating the break-the-glass access control model to our SSBAC model would provide the ability to override the strict security restrictions of our model in some situations. Since the data in the HIT systems are private and sensitive, we need to pay more attention to how to prevent misuses of break-the-glass access.

CONCLUSION

This chapter has presented the Spatio-Situation-Based Access Control (SSBAC) model that is based on Spatio-temporal Access Control (STRBAC) (Ray & Toahchoodee, 2007) that leverages time and location to determine and check permissions and the Situation-Based Access Control model (SitBAC) (Peleg et al., 2008) that defines the various situations for permissions. SSBAC combines STRBAC and SitBAC and extends the result with the Zone concept to characterize the geographic location within which the situation is occurring. To organize the presentation, the *Background* section examined work on RBAC, STRBAC, and SitBAC. Using that as a basis, the *Spatio-Situation-Based Access Control (SSBAC) Model* section contained: a formal definition of the SSBAC model that includes zones, situation schemas, users, roles, and permissions; a set of functions that define Zones and their relationship to users, roles, and permissions, and constraints that define user-role and role-permission, and three different types of separation of duties (SoD); and, a software architecture with three major modules. To illustrate SSBAC in practice, the *Scenario of Usage for SSBAC in Healthcare* section demonstrated SSBAC for a physician utilizing a mobile health (mHealth) application as he/she moves among various locations at different times. The *Future Literature Reviews* section reviewed efforts on Geo-Social RBAC (Baracaldo, Palanisamy & Joshi, 2014), users performing tasks in conjunction with the role (Mallare & Pancho-Festin, 2013), context-aware and associated workflow behavior (Yarmand, Sartipi & Down, 2008), and access control extended with break-the-glass (Bruker and Petritsch, 2009). SSBAC provides an important first step to allowing mobile applications to be able to dynamically adapt based on time and location, as a user moves among many different Zones that have unique and specific permissions by situation.

REFERENCES

Abdunabi, R., Sun, W., & Ray, I. (2014). Enforcing spatio-temporal access control in mobile applications. *Computing*, *96*(4), 313–353. doi:10.1007/s00607-013-0340-2

Aitken, M. (2013, October). Patient Apps for Improved Healthcare: from Novelty to Mainstream. *imshealth*. Retrieved from http://www.imshealth.com/en/thought-leadership/ims-institute/reports/patient-apps-for-improved-healthcare#ims-form

Baracaldo, N., Palanisamy, B., & Joshi, J. (2014). Geo-Social-RBAC: A Location-Based Socially Aware Access Control Framework. In Network and System Security (pp. 501-509). Springer International Publishing. doi:10.1007/978-3-319-11698-3_39

Beimel, D., & Peleg, M. (2011). Using OWL and SWRL to represent and reason with situation-based access control policies. *Data & Knowledge Engineering*, *70*(6), 596–615. doi:10.1016/j.datak.2011.03.006

Brucker, A. D., & Petritsch, H. (2009, June). Extending access control models with break-glass. In *Proceedings of the 14th ACM symposium on Access control models and technologies* (pp. 197-206). ACM. doi:10.1145/1542207.1542239

Caine, K., & Hanania, R. (2013). Patients want granular privacy control over health information in electronic medical records. *Journal of the American Medical Informatics Association*, *20*(1), 7–15. doi:10.1136/amiajnl-2012-001023 PMID:23184192

Cryderman, J. (2011). *Anywhere Computing: How Mobile Apps Are Changing the World.* Pipeline Publishing. Retrieved from http://www.pipelinepub.com/0111/Anywhere-Computing-Mobile-Apps1.html

Epic. (1979). *Epic EHR*. Retrieved from http://www.epic.com/about-index.php

FDA. (2015, February 9). *FDA Guidelines for Mobile Medical Applications*. Retrieved from http://www.fda.gov/downloads/MedicalDevices/.../UCM263366.pdf

Ferraiolo, D. F., Barkley, J. F., & Kuhn, D. R. (1999). A role-based access control model and reference implementation within a corporate intranet. *ACM Transactions on Information and System Security*, *2*(1), 34–64. doi:10.1145/300830.300834

Ferraiolo, D. F., & Kuhn, D. R. (2009). *Role-based access controls.* arXiv preprint arXiv:0903.2171

Ferraiolo, D. F., Sandhu, R., Gavrila, S., Kuhn, D. R., & Chandramouli, R. (2001). Proposed NIST standard for role-based access control. *ACM Transactions on Information and System Security*, *4*(3), 224–274. doi:10.1145/501978.501980

Haberle, C. (2014, August 27). Changing Healthcare Through Mobile Technology. *phx a cost management company*. Retrieved from http://www.phx-online.com/ecudednews/changing-healthcare-through-mobile-technology/

Harrison, R., Flood, D., & Duce, D. (2013). Usability of mobile applications: Literature review and rationale for a new usability model. *Journal of Interaction Science*, *1*(1), 1–16. doi:10.1186/2194-0827-1-1

Healthcare, G. E. (2002). *GE Healthcare Centricity EMR.* Retrieved from http://www3.gehealthcare.com/en/products/categories/healthcare_it/electronic_medical_records/centricity_emr

HHS. (1996). *Health Insurance Portability and Accountability Act of 1996.* U.S. Department of health & Human Services. Retrieved from http://www.hhs.gov/hipaa/index.html

Himss. (2014, June 17). How #mHealth is Changing Health and Healthcare. *Himss transforming health through IT*. Retrieved from http://www.himss.org/how-mhealth-changing-health-and-healthcare

Kirkpatrick, M. S., Damiani, M. L., & Bertino, E. (2011, November). Prox-RBAC: a proximity-based spatially aware RBAC. In *Proceedings of the 19th ACM SIGSPATIAL International Conference on Advances in Geographic Information Systems* (pp. 339-348). ACM.

Le, X. H., Doll, T., Barbosu, M., Luque, A., & Wang, D. (2012). An enhancement of the role-based access control model to facilitate information access management in context of team collaboration and workflow. *Journal of Biomedical Informatics*, *45*(6), 1084–1107. doi:10.1016/j.jbi.2012.06.001 PMID:22732236

Lee Ventola, C. (2014, May). *Mobile Devices and Apps for Health Care Professionals: Uses and Benefits.* PMC US National Library of Medicine National Institutes of Health. Retrieved from http://www.ncbi.nlm.nih.gov/pmc/articles/PMC4029126/

Lodderstedt, T., Basin, D. A., & Doser, J. SecureUML: A UML-Based Modeling Language for Model-Driven Security. In *Proceeding UML '02 Proceedings of the 5th International Conference on The Unified Modeling Language* (pp. 426-441). Springer-Verlag. doi:10.1007/3-540-45800-X_33

Mallare, I. J. G., & Pancho-Festin, S. (2013, December). Combining Task-and Role-Based Access Control with Multi-Constraints for a Medical Workflow System. In *IT Convergence and Security (ICITCS), 2013 International Conference on* (pp. 1-4). IEEE. doi:10.1109/ICITCS.2013.6717814

Microsoft. (2007). *Microsoft HealthVault.* Retrieved from https://www.healthvault.com/us/en

Mossakowski, T., Drouineaud, M., & Sohr, K. (2003, July). A temporal-logic extension of role-based access control covering dynamic separation of duties. In *Temporal Representation and Reasoning, 2003 and Fourth International Conference on Temporal Logic. Proceedings. 10th International Symposium on* (pp. 83-90). IEEE. doi:10.1109/TIME.2003.1214883

NEXTGEN Healthcare. (2009). *Nextgen EHR.* Retrieved from https://www.nextgen.com/Products-and-Services/Ambulatory/Electronic-Health-Records-EHR

NIST. (2015, January 15). *The NIST RBAC Standards.* NIST. Retrieved from http://csrc.nist.gov/groups/SNS/rbac/

OASIS. (2013, January 22). *eXtensible Access Control Markup Language (XACML).* OASIS. Retrieved from http://docs.oasis-open.org/xacml/3.0/xacml-3.0-core-spec-os-en.html

OpenEMR. (2012). *Open Source EMR.* Retrieved from http://www.open-emr.org/

Park, J., & Sandhu, R. (2004). The UCON ABC usage control model. *ACM Transactions on Information and System Security*, *7*(1), 128–174. doi:10.1145/984334.984339

Peleg, M., Beimel, D., Dori, D., & Denekamp, Y. (2008). Situation-based access control: Privacy management via modeling of patient data access scenarios. *Journal of Biomedical Informatics*, *41*(6), 1028–1040. doi:10.1016/j.jbi.2008.03.014 PMID:18511349

Posada, M. (2014, June 23). *The Evolving Landscape of Medical Apps in Healthcare.* HIT Consultant. Retrieved from http://hitconsultant.net/2014/06/23/the-evolving-landscape-of-medical-apps-in-healthcare/

Progno C. I. S. (2010). *The PrognoCIS EHR.* Retrieved from http://prognocis.com/

Ramey, K. (2013, December 14). Mobile Technology – 7 Location Based Apps Changing The Mobile Industry. *Use of Technology.* Retrieved from http://www.useoftechnology.com/7-location-based-apps/

Ray, I., & Toahchoodee, M. (2007). A spatio-temporal role-based access control model. In *Data and Applications Security XXI* (pp. 211–226). Springer Berlin Heidelberg. doi:10.1007/978-3-540-73538-0_16

Russello, G., Dong, C., & Dulay, N. (2008, March). A workflow-based access control framework for e-health applications. In *Advanced Information Networking and Applications-Workshops, 2008. AINAW 2008. 22nd International Conference on* (pp. 111-120). IEEE. doi:10.1109/WAINA.2008.131

Sandhu, R., Ferraiolo, D., & Kuhn, R. (2000, July). The NIST model for role-based access control: towards a unified standard. In ACM workshop on Role-based access control (Vol. 2000). doi:10.1145/344287.344301

Savitz, E. (2012, June 4). 5 Ways Mobile Apps Will transform Healthcare. *Forbes*. Retrieved from http://www.forbes.com/sites/ciocentral/2012/06/04/5-ways-mobile-apps-will-transform-healthcare/#7490182d6509

Simon, R. T., & Zurko, M. E. (1997, June). Separation of duty in role-based environments. In *Computer Security Foundations Workshop, 1997. Proceedings., 10th* (pp. 183-194). IEEE.

Torgan, C. (2009, November 6). The mHealth Summit: Local & Global Converge. *Kinetics: From Lab Bench to Park Bench*. Retrieved from http://caroltorgan.com/mhealth-summit/

Venturebeat Staff. (2014, September 29). Mobile Apps 2015: 5 Predictions of How Apps Will Change. *VentureBeat*. Retrieved from http://venturebeat.com/2014/09/29/mobile-apps-2015-5-predictions-of-how-apps-will-change/

Yarmand, M. H., Sartipi, K., & Down, D. G. (2008, June). Behavior-based access control for distributed healthcare environment. In *Computer-Based Medical Systems, 2008. CBMS'08. 21st IEEE International Symposium on* (pp. 126-131). IEEE. doi:10.1109/CBMS.2008.14

Zhang, X., Parisi-Presicce, F., Sandhu, R., & Park, J. (2005). Formal model and policy specification of usage control. *ACM Transactions on Information and System Security*, *8*(4), 351–387. doi:10.1145/1108906.1108908

KEY TERMS AND DEFINITIONS

Access Control: In information security, access control is a restriction of access to a resource.

Electronic Medical Record (EMR): An electronic copy of health information on a patient to support patient care by medical providers in a health care setting.

Global Position System (GPS): A satellite-based navigation system that provides location and time information.

Health Information Technology: (HIT): An abbreviation for Health Information Technology that are the systems used to support the delivery of healthcare.

Health Insurance Portability and Accountability Act (HIPAA): A law to protect the confidentiality and security of healthcare information.

mHealth: An abbreviation for mobile health, a term used for the applications related to health care utilized by patients and medical providers on mobile devices.

Mobile Application: Applications designed to run on mobile devices, like smartphones and tablets.

Role: The representation of a set of tasks or a job responsibility to which permissions can be assigned.

Role-Base Access Control (RBAC): An access control model where permissions are assigned directly to roles, which are assigned to users.

Separation of Duties (SoD): Places a restriction on the users that a role can play, so that a user authorized to one would be prohibited from another role.

Chapter 8
Preserving User Privacy and Security in Context-Aware Mobile Platforms

Prajit Kumar Das
University of Maryland – Baltimore County, USA

Pramod Jagtap
University of Maryland – Baltimore County, USA

Dibyajyoti Ghosh
University of Maryland – Baltimore County, USA

Anupam Joshi
University of Maryland – Baltimore County, USA

Tim Finin
University of Maryland – Baltimore County, USA

ABSTRACT

Contemporary smartphones are capable of generating and transmitting large amounts of data about their users. Recent advances in collaborative context modeling combined with a lack of adequate permission model for handling dynamic context sharing on mobile platforms have led to the emergence of a new class of mobile applications that can access and share embedded sensor and context data. Most of the time such data is used for providing tailored services to the user but it can lead to serious breaches of privacy. We use Semantic Web technologies to create a rich notion of context. We also discuss challenges for context aware mobile platforms and present approaches to manage data flow on these devices using semantically rich fine-grained context-based policies that allow users to define their privacy and security need using tools we provide.

INTRODUCTION

Smartphones or mobile devices that run advanced mobile operating systems are transforming how we communicate with people and connect with the world. Modern mobile operating system platforms like Android and iOS provide applications or "apps" through their "marketplaces". Combining computing ability with apps allows a "smart" phone to accomplish tasks that would either require a personal computer or special hardware components. For example a user can take pictures, record a video, connect to the Internet, navigate using GPS, prepare a presentation and accomplish many other day-to-day tasks, on smartphones.

DOI: 10.4018/978-1-5225-0945-5.ch008

However, with great power that comes with substantial computing and special hardware based sensing ability of smartphones, comes with added risks to user data. Advanced sensing abilities on smartphones have given rise to a new generation of intelligent applications. Smart assistants like Siri, Google Now and Microsoft Cortana are just a few examples of intelligent applications that are context-aware. All such apps exploit a user's location context to deliver personalized services. They do this by leveraging the user's location at the level of position, i.e., geospatial (latitude-longitude) coordinates. Integrating this with readily available background knowledge allows such systems to identify the location with a known place (e.g., Baltimore), facility (e.g., the BWI airport) or an organization (e.g., UMBC). As a result, location becomes an important aspect of a user's context but there are additional contextual information that includes a user's activity, identity and temporal information (Dey & Abowd, 1999). Naturally, protecting the security and privacy of user data now includes the critical task of protecting contextual data. In this chapter, we will discuss access control issues that need to be focused on and discuss solutions that have been proposed by researchers in the domain.

BACKGROUND

Access control generally refers to the process of determining what actions are allowed by a given subject upon objects and resources (Sandhu & Samarati, 1996). The security domain has seen the emergence of various access control models over the years. The most popular models include Discretionary Access Control (DAC), Mandatory Access Control (MAC), Role Based Access Control (RBAC) and Attribute Based Access Control (ABAC). DAC refers to access control mechanisms where it is at the "discretion" of the owner of an object. On the other hand, MAC "mandates" control based on security labels assigned to an object. RBAC is a model that uses "roles" to determine access control and in this model permissions are associated with roles, and users are made members of appropriate roles. RBAC suffers from issues of setting up initial role structure and inflexibility in dynamic domains (Kuhn, D. R., Coyne, E. J., & Weil, T. R., 2010). A pure RBAC solution will not consider dynamic attributes like time of day, which could be critical for determining user permissions. Essentially, it does not take into consideration the context aspect that we so often see, especially in the mobile domain. ABAC models are better equipped in handling access control for such dynamic systems. When it comes to using ABAC models one of the standard system implementations created by (Godik, S., Anderson, A., Parducci, B., Humenn, P., & Vajjhala, S., 2002) is XACML. The XACML standard defines a declarative access control policy language implemented in XML and provides a processing model on how to evaluate access requests. The access control mechanisms that we will discuss are modeled on ABAC.

In this chapter, we focus on the work done in the Platys project and various solutions suggested in it. The Platys project has developed a high-level abstraction of context. Context in Platys is generated by leveraging capabilities of smartphones, discussed in the introduction. Today a significant portion of the human population owns a smartphone and such devices are always on their person. This allows an app on the phone to capture key elements of context: like the user's location and, through localization, characteristics of the user's environment, etc. This leads to a deeper contextual understanding that comes from semantics associated with the location coordinates that are captured. By semantics of a location we mean the notion of a *Place*, i.e., a location in conceptual terms. For example a place could be "at a study group meeting," "out for jogging" or "shopping for grocery at the local market" – descriptions that combine a set of positions with user's activity, properties of user's environment, and activities of people surrounding the user or interacting with the user. Context extraction using sensors on mobile devices has received great attention in the field. The techniques proposed can be broadly classified into machine learning based models (Lane, N. D., Miluzzo, E., Lu, H., Peebles, D., Choudhury, T., & Campbell, A. T., 2010) or context modeling based models. Context modeling is carried out using ontologies and rules, and reasoning to infer high level context information from low level sensor data (Gu, T., Wang, X. H., Pung, H. K., & Zhang, D. Q., 2004). The Platys project builds on previous work where strong support for context reasoning using ontologies for explicit semantic representation of context (Chen, H., Finin, T., & Joshi, A., 2003) has been developed. Platys uses Semantic Web technologies to specify high-level, declarative policies in the form of Jena rules, for defining information sharing constraints using semantic context model. Apache Jena or Jena in short is a free and open source Java framework for building Semantic Web and Linked Data applications.

In other related work, Rein, a decentralized framework for representing and reasoning over distributed policies (Kagal, L., & Berners-Lee, T., 2005), is an extension of the Rei policy specification language, developed as part of research done in the Platys core group (Kagal, L., Finin, T., & Joshi, A., 2003). The Rei language is based on OWL-Lite and allows policies to be specified as constraints over allowable and obligated actions on resources in the environment. Rein, on the other hand, permits policies to be represented in different policy ontologies and requires the use of Semantic Web rules encoded in N3.

KAoS (Uszok, A., Bradshaw, J., Jeffers, R., Suri, N., Hayes, P., Breedy, M., & Lott, J., 2003) is an early research work in this domain that used the DARPA Agent Markup Language (DAML) language and Description Logic based ontology of the environment, app context and policy to determine access control at different levels of abstraction.

The final policy specification language we will look at is the Ponder policy specification language created by (Damianou, N., Dulay, N., Lupu, E., & Sloman, M., 2001). The Ponder language defines security policy specification for event triggered condition-action rules that define obligations and can be used for auditing access events to critical resources.

MAIN FOCUS OF THE CHAPTER

As we go forward, we need to ask an important question that motivates a need for strong access control: *What sort of data can be found on a user's mobile device?* These devices are capable of collecting and/or storing extremely private data. For example they can be used to generate a comprehensive digital profile of its users, through social media identities on the phone, photos, financial information, information about books users are reading or movie/TV shows they are watching, Web search history and users' contacts. At times mobile devices are used for authenticating logins into various services through secret shared key, thus acting as a substitute user identity. Mobile devices can also collect and store professional data due to their use in keeping track of work emails, messages, voicemails, calendar appointments and even accessing and storing digital files through cloud storage services. All of these features are available to users through apps that they may install from various app stores. Apps are capable of providing all these variety of services and at the same time they are capable of transmitting stored data on these devices. As a result, improper access controls may enable an app to steal a user's data.

Another reason why access control is becoming a critical requirement on mobile devices stems from the trend of corporations like IBM and many others (BYOD adoption rate is 74% among surveyed companies in January of 2015 as per the Tech Pro report) including healthcare companies adopting

Bring-Your-Own-Device (BYOD) as a corporate policy. Naturally, such a policy leads to a potential for data breach and thus to a need for stronger access control on mobile devices, one that is not necessarily available. A 2012 study of medical professionals (Fonseca, 2012), showed that 84% use the same device for personal and professional activities. Surprisingly 49% of the users' stated that their IT departments had not discussed mobile security issues with them. Such a glaring oversight in healthcare domain has happened due to a preference towards convenience over security, according to the study. Through the above-mentioned scenarios, it is evident that a mobile device has not only become an integral part of a user's life, but essentially it has also become a digital representation of that person. At the same time they are a critical part of a users' professional world.

What are the default security mechanisms that are in place on users' mobile devices? On one of the most popular mobile platforms in the world (i.e. Android), the security model deployed takes advantage of the features provided by the Linux kernel. In the Linux system one user cannot access files of another user. On mobile device users are traditionally implicit because they are an individual's device. As a result the UID associated with users in a Linux system are replaced by appID(s). Thus Android has the apps isolated at a process level and data level through the app's private directory. Access to components on the device, for example hardware like Camera or other OS services or to the Internet is controlled through permissions. Permissions are obtained at install time, in pre-Android Marshmallow era, or at run time starting from Android Marshmallow. Once obtained the OS enforces the permissions.

Why care then? We already have permissions! Unfortunately permissions defined in Android or iOS are too coarse-grained to adequately manage access control on mobile devices. Although the user may be able to use the device permission settings to manage access to say Internet for an app, they are not able to control what domains an app may connect to. They can control whether an app has access to the camera or not but they still do not have any way of stating that an app might only access the flashlight and not the camera. Users do not have the option to block an app from accessing ad api(s). On top of that none of the mobile platforms have provisions for fine-grained access control which are dependent on the context of the user. For example, a user might prefer to disallow camera access to social media apps at a work location or they might want to disable camera completely during a certain time period of the day.

Therefore, there is a need for fine-grained context driven access control mechanism for apps. This requires a rich notion of context for usage in policy execution. Context may be represented using Semantic Web technologies which will allow handling of various data flow scenarios from and through users' mobile devices. Understanding the factors that impact users' privacy and security is an important part of this work. Only then it will be possible to design policies to mitigate such issues. The challenges in achieving the goal of strong access control on users' mobile devices can thus broadly be broken down into three parts listed below.

- Collaborative context modeling,
- Access control policies,
- Rule specification.

In the following sections we delve into each of these sub problems.

COLLABORATIVE CONTEXT MODELING

In this section we are going to present techniques used by the Platys project for collaborative context modeling. We have seen applications for smartphones evolve to take advantage of features beyond localization like environmental data for example, ambiance, nearby people and resources, and the activities in which they are engaged. An ontology presented by (Chen, Finin, & Joshi, 2003) represents various categories of contextual information in pervasive computing environments, specifically, smart meeting rooms. Further generalization of the model to a lightweight, high-level context ontology, in form of the Place ontology, is carried out by (Zavala, Dharurkar, Jagtap, Finin, & Joshi, 2011). This ontology is used to reason about a general notion of context, as well as to share contextual knowledge. The vision is to generate a collective context using various users' devices and integrating the shared context knowledge from them.

Semantic Context Model

In the Platys project, context is modeled to include a semantic notion of a *Place*. Such elements of context are used in mobile devices to serve intelligent functionality by personal agents. Personal agents are used to proactively control activities on the phone such as switching off during a scheduled meeting or enforcing relevant privacy policies. User's location captured at the level of position, i.e., geospatial (latitude-longitude) coordinates is mapped to a *Place* or geographic entity, such as a region, political division, populated place, locality, and physical feature. Although position and geographic place information are potentially valuable on their own, from the standpoint of context, *Place* is a more inclusive and a higher-level abstraction.

A *User* is associated with a *Device* whose *Position* maps to a geographic place (*GeoPlace*) such as "ABC University" and to a conceptual place (*Place*) such as "*At Work*". Some *GeoPlaces* are part of others due to spatial containment and such relationship (*part_of*) is transitive. The mapping from *Positions* to *GeoPlaces* is many to one and the mapping from *Positions* to *Places* is many-to-many (the same *Position* may map to multiple *Places*, even for the same *User*; and, many *Positions* map to the same *Place*). Mapping from *Positions* to *Places* is done through *GeoPlaces* (*maps_to* is a transitive property). An *Activity* involves *Users* under certain *Roles*, and occurs at a given *Place* and *Time*. *Activities* have a compositional nature, i.e., fine-grained activities make up more general ones. This approach reflects the pragmatic philosophy that the meaning of a place depends mainly upon the activities that occur there, especially the patterns of lower-level activities. The idea applies at both the individual and collaborative level.

The Place ontology, mentioned earlier is a lightweight, high-level ontology that models the concept of place in terms of activities that occur at a geo-location. Description logics (Baader & Sattler, 2003) is adopted, specifically the Web Ontology Language OWL (Bechhofer et al., 2004), and associated inference mechanisms to develop the model. OWL supports the specification and use of ontologies that consist of terms representing individuals, classes of individuals, properties, and axioms that assert constraints over them. Figure 1 shows the core classes in the ontology and their relationships.

Figure 1. The place ontology models the concept of place in terms of activities that occur there

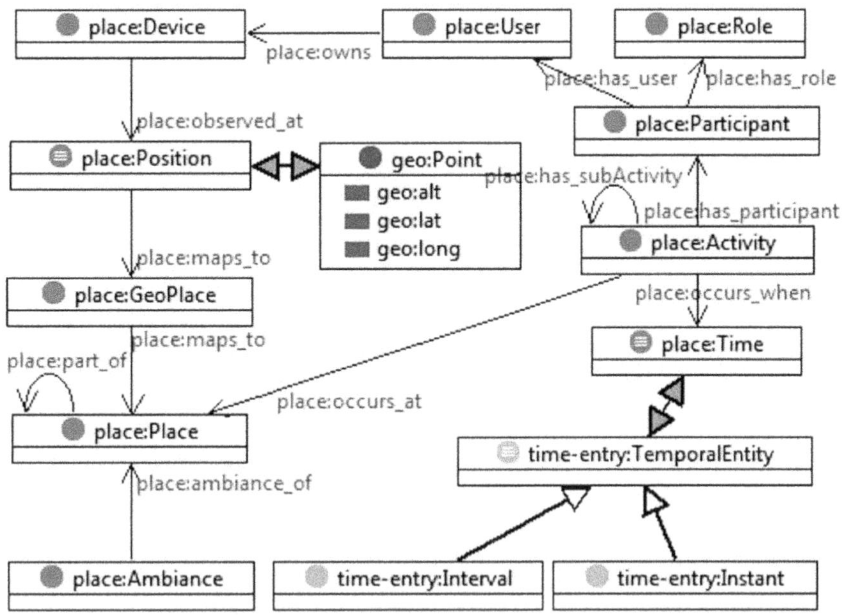

Information Sharing Policies

Users require appropriate levels of privacy control to protect the personal information their mobile devices are collecting including the inferences that can be drawn from the information. For example, in a healthcare scenario, if a user has an accident, it might be right to disclose relevant information (medical records, history, etc.) to the paramedics on the scene and only while they are providing their services. Semantic Web technologies are adopted due to two primary purposes:

1. Creation of models for representing and reasoning about a high-level notion of context.
2. Specification of expressive policies to control the sharing of contextual information.

Policies involve attributes of the subject (i.e., information recipient), target (i.e., the information) and their dynamic context (e.g., are the parties co-present). A prototype system in a university environment is used to demonstrate the workings of the system. Information aggregated from sensors on an Android phone, online sources, as well as sources internal to the campus intranet are used to individually infer the dynamic user activity using existing machine learning algorithms. The system allows sharing of contextual information directly between devices or through a server. Each device in the system contains a knowledge base (KB) that aligns with the Place ontology. The system also implements a model for specifying and enforcing privacy through declarative policies. The policies allow users to specify situations under which they allow sharing of their context information as well as the level of accuracy at which such information should be shared.

General Interaction Architecture

A general interaction architecture for mobile context-aware systems which share and integrate knowledge about their context is depicted in Figure 2. Sensors on devices sense the local context of the user, using mobility tracking and ambient sensing such as light, sound, and motion. The network component opportunistically gathers and disseminates local context information to neighboring fixed or mobile wireless devices. Its policy engine verifies the release policies to ensure context dependent release of information in accordance to the user preferences. Devices interact directly or through services on the Internet. Inferences such as current activity can be drawn from information collected by the sensors, the context information gathered, and additional resources (e.g., the user calendar and open geo-location KBs). The sensor's raw data as well as the inferred context knowledge is stored in a local knowledge base on the device. Context-aware applications and network components use this context knowledge to enhance their functionality. The locally inferred context knowledge may be sent to context-aware services located on the Internet. These services on the Internet, verify, if needed, the statements (proof) of the clients against the access policies. Depending on their functionality, these services provide context information of the user to other users.

Users' information sharing policies provide appropriate levels of privacy to protect the personal information their mobile devices collect, need to be expressive, flexible, and allow for context-dependent release of information. Semantic Web technologies represent a key building block for supporting expressive context policy modeling, reasoning and adaptation (Weitzner, Hendler, Berners-Lee, & Connolly, 2006). As a result, they are used to model a high-level notion of context and to specify high-level, declarative policies that describe users' information sharing preferences under given contextual situations.

Figure 2. Interaction among entities in a collaborative information sharing, context-aware system

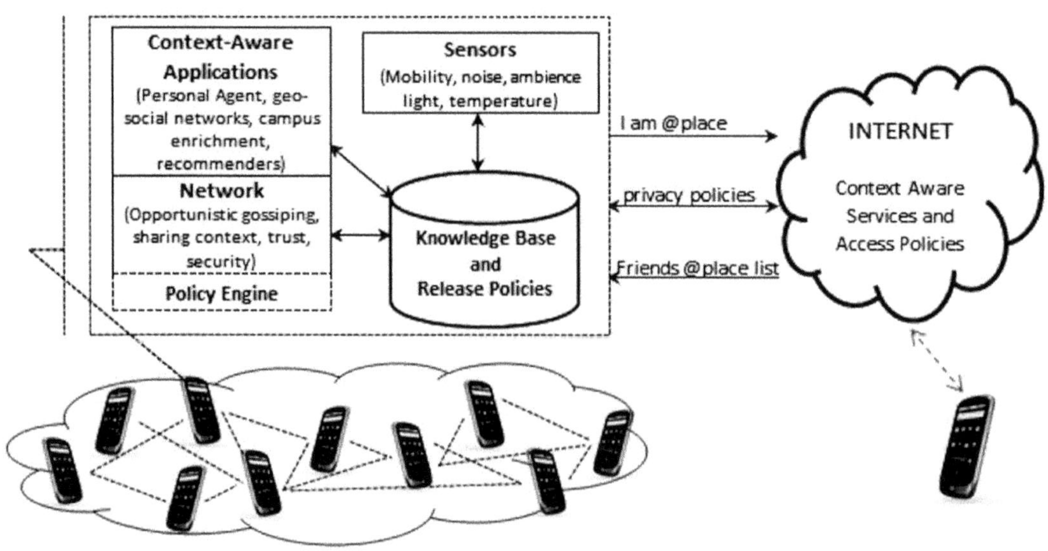

- **Knowledge Base:** The knowledge base (KB) on each device aligns with the Place ontology. Using the Place ontology, devices can share information about their context. The Android Location APIs are used to obtain the *position* of the device. Given the *Position* of the users' device, assertions are made and stored as triples into the KB (see Figure 3). Additional online resources are used, specifically GeoNames spatial KB (RDF version) and its associated services, to infer the user's *GeoPlace* using the following process:
    - Using reverse geocoding services to find the closest GeoNames entity to the current position.
    - Querying GeoNames through SPARQL to get further information about that entity.
    - Applying transformation rules to the data obtained from GeoNames (see Figure 3).
    - Using OWL inference to obtain the triples corresponding to the spatial containment of entities (transitivity of the *part of* relationship).
    - Using ad-hoc property chains (Figure 4) to infer knowledge about a user's *GeoPlace* based on the places her associated device is observed.
- **Activity and Place Inference:** The system uses machine learning algorithms to recognize activity (e.g., "sleeping", "walking", "sitting", "cooking"), coarse-grained geographic place, and conceptual place (e.g., "at work", "at home") at different levels of granularity.

Figure 3. An excerpt of the assertions made to the KB (left) in Turtle syntax and an example of a Jena rule used to integrate knowledge from GeoNames (right)

```
@prefix rdf: <http://www.w3.org/1999/02/22-rdf-syntax-ns#>.
@prefix place: < http://ebiquity.umbc.edu/ontologies/platys#>.
@prefix geo: <http://www.w3.org/2003/01/geo/wgs84_pos#>.
@prefix kb: <http://example.org/kb/device/>.
@prefix gn: < http://www.geonames.org/ontology#>.
```

```
kb:droid1 place:observed_at kb:anon01f
kb: anon01f rdf:type place:Position
kb: anon01f geo:lat 39.253525
kb: anon01f geo:long -76.710706
```

```
[partof:
(?a gn:parentFeature ?b)
->
(?a platys:place_part_of ?b)
(?b platys:spatially_contains ?b)
]
```

Figure 4. Property chain axioms to assert knowledge about a user's location: a) device is observed at the place whose location maps to b) user's location is the place where her associated device has been observed at c) Generalization of user location based on spatial containment (part_of)

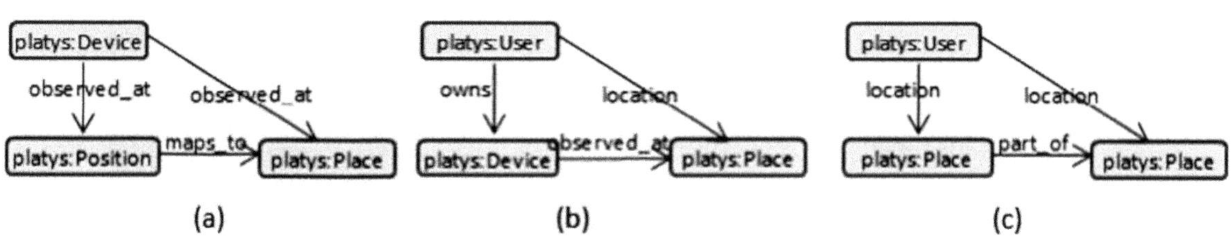

Privacy Reasoning and Enforcement

In the prototype system, the context is shared among devices by means of queries sent directly between them or through a server. The integration happens on individual device and is a simple operation where the results are added to the knowledge base. For privacy enforcement, users specify privacy policies that regulate the disclosure of Sensor and Inferred context information to server or Inferred context information to other users.

A user defines groups of contacts such as friends and family which are stored in the KB too. The user also specifies context dependent privacy policies and sharing preferences for each group. Privacy policies are expressed as Jena rules over the KB. The focus is not on the information exchange protocol, but on the privacy control mechanisms. Requests are simple messages with required information embedded in them. Whenever a request is received, either at the server or at a device, the privacy control module fetches static knowledge about a user (e.g. personal information and defined groups), the dynamic context knowledge and the user-specified privacy preferences. Access rights obtained by performing backward reasoning confirms conclusions by verifying conditions. Additionally, when access is allowed and according to the user defined sharing preferences, certain pieces of the information might be obfuscated in order to protect user privacy. The implementation used Jena Semantic Web framework (Carroll et al., 2004). Privacy rules are defined as Jena rules and the Jena reasoning engine is used to perform the reasoning. AndroJena, a porting for Jena to the Android platform (Lorecarra, 2009) is used for drawing inferences on devices.

- **Policies for Information Sharing:** These policies can describe which information a user is willing to share, with whom, and under what conditions. Conditions are defined based on attributes like a user's current location, current activity or any other dynamic attribute. Since users can have different networks of friends, a variety of group level privacy preferences are employed. For example: "share detailed contextual information with family members all the time" and "do not share my sleeping activity with Teachers on weekdays from 9am to 5pm". Figure 5 shows the representation of the first rule as a Jena rule (left) and the results on a test screen are provided using the results of the reasoning engine (right).

Figure 5. Left: Jena rule for expressing the policy "share detailed contextual information with family members all the time"; right: Android device screen with reasoning results. It shows access levels for requester "Ron" who is a member of the group family.

```
@prefix kb: <http://example.org#device>.
@prefix rdf: <http://www.w3.org/1999/02/22-rdf-syntax-ns#>.
@prefix foaf: <http://xmlns.com/foaf/0.1/>.

[AllowFamilyRule:
(?requester kb:contextAccess kb:userPermitted)
<-
(?requester rdf:type kb:requester)
(?groupFamily foaf:member ?requester)
(?groupFamily foaf:name "Family")
]
```

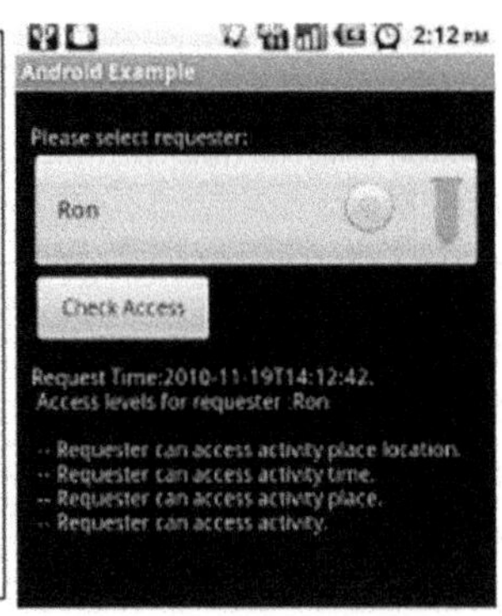

- **Policies for Obfuscating Shared Information:** These policies can depict what information a user is willing to disclose with different accuracy levels. For instance, she may be willing to reveal to her close friends the exact room and building on which she is located, but only the vicinity or town to others. Furthermore, a user may decide not to disclose her location to advertisers. For these purposes generalization models maybe used, which are discussed in detail in the Rule Specification section. These models are simple subsumption hierarchies over location and activity entities (e.g., *City* is subclass of *State* which is a subclass of *Country*).

ACCESS CONTROL POLICIES

In this section, the two high level aspects of access control that we will learn about, involves the information sharing between a device and a server or another device, a device and the apps on the device. In the previous section we have discussed the generation of a collaborative context model and the use of information sharing policies that control the context data that will be shared to generate such a model. We further examine this scenario with respect to a BYOD policy set (as system level policy described in Rule Specifications section) and a user policy set in a social media application scenario.

Inter-Device Information Sharing

In Figure 6 we present the major components of the system. The system consists of client devices, server side modules and the Internet services that provide social media requests. The client devices are context-aware smartphones. Client devices as well as the server side modules contains a user profile repository, a privacy control module and content preferences. The server side also contains content aggregator, *learn & share* module and *privacy control* module. The content aggregator combines social media data, photos, and videos from Internet services or other sources like university information portals. *Learn & share* module inferred a user's dynamic context using sensor data collected on phone, information from the content aggregator and online sources such as user's calendar. The inferred context is shared with corresponding client device so that the device along with server can handle further context sharing queries from other clients. The requester queries are passed through the *privacy control* module to constrain the information flow and hence to protect the user privacy. The *privacy control* module provides access control mechanisms and aids in controlling information flow within the system. On the client device, it enables privacy sensitive and resource sensitive reasoning over sensed data along with privacy enforcement between peer devices sharing contextual information. The interaction between various components of the system can be described as follows:

- The user has a client device to collect the sensor data periodically. This data is passed to the "learn & share" module on the server as allowed by the privacy control module on client device. The privacy control module decides which specific sensor data can be shared with the server based on user-specified privacy policies.
- The "learn & share" module infers the user's context using sensor data and information from content aggregator and other online sources. The context consists of current location, activity and additional surrounding information like nearby people. The inferred knowledge is passed to the corresponding client device so that it can handle context access queries from other clients.

Figure 6. Server-client data flow architecture

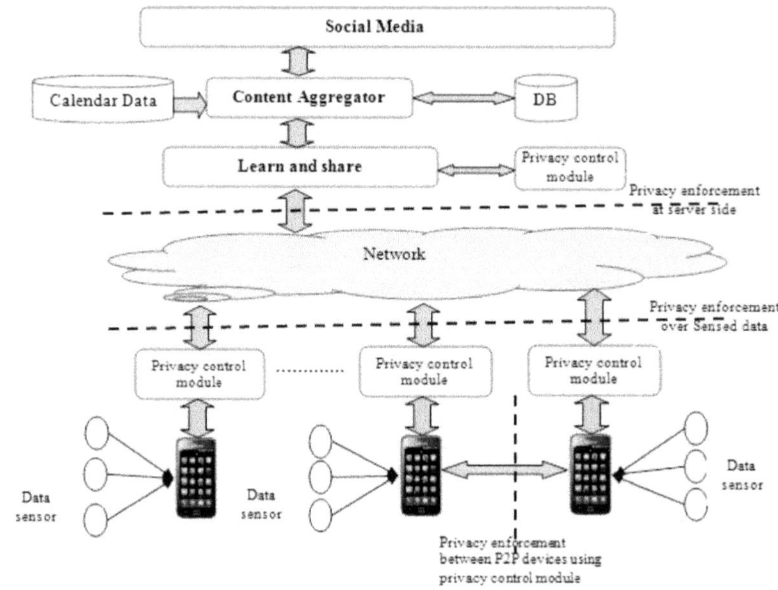

- Access requests are passed through the privacy control module which in turn decides whether to allow or deny access. If the requester is granted the access then it determines a set of information to be shared by performing reasoning over the context information and user's privacy preferences. These requests can be made by one client device to another or from a client device to the server.

Figure 6 shows the three different ways in which information can be shared in the system, namely: context information sharing between the client devices, sensor data sharing between a client device and server, and context information sharing between a client device and the server.

- **Access Control:** The user's personal information can be shared between a client device and the server side application or between two client devices. To constrain the information flow, privacy enforcement can be done between peer client devices and at server side for contextual information.
- **Privacy Enforcement Between Peer Client Devices:** Learn and share module from server side shared the owner's contextual information with corresponding client device. The client device further keeps track of the context and responds to queries made by other peer devices. Code 1 shows the sample contextual information for user "Alice". The contextual information is to be protected and should be shared only with requesters having sufficient privileges. The user can provide detailed privacy policies specifying what context information can be shared with whom, when, and under what conditions. If users are reluctant to provide any specific policies then they can opt for either default models of the system viz.
    - **Optimistic Model:** Where the system can provide response to any query with all possible relevant information associated with a user's activity such as associated place, location and the timing details, or
    - **Pessimistic Model:** Where the system can refrain from revealing activity associated information.

Code 1. Contextual information represented in N3; it consists of activity, associated place, location, time and nearby users.

```
ex:Alice a foaf:Person ;
foaf:name "Alice" ;
platys:has role platys:Student .
platys:Sleeping a platys:Activity ;
platys:is performed by ex:Alice ;
platys:has participant ex:Alice, ex:John ; platys:occurs at
platys:Class LH1  ;
platys:occurs when "2010-11-19T14:12:42".
platys:Class LH1 a platys:Place  ;
platys:has location "39.253525, -76.710706".
```

Apart from these default system settings the user can define her privacy rules with various degrees of accuracy levels. She can also use the system to obfuscate certain pieces of information to protect the context information. This way the system can protect the users' privacy by varying accuracy levels of activities, associated locations and timestamps.

Whenever any participant in the systems tries to access any protected resource (activity, place, location or any additional information) the query has to be sent to the privacy control module. This module fetches the user knowledge, dynamic knowledge and user-specified privacy preferences to evaluate the query. As a result it will decide whether participant is allowed to access the *protected* resource or not. In the former case, it might obfuscate certain pieces of information as per user-specified privacy policies to protect user's privacy. These policies are represented as Jena rules in Code 2, Code 3 and Code 4 respectively. When a request would be made by "Ron" who is a family member of the user then he should be able to access the user's detailed contextual information. If the request came from "Bob" who is a member of the friend group and the user's current activity is "Sleeping" then the requester is allowed to access a user's activity information excluding the associated place and location. Figure 8 shows the access level for requester "Ron" and Figure 7 shows allowed access for user "Bob" after performing reasoning on the device using user information, dynamic knowledge and privacy policies mentioned in Code 2, Code 3 and Code 4.

Privacy Enforcement at the Server Side

At the server side learn and share module, infers the user's dynamic context such as current activity, associated place and location, and nearby people. This contextual information needs to be protected and should only be shared with requesters with sufficient privileges. The server has information about all the system users whereas a client device has information about its owner. Due to this, the server can handle requests for all the users whereas the client device can handle requests about its owner only. The main distinction between the access requests made by a client device to a peer device and to a server is that the latter request contains a specific userId. This userId is used to retrieve specific users' information. Consider a privacy policy as shown in Code 4, which states "allow location access to teachers on weekdays only between 9am and 6pm". The system uses the userId to retrieve the related information

Figure 7. Android device screen with reasoning results; it has access levels for requester "Bob" who belongs to friend group.

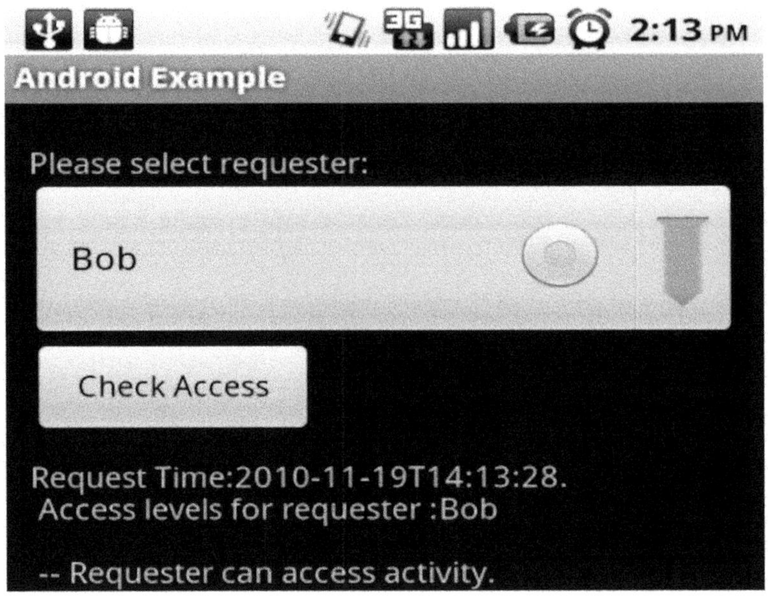

Figure 8. Android device screen with reasoning results; it has access levels for requester "Ron" who belongs to family member group.

and then checks whether the requester is a member of the group by verifying the requesters' userId. The example explained above involves representation of a user's personal resources such as list of friends, group's information, contextual attributes like current location and current activity.

Reasoning Engine

The reasoning engine handles the requester queries and performs reasoning for access control decisions. The system uses Jena Semantic Web framework (Lorecarra, 2009) for performing the reasoning over context data. Jena inference system allows the support of various inference engines or reasoners. These reasoners are used to infer additional facts from the existing knowledge base coupled with ontology and rules. In particular, Jena uses the generic rule reasoner which is included in Jena2 as a general purpose rule-based reasoner. It is used to implement both the RDFS and OWL reasoners. It needs at least a rule set to define its behavior. In the system, the reasoner uses the context ontology, static user facts like identity and group information along with the user-specified privacy rules to generate an inference model. This inference model is used for responding to the requester queries. This process is shown in the Figure 9 and works as follows:

- Create the instance of OWL reasoner specialized for context ontology and then apply that to the users' static information to generate an inference model. This inference model consists of additional statements inferred from static knowledge and ontology. As the user information and ontol-

Figure 9. Reasoning flow

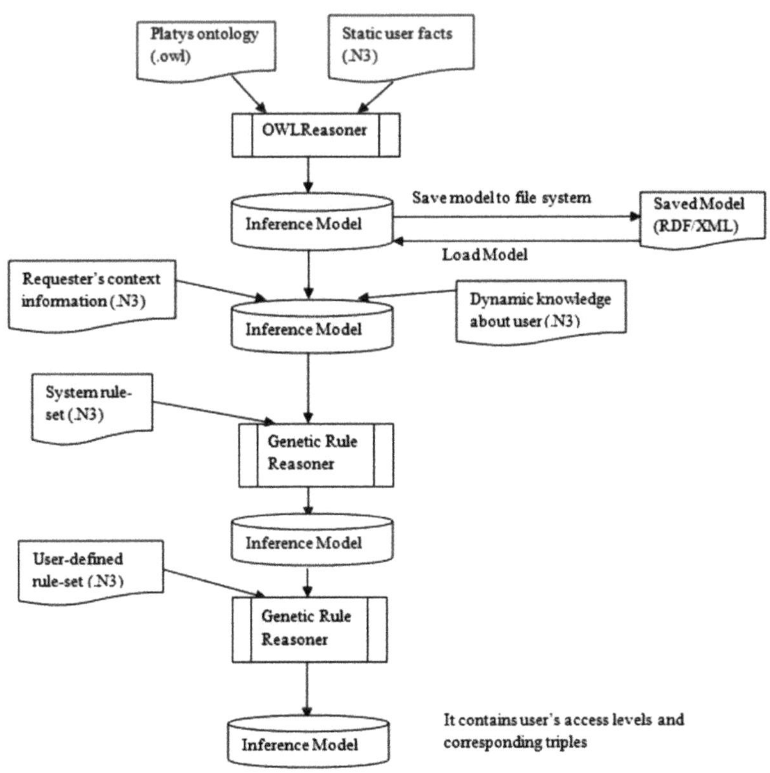

ogy are not changed often, it is quite safe to save the model on external storage and reload it for subsequent queries rather than generating it each time.

- The requester's contextual information is extracted from requester query and along with user contextual information it is added to the inference model to generate a new model.
- The system-level polices are executed against the inference model using an instance of generic rule reasoner. It is an optional feature and it's used to enforce certain organization level policies. It will create a new model having SystemPermitted and SystemProhibited statements to enforce system policies over the users' contextual information. If the user is a sole owner of client device then this step can be skipped. The detailed description of this feature is provided in the next section.
- The user-specified privacy rules are executed against the inference model from previous step to generate a new inference model having requester access levels.

The system will use the new model to decide what can be shared with requester and respond accordingly.

System Level Policies

The context-aware systems are used by individuals to organization and from social-networking application to military domains. In case of military domains or organizations, the user may not be the sole owner of client device and there is a strong need of robust security mechanisms. It can be in the form of multi-level secure systems where the system-level policies must override user-level policies. This highlights

Code 2. Policy to share detailed contextual information with family members

```
[AllowFamilyRule:
        (?requester ex:memberOf ?groupFamily)
        (?groupFamily foaf:name "Family")
->
        (?requester ex:canAccessActivity "True")
        (?requester ex:canAccessActivityPlace "True")
        (?requester ex:canAccessActivityTime "True")
        (?requester ex:canAccessPlaceLocation "True")]
```

Code 3. Policy to share activity information with friends all the time except when a user is attending lecture

```
[ShareActivityWithFriendsRule:
        (?requester ex:memberOf ?groupFriends)
        (?groupFriends foaf:name "Friends")
        (?someActivity platys:is performed by ex:Alice)
        notEqual(?someActivity, platys:Listening To Lecture)
->
        (?requester ex:canAccessActivity "True")]
```

Code 4. Policy to not share sleeping activity with teachers on weekdays from 9am - 9pm

```
[ShareActivityWithTeachersRule:
        (?requester ex:memberOf ?groupTeachers)
        (?groupTeachers foaf:name "Teachers")
        (?requester ex:requestTime ?localTime)
        (?localTime time:dayOfWeek ?day)
        ge(?day, 1) le(?day, 6)
        (?localTime time:hour ?hour)
        ge(?hour, 9) le(?hour, 21)
        (?someActivity platys:is performed by ex:someUser)
        equal(?someActivity, platys:Sleeping)
->
        (?requester ex:canAccessActivity "False")]
```

the need of system-level policies along with user-specified policies. The system-level policies should be defined by the system-administrator to ensure that sensitive resources are always protected from illegitimate access. Consider a system-level policy as "Do not share the user's context if she is inside a military building BuildingXYZ" and a user-specified policy as "Share my context with family members all the time". The system- level policy states that the user context won't be shared with anyone if she is inside BuildingXYZ whereas in latter policy user specifies to share her context with family members all the time. In this case the system-level policy should override user- specified policy and hence, if the user is inside BuildingXYZ then her context will not be shared to anyone including her family members.

Intra-Device Information Sharing

In this use-case scenario intra-device information sharing policies consider the data flow from the device's sensors to requester of said data. However, the requester in this case, is an entity which resides on the phone itself, i.e. an app. Data flow control, in this case, will have obvious user implications. When the data is leaving the device and is being shared with an external entity it makes sense that the user might want to control what information is shared. On the other hand when the data is being shared with apps on a user's device an implicit trust assumption should not be the norm. The reason being, most users install apps from a variety of developers and sometimes a variety of sources. It may be presumed that standard app market places are monitored by their respective owners but such a presumption may not necessarily be true (Lindorfer et al., 2014). There are examples of legitimate apps stealing user sensor data and sending them over the internet for ad purposes (Android Flashlight App Developer Settles FTC Charges It Deceived Consumers). Under such circumstances it's important to control what data flows from the sensors on a mobile device to the apps.

As discussed in the introduction, Android's security model is based on the Linux kernel's security features and application sand-boxing. Android application packages execute in their individual sand box and are built from multiple components that provide various functionality. To understand the intra device access control let's take a look at a prototype system. In this system few of the important device services on Android platform have been modified to serve as a way to control data flow. Specifically the LocationManagerService, AudioService and WifiService are modified. The altered system runs a reasoner on top of user-context and access control policies to determine response for an app's request to system resources. The system architecture of such a controlled system can be seen in Figure 10

Access Control

A significant point of failure for such a system, would be that apps do not expect to be blocked from accessing the data on a mobile device. As a result, apps would simply crash if such a block is put in place. Therefore, as an alternative solution, obfuscation is used. As part of the obfuscation solution a location randomization module is created. This module is used to generate fake coordinates for location of the device. The algorithm used for generation of a new set of coordinates from device's current location is similar to the ones deployed in apps reporting nearby places of interest.

Given a location L and a radius R location randomization module generates L' where L' ∈ *{*l: l is in the bounded circle with radius R and origin L*}*. This technique is used to find points within a distance of a latitude/longitude using bounding coordinates. The shortest geodesic distance between two given points $P1=(lat_1,lon_1)$ and $P2=(lat_2,lon_2)$ on the surface of a sphere with radius R can be calculated using the formula:

$$d=\arccos(\sin(lat_1)\times\sin(lat_2)+\cos(lat_1)\times\cos(lat_2)\times\cos(lon_1-lon_2))\times R$$

Figure 10. Architecture of the intra-device access control prototype Android system

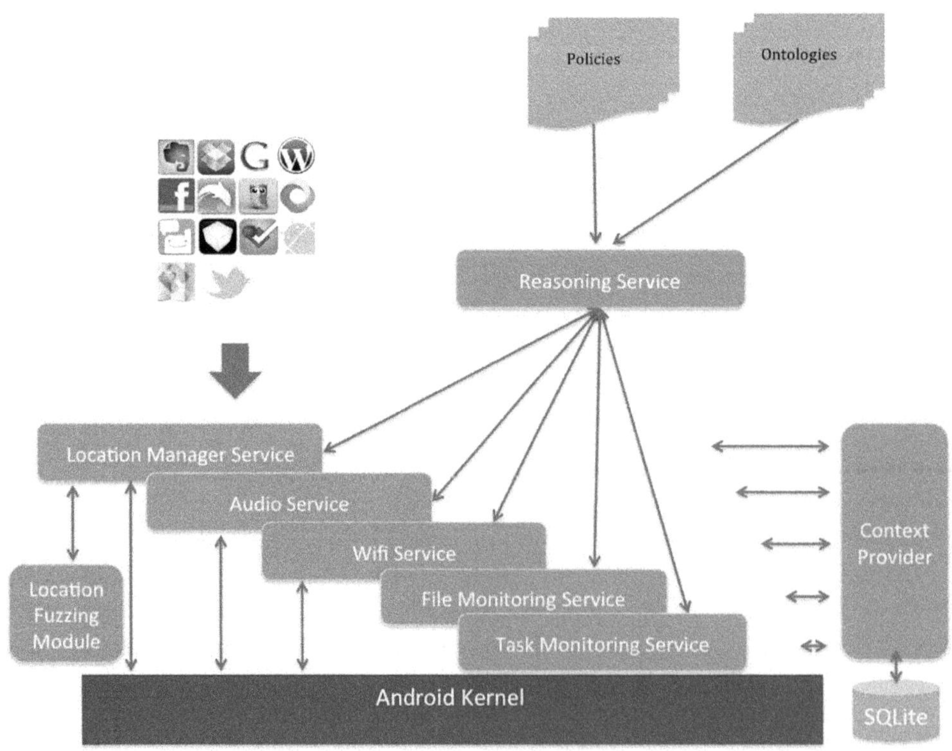

It computes the bounding coordinates of all points on the surface of a sphere that have a great circle distance to the point represented by this GeoLocation instance that is less or equal to the distance argument. Once these coordinates are obtained one maybe randomly selected and returned to the calling app instead of a failure response. This ensures that apps do not crash and user data is protected at the same time. Similar obfuscations can be generated for other services too. For example the camera component can be interrupted to provide fake images to a requesting app or fake contacts could be returned to an app which requests user's contact lists. The reasoning process remains the same as before. A set of static user data, an ontology defining user context and user's dynamic context information is combined with a list of system and user defined policies to reason over and decide whether the data should be shared or not.

Reasoning Engine

The heart of the access control system is the reasoning engine. Running as a system service it uses the policies that are stored on the device. The policies may be downloaded from a server or maybe selected by the user. Details of such a process will be discussed in the section Rule Specifications. As seen in the architecture diagram in Figure 10 upon a data/resource request made by an app, the various services on the device queries the reasoning service for an access control decision. The reasoner then uses policies on the device and based on the current context returns a decision asynchronously to the requesting service. The context string is forwarded to the reasoner by the requesting service.

RULE SPECIFICATIONS

Till now we have discussed context generation and presented use-cases for access control implementations. The last and equally important aspect of managing data flow in a smartphone requires a process to specify rules to be implemented. For this purpose we discuss the use of system level access control rules that might be specified by administrators. We also discuss the use of user policies for added protection. The two scenarios effectively take care of BYOD use-case organizational policies and users' personal privacy and security policies. An initial set of default policies may be obtained through a trusted third party. In case of a rule conflict between system and user policies the following resolution process is used. Access predicate of the rules can take the following values: SystemPermitted, SystemProhibited, UserPermitted and UserProhibited. As default deny policy is followed to determine access control, if SystemProhibit is present in the set of conflicting rules, the access is denied. Additionally system policies trump User policies and therefore if SystemPermitted and UserProhibited are present in rule set, the access is granted. Finally, there are UserPermitted and UserProhibited values in the access predicate of the rules, data access is denied.

User Policy Editor

As described by (Patwardhan, Korolev, Kagal, & Joshi, 2004), policies may be enforced indefinitely or for a certain time period based on a policy certificate validity period or a combination of timeout or loss of contact with an assigned network. However, the user has the option of modifying or adding rules to the policy through the interface shown in Figure 11. The Requester option allows choice of permitting

Figure 11. Policy editor settings app

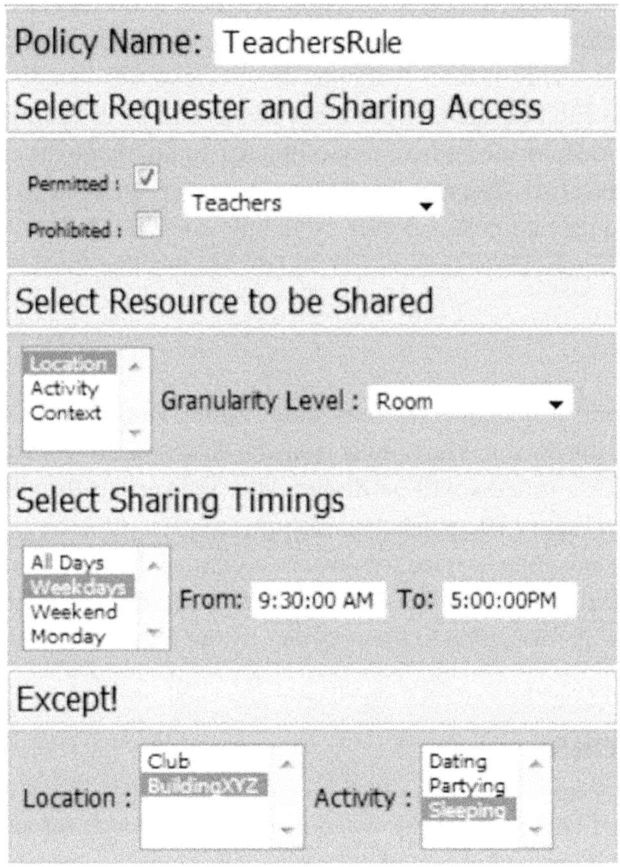

or prohibiting access to specific entity. This entity may be outside the user's mobile device like a friend or family member's phone. It may also be an app on the user's own mobile device. The second clause defines what data/resource may be shared. The third clause defines a timing contextual constraint within which the rule applies and finally the system has exception clauses that may negate the rule if necessary. The generalization and specialization options for context constraints are defined using the Place ontology discussed in Section on Collaborative Context Modeling.

Generalization

Generalization involves replacing a value with a less specific but semantically consistent value in order to protect user data privacy (Sweeney, 2002). The system uses context-data generalization to allow information sharing on different levels of granularity.

- **Location Generalization:** In order to support location generalization, the ontology uses hierarchical model for location. Location is a super class of *Point, Room, Building, City* and *State* classes. The *Point* class is used for denoting GPS coordinates whereas *Room* and other subclasses are used to denote different levels of abstractions for location. The transitive "Part Of" property creates a location hierarchy based on some simple axioms like "Room is a part of Building". The reasoning engine uses this ontology to infer different relations existing between instances of these subclasses (see Figure 12).

- **Activity Generalization:** Along the lines of location generalization, let's look at activity generalization for allowing users to share different descriptions of their current activity to different set of requesters. In many cases, the user is willing to share more generalized activity rather than a precise one. For instance, if a user is attending a confidential "project meeting" then she might want to share it in a more generalized way as "working" or simply as a "meeting" (see Figure 13).

Figure 12. Location hierarchical model

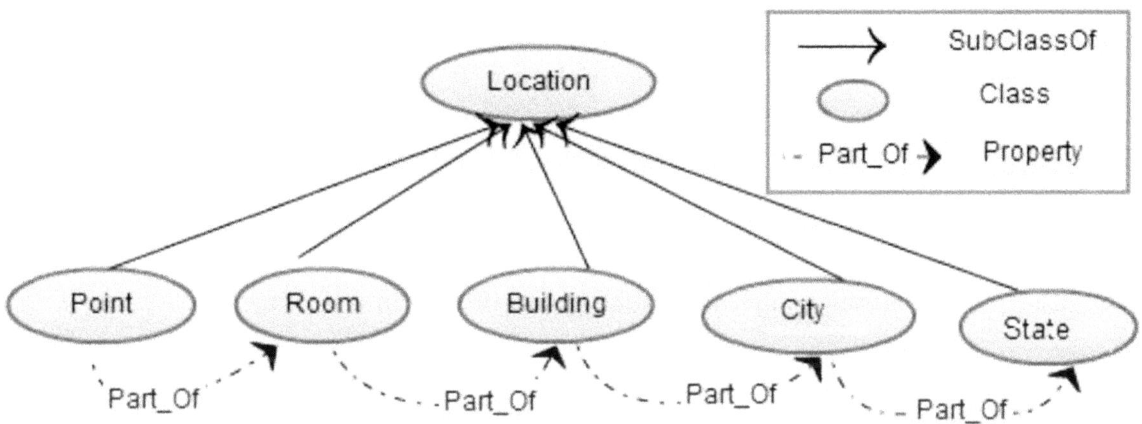

Figure 13. Activity hierarchical model

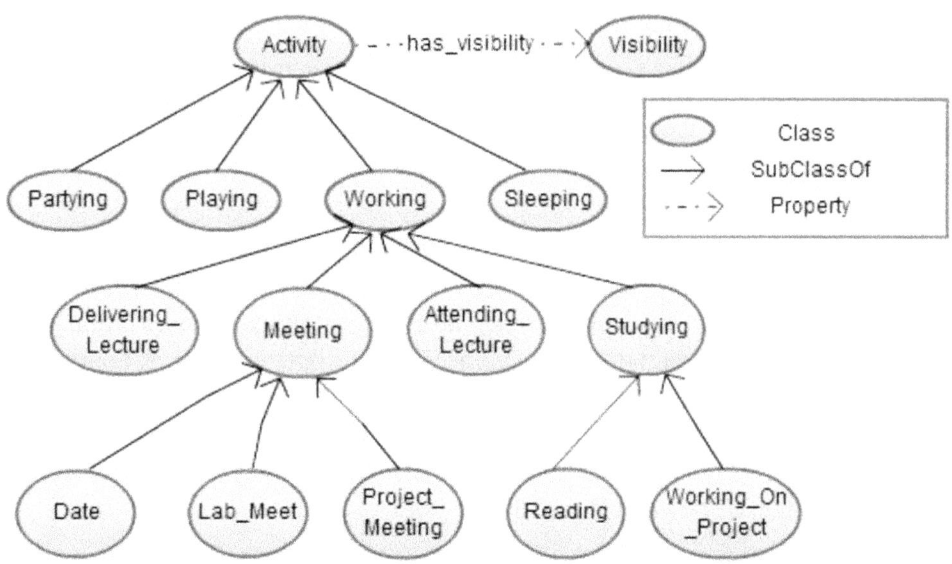

Policy Determination

Selecting policy that needs to be implemented cannot always be driven by human users or administrators. A certain level of automation is desired in this process. *Why?* Studies have shown that people need to wade through a sea of information in order to determine the right privacy preference (Lin, Liu, Sadeh, & Hong, 2014). However, such information might not necessarily enable them to choose the best possible policies (Lin, 2013). Allen Westin's perspective on privacy defines privacy as the ability for people to determine for themselves "when, how, and to what extent, information about them is communicated to others" (Westin, 1970). However, these approaches on information access control have been found to be flawed (Kagal, L., & Abelson, H., 2010). According to (Weitzner et al., 2008) information accountability is the ability to determine whether usage of information is appropriate and the ability to identify the violator.

Considering data being accessed by apps on a user's mobile device, app provenance maybe used as a way to assign accountability. It should be noted that the notion of app provenance is not restricted merely to geographical origin; of the developer of the app. It can also refer to online repository from where the app is downloaded to the mobile device. Availability of such information opens up possibility of capturing a whole new set of access controls on device; apps can be restricted from being installed in first place based on geographical origin or online origin. Policies can be pre-written to restrict access to only a subset of device resources for apps, which takes into account app origin and other contextual information of the device user and the app itself.

Automatic policy generation/implementation can thus be done in three stages.

Stage 1 - App Data Gathering Phase: Apps are searched in android marketplace by name. Thereafter, package name, app version, app rating, app developer information, app permission data, app descriptions are collected. Once app developer information is available app's organization data from DBpedia is retrieved and location information of the organization is extracted. In (Ghosh, 2012) the author collected developer information from DBpedia.

Stage 2 - Pre-Decision Making Phase: Triples are generated which represent facts about the app and stored on the mobile device.

Stage 3 - Decision Phase: The reasoner upon a receipt of an access control request uses the triples stored about the app, the policy stored on the phone and the context information, generates a grant or deny response.

For this purpose, take a look at the *PlatMob* (Ghosh, Joshi, Finin, & Jagtap, 2012), ontology for mobile device applications to capture application provenance data from heterogeneous sources. *PlatMob* is an extension of the *Platys* ontology presented earlier. The Place ontology represents a high level context ontology by (Zavala, Dharurkar, Jagtap, Finin, & Joshi, 2011). The concept of a requester of data is defined in the Platys ontology (Jagtap, Joshi, Finin, & Zavala, 2011b, 2011a). The *PlatMob* ontology allows modelling of richer notion of an application's context and developing privacy preservation policies far more complex than that of Android's default all or nothing policy. *PlatMob* is an aggregation of four domains of knowledge as shown in Figure 14.

- **PlatMobile:** Representative application data and context on the device the app is installed.
- **Platys:** Representative device users' context.

Figure 14. PlatMob ontology

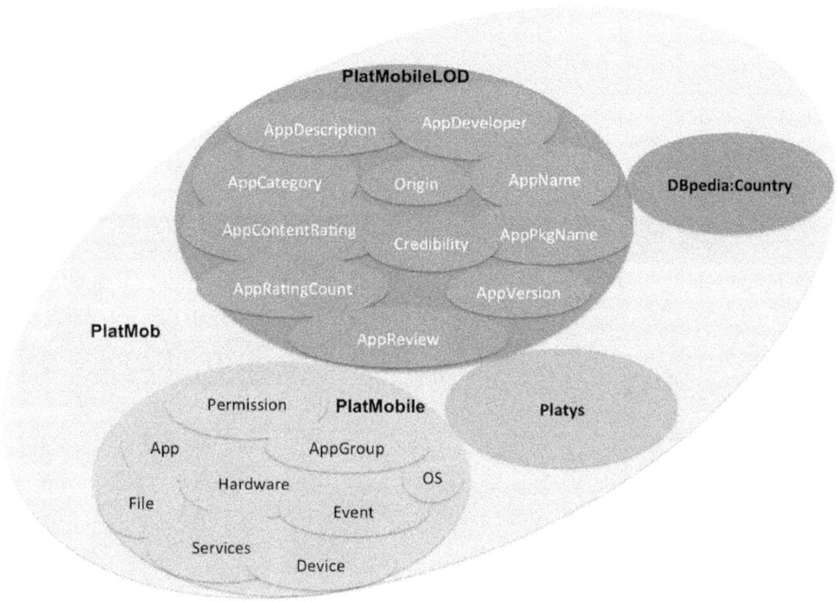

- **PlatMobileLOD:** Representative application context sourced from marketplace and linked open data cloud.
- **DBpedia:** Country.

PlatMobileLOD models data available for an app from external sources whereas *PlatMobile* models on-device runtime information about the application. *PlatMobile* represents on-device resources using the following entities. *AppGroup* defines an application group; can be apps accessing Internet or apps which deals with location data or apps which does media capture etc., *Permissions* defines classes of permissions a specific app has; and is represented as a collection of String literals which can be collected from apps *AndroidManifest.xml* file, *File* represents the file system resource of the device; it is further divided into *ImageFile*, *AudioFile*, *VideoFile*, *BinaryFile*, *TextFile* subclasses, *Hardware* abstracts devices hardware resources e.g. *Camera*, *Keyboard*, *Microphone* and *Storage*, *Service* is representative of system and third party services. *PlatMobileLOD* represents different entities that captures app context from external data sources like app marketplace and linked open data sources like DBpedia. *PlatMobileLOD* comprises of *AppCategory*, *AppContentRating*, *AppDescription*, *AppDeveloper*, *AppPkgName*, *AppName*, *AppRatingCount*, *AppReview*, *AppVersion*, *Origin*, *BlackList* and *Credibility* entities. *PlatMobileLOD* uses DBpedia Country ontology to describe *Origin* and *Blacklist* entities, e.g.

```
http://dbpedia.org/resource/Iran a platmob:BlackList .
http://dbpedia.org/resource/Iran platmob:origin ex:com.farsitel.bazaar .
```

Using the *dbpedia-owl:location* property, origin country information can be extracted for a developer name. Figure 15 displays Google Play's entry of a popular photography app, Instagram. Google Play displays a list of attributes, including but not limited to the ones listed below:

Figure 15. Google Play entry for Instagram Android app

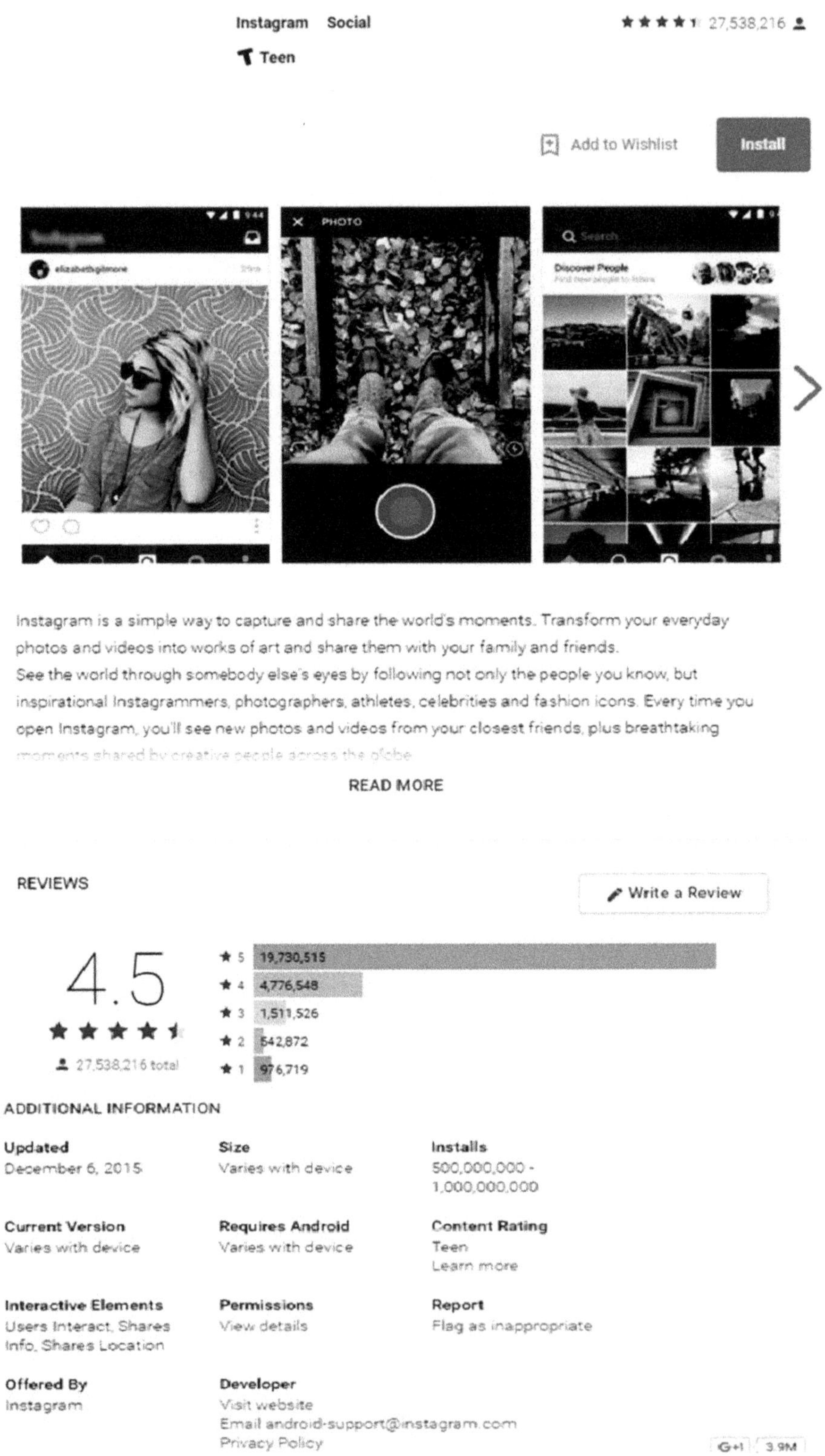

- App description,
- User reviews,
- User rating and rating count,
- Developer information:
  - Developer website,
  - Developer email.
- App version,
- Download count,
- User review count,
- Category, etc.
- **Complex Policy Generation:** Using the app provenance information it is possible to represent highly rich context-based policies. An example scenario could be as follows. Imagine a person with a top security clearance who might be the target of a foreign entity for perpetrating acts of espionage. The targeted individual has access to sensitive information and is an avid smartphone user. The foreign entity has considerable reach to this person and has successfully installed a backdoor on this person's device. The targeted user visits a secure data facility on a regular basis due to his/her nature of job. The malware residing on this individuals compromised device does a string of activities it starts audio/video capture at stipulated time of the day or day of the week based on this individuals calendar schedule and only if device screen is not active or the user is not in the middle of any calls at that point of time; these media files are then uploaded to remote servers when connected to Wi-Fi and not using the data plan to avoid raising suspicion; once uploaded the malware removes these files to cover its track. Accessing user calendar is possible for any user land app once it is given access at install time. And the use case delineated here is fairly generic since all the associated activities are straightforward. Existence of privacy policies capable of controlling application specific resource access based on dynamic requester app and device context can address breach of privacy under such circumstances. *PlatMob* may be used to generate privacy policies for situations outlined below:

Disable camera recording on weekdays between 9 AM – 5 PM if device user is in a meeting

OR

Disable audio / video capturing by some app A belonging to app group Internet and with origin country O, where O is Blacklisted between 9 AM - 6 PM if device user is at location L

OR

Do not allow some app A to run on device between time t on days d, when app A belongs to app group G, app A has origin country O, app rating > R, no of downloads > N, app category is P, app is developed by group X, device user is at location L engaged in an activity Q, and location L belongs to country C

For the first and second scenario above, all context information are available from the device. However, that changes in the third scenario. Applications rating, download count, category, developer and app origin information is harnessed from apps market place and DBpedia.

DISCUSSION

We have discussed two prototypes one for inter-device privacy and security implementation and the other for intra-device privacy and security control. The first prototype uses a client server model with the requester being able to send requests to another device directly or through a server. The access request is processed by the policy framework. For the intra-device privacy and security the prototype implementation has two major components: a privacy control module and a device operating system. The privacy control module aims to protect user privacy by performing reasoning over the context. It deals with the resource to be protected, the owner of a resource and the requester who wants to access it. More abstractly, it accepts an RDF triple (U, C, Q), where U is the identity of the requester, C is the requester's context (expressed as RDF triples in the Platys ontology), and Q is the query pertaining to context information. Both prototypes consider contextual information and sensor information as the resources that changes dynamically for the user, and provide mechanisms to specify more expressive policies to control its sharing. The users can create policies by using Policy Editor Interface.

Context-aware systems have been studied for a long time, though the focus has been mainly on the location and activity inference. The research project MyCampus (Sadeh, Chan, Van, Kwon, & Takizawa, 2003) presented a mobile application that involved a collection of customizable agents capable of (semi-) automatically discovering and accessing Intranet and Internet services during the process of assisting their users in carrying out various tasks. During the past decade a body of work on rule-based policy frameworks and access control systems has emerged. Rei (Kagal, Finin, & Joshi, 2003) is a policy language designed for pervasive computing applications. It has been used to build a security framework that addresses the issues of security for web resources, agents and services in Semantic Web. Rein (Rei and N3) (Kagal & Berners-Lee, 2005) is a distributed framework for describing and reasoning over policies in Semantic Web. It supports N3 rules for representing interconnections between policies and resources. Taintdroid (Enck et al., 2010) uses a taint tracking mechanism to detect sensitive data flow inside the device. CRePE(Conti, Nguyen, & Crispo, 2011) is the first to introduce a policy based Android extension but the user context model and the user level CRePE assumes, is trivial when one considers the extent of granularity of user context and user role possible in real life circumstances.

FUTURE WORK AND CONCLUSION

As mobile devices become the dominant communication and information access medium these devices will model our interests, activities and behavior. When appropriate, aspects of this learned information which includes context may be shared with other devices in order to collaborate and provide enhanced service. This development introduces a need for a stronger flow control. However, leaving this control solely in hand of users might not be safe. Eventually we will have to think of systems that are capable of making suggestions to users about how to protect their privacy and security. Such systems could provide understandable policies to users or even enforce generic corporate or certified policies from trusted authorities. One way of going forward is to work on logic based and learning based systems which are capable of determining security vulnerabilities on devices and make such suggestions.

Another problem lies on increasing the computing ability of mobile devices. Executing complex inference mechanisms or generating context models can be a compute intensive task. We need to make sure these tasks are efficient in order to ensure it does not consume all the resources on an already limited-resource device. Therefore, we need to work on algorithms to ensure efficient executions of privacy and security rules which consume less resources for generating context.

REFERENCES

AppBrain. (2015). *Number of android applications.* Retrieved from http://www.appbrain.com/stats/number-of-android-apps

Baader, F., & Sattler, U. (2003, December). Description logics with aggregates and concrete domains. *Inf. Syst., 28*(8), 979–1004. Retrieved from doi:10.1016/S0306-4379(03)00003-6

Bechhofer, S., Van Harmelen, F., Hendler, J., Horrocks, I., McGuinness, D. L., Patel-Schneider, P. F., (2004). Owl web ontology language reference. *W3C recommendation, 10.*

Carroll, J. J., Dickinson, I., Dollin, C., Reynolds, D., Seaborne, A., & Wilkinson, K. (2004). Jena: Implementing the semantic web recommendations. In *Proceedings of the 13th international World Wide Web conference on alternate track papers & posters* (pp. 74–83). New York, NY: ACM. doi:10.1145/1013367.1013381

Chen, H., Finin, T., & Joshi, A. (2003, September). An ontology for context-aware pervasive computing environments. *The Knowledge Engineering Review*, *18*(3), 197–207. doi:10.1017/S0269888904000025

Conti, M., Nguyen, V. T. N., & Crispo, B. (2011). Crepe: Context-related policy enforcement for android. In M. Burmester, G. Tsudik, S. Magliveras, & I. Ilic (Eds.), *Information security* (Vol. 6531, pp. 331–345). Springer Berlin Heidelberg. doi:10.1007/978-3-642-18178-8_29

Damianou, N., Dulay, N., Lupu, E., & Sloman, M. (2001). The ponder policy specification language. In *Policies for Distributed Systems and Networks* (pp. 18–38). Springer Berlin Heidelberg. doi:10.1007/3-540-44569-2_2

Dey, A. K., & Abowd, G. D. (1999). Towards a better understanding of context and context-awareness. In *First int. symposium on handheld and ubiquitous computing (HUC).*

Enck, W., Gilbert, P., Chun, B.-G., Cox, Jung, J., & Sheth, A. N. (2010). Taintdroid: an information-flow tracking system for real-time privacy monitoring on smartphones. In *Proceedings of the 9th usenix conference on operating systems design and implementation* (pp. 1–6).

Fonseca, O. (2012, November). *Byod leads to data breaches in the workplace.* Retrieved from https://github.com/lencinhaus/androjena

Ghosh, D. (2012). *Context based privacy and security in smartphones.* (Unpublished master's thesis). University of Maryland, Baltimore County.

Ghosh, D., Joshi, A., Finin, T., & Jagtap, P. (2012). Privacy control in smart phones using semantically rich reasoning and context modeling. In Security and privacy workshops (spw), 2012 IEEE symposium on (pp. 82-85). doi:10.1109/SPW.2012.27

Godik, S., Anderson, A., Parducci, B., Humenn, P., & Vajjhala, S. (2002). *OASIS eXtensible access control 2 markup language (XACML) 3. Tech. rep*. OASIS.

Gu, T., Wang, X. H., Pung, H. K., & Zhang, D. Q. (2004, January). An ontology-based context model in intelligent environments. In *Proceedings of communication networks and distributed systems modeling and simulation conference* (pp. 270-275).

Jagtap, P., Joshi, A., Finin, T., & Zavala, L. (2011a, Sept). Preserving privacy in context-aware systems. In *Semantic computing (ICSC), 2011 fifth IEEE international conference on* (p. 149-153). doi:10.1109/ICSC.2011.87

Jagtap, P., Joshi, A., Finin, T., & Zavala, L. (2011b). Privacy preservation in context aware geosocial networking applications. *Organization.*

Kagal, L., & Abelson, H. (2010). Access control is an inadequate framework for privacy protection. In *W3C Privacy Workshop.*

Kagal, L., & Berners-Lee, T. (2005). *Rein: Where policies meet rules in the semantic web.* Cambridge, MA: Computer Science and Artificial Intelligence Laboratory, Massachusetts Institute of Technology.

Kagal, L., Finin, T., & Joshi, A. (2003, October). A policy based approach to security for the semantic web. In *International Semantic Web Conference* (Vol. 2870, pp. 402-418). doi:10.1007/978-3-540-39718-2_26

Kuhn, D. R., Coyne, E. J., & Weil, T. R. (2010). Adding attributes to role-based access control. *IEEE Computer*, *43*(6), 79–81. doi:10.1109/MC.2010.155

Lane, N. D., Miluzzo, E., Lu, H., Peebles, D., Choudhury, T., & Campbell, A. T. (2010). A survey of mobile phone sensing. *Communications Magazine, IEEE*, *48*(9), 140–150. doi:10.1109/MCOM.2010.5560598

Liebowitz, M. (2011). *Developer sneaks fake apps into android market.* Retrieved from http://www.nbcnews.com/id/45641853/ns/technology_and_science-security/t/developer-sneaks-fake-apps-android-market/

Lin, J. (2013). *Understanding and capturing people's mobile app privacy preferences* (Unpublished doctoral dissertation). Carnegie Mellon University, Pittsburgh, PA, USA. (AAI3577905)

Lin, J., Liu, B., Sadeh, N., & Hong, J. I. (2014, July). Modeling users' mobile app privacy preferences: Restoring usability in a sea of permission settings. In Symposium on usable privacy and security (soups 2014) (pp. 199–212). Menlo Park, CA: USENIX Association. Retrieved from https://www.usenix.org/conference/soups2014/proceedings/presentation/lin

Lindorfer, M., Volanis, S., Sisto, A., Neugschwandtner, M., Athanasopoulos, E., Maggi, F., & Ioannidis, S. (2014). Andradar: Fast discovery of android applications in alternative markets. In S. Dietrich (Ed.), Detection of intrusions and malware, and vulnerability assessment (Vol. 8550, pp. 51-71). Springer International Publishing. doi:4 doi:10.1007/978-3-319-08509-8

Lorecarra. (2009). *Androjena: Jena android porting.* Retrieved from http://www.experian.com/blogs/data-breach/2012/05/02/medical-and-mobile-convenience-trumps-security/

Patwardhan, A., Korolev, V., Kagal, L., & Joshi, A. (2004, August). Enforcing policies in pervasive environments. In *Mobile and Ubiquitous Systems: Networking and Services, 2004. MOBIQUITOUS 2004. The First Annual International Conference on* (pp. 299-308). IEEE. doi:10.1109/MOBIQ.2004.1331736

Sadeh, N. M., Chan, T.-C., Van, L., Kwon, O. B., & Takizawa, K. (2003). A semantic web environment for context-aware m-commerce. In *Proceedings of the 4th ACM conference on electronic commerce* (pp. 268–269). New York, NY: ACM. doi:10.1145/779928.779992

Sandhu, R., & Samarati, P. (1996, March). Authentication, access control, and audit. *ACM Computing Surveys*, *28*(1), 241–243. doi:10.1145/234313.234412

Security, L. M. (2015). *App vetting API.* Retrieved from https://www.mylookout.com/app-vetting-api

Statista. (2015). *Number of apps available in leading app stores as of July 2015.* Retrieved from http://www.statista.com/statistics/

Sweeney, L. (2002). k-anonymity: A model for protecting privacy. *International Journal of Uncertainty, Fuzziness and Knowledge-based Systems*, *10*(05), 557–570. doi:10.1142/S0218488502001648

Uszok, A., Bradshaw, J., Jeffers, R., Suri, N., Hayes, P., Breedy, M., & Lott, J. (2003, June). KAoS policy and domain services: Toward a description-logic approach to policy representation, deconfliction, and enforcement. In *Policies for Distributed Systems and Networks, 2003. Proceedings. POLICY 2003. IEEE 4th International Workshop on* (pp. 93-96). IEEE.

Weitzner, D. J., Abelson, H., Berners-Lee, T., Feigenbaum, J., Hendler, J., & Sussman, G. J. (2008). Information accountability. *Communications of the ACM*, *51*(6), 82–87. doi:10.1145/1349026.1349043

Westin, A. F. (1970). *Privacy and freedom*. Academic Press.

Chapter 9
Mobile Location Tracking:
Indoor and Outdoor Location Tracking

Sima Nadler
IBM Haifa Research Lab, Israel

ABSTRACT

One of the key things that differentiate mobile devices from static computing platforms is the ability to provide information about the device user's location. While the raw location is often useful, it is the ability to understand the user's context that makes this capability so powerful. This chapter will review the technologies used today to provide location tracking of mobile devices and which are best for different types of use cases. It will also address challenges associated with location tracking, such as accuracy, performance and privacy.

INTRODUCTION

Mobile phones initially disrupted the market by enabling voice communication literally anywhere and anytime. As they evolved though, mobile devices became powerful computing platforms with new capabilities not available on traditional desktop and server computers. One of the first differentiators was the ability to track the mobile device user's location. This opened the door to new types of capabilities and apps previously inconceivable to the general public. The ability to identify and track a mobile user's location, whether inside or outside, has changed the way we live and work. This chapter describes the different types of location tracking technologies, their advantages and disadvantages as they relate to different use cases, as well as the challenges associated with the technologies and location tracking in general. Among the challenges highlighted will be the issue of data privacy as it relates to mobile location data.

DOI: 10.4018/978-1-5225-0945-5.ch009

BACKGROUND

Location-based services have become very popular both in the general public and enterprises, with the advent of smartphones and other sophisticated mobile devices. While many people are accustomed to using such services, knowledge of how they work, and which underlying technologies are most appropriate for different types of use cases, remains limited even to experienced developers.

MAIN FOCUS OF THE CHAPTER

Mobile Device Based Tracking Technologies

To fully understand the options available for tracking a mobile device, it is important to understand how mobile devices work and the capabilities embedded in them. Voice and data communication with the mobile device are enabled by the telecommunications provider's (telco) infrastructure, together with the mobile device. The mobile device sends out signals including its unique IMEI number via the 3G, 4G, and/or LTE protocols, which are picked up by the telco's nearest antennas. These signals are correlated by the telco's backend systems. When communication with the device is desired (for voice, data, and/or video), the telco knows which base stations and antennas to send the data to so it reaches the designated device. As such, the telco knows which cellular antenna is closest to the device. Since it knows the location of its antennas, it thus knows the location of the mobile device – at least to the resolution of the area covered by the given antenna. This can be anywhere from a hundred meters up to several kilometers, depending on the density of the antenna coverage in the given area. The telco may put in place a system to triangulate the location of a mobile device. Since the mobile device's signal is usually captured by multiple cellular antennas, it is possible to compare parameters such as signal strength, time difference of arrival, and bearing, and then calculate a more accurate location of the mobile device. Not all telcos implement such triangulation systems. For those that do, external developers can only access such information if the telco provides a public API to it. The accuracy of location for cellular triangulation is between 100 meters and several kilometers, depending on the topology of the cell towers and antennas as well as the general topography and buildings nearby. Much research has been done in developing algorithms to improve cellular-based location tracking. One such example is *Mixing and Combining with AOA and TOA for the Enhanced Accuracy of Mobile Location* (Chen, Chiu & Tu 2003).

WIFI tracking works very similar to the cellular method just discussed, and is typically used indoors. Rather than using cellular antennas and base stations, WIFI deployments consist of WIFI access points that are installed throughout the building and connected to controllers. Modern mobile devices are equipped with WIFI radios that enable them to send/receive data via the internet using the WIFI protocol IEEE 802.11 ("IEEE Standard for Information Technology", 2012); this is possible if WIFI is activated on the device and the WIFI infrastructure is connected to the internet. Where cellular data services are expensive with limited or low bandwidth, many people tend to take advantage of WIFI networks where they are available. As a person moves around a venue with their device, the device connects to the WIFI router that is closest or has the strongest signal. Thus, someone with access to the WIFI infrastructure can know where the device is located. If the appropriate WIFI infrastructure is in place, it is possible to more accurately identify the location of the device since the mobile device's signal together with its unique MAC ID is captured by multiple WIFI routers. This information is calculated by the backend

Table 1. Indoor cellular tracking: advantages and disadvantages

Advantages	Disadvantages
Cellular is always on, unlike with WIFI which must be turned on for location tracking to work.	Because there are multiple telcos in each country, the visitors to a single venue will likely be from different telcos. The infrastructure must support all the different telcos and cellular protocols, and business agreements must be in place with all parties.
All mobile phones support cellular.	More work has been done on indoor location tracking for WIFI than has been done on cellular-based indoor location tracking.
	Not all tablets, wearables, and other devices are connected to the cellular network.

WIFI infrastructure and can be accessed by developers and others if it is made available through APIs, as is done with solutions such IBM Presence Insights, Motorola Air Defense and Cisco Unified Wireless Location Based Services.

There are now options for cellular tracking indoors as well, because the telcos are providing indoor antennas, known as small cells, for improved indoor communications. There are both advantages and disadvantages to this approach shown in Table 1.

The approaches mentioned so far have been of the type where the environment senses the mobile device. However, the mobile device can also identify its own location using sensors on the device. The most prevalent approaches are as follows:

Global Positioning System (GPS)

Most modern smartphones have a GPS receiver in them. GPS technology relies on satellites. Satellites circling the earth transmit signals that can be captured by the GPS chip in the mobile device. In order to determine its location, the mobile device needs to be in "line of sight" with at least 4 different satellites. Thus the device must be outdoors, making GPS a technology that is not appropriate for indoor location tracking. Very tall buildings, tunnels, and other large objects can obstruct the ability of the device to capture the satellite signals. However, excluding those limitations, GPS typically provides an accuracy of 3 to 15 meters. In addition, any application running on the mobile device can access the GPS data, assuming location tracking is activated by the user. Assisted GPS, also known as A-GPS, augments location information based on satellites with cell tower data to improve accuracy.

WIFI

As mentioned previously, most mobile devices have the ability to connect to WIFI networks. If WIFI is activated, the device constantly scans to see what WIFI networks are available in the vicinity. The device may or may not connect to the WIFI networks it senses (based on user settings and WIFI network permissions) but Apple, Google, and other companies have mined and mapped the location of WIFI networks across the world. This enables them to provide increased location accuracy by combining data from the different sources – GPS, cellular, and WIFI. For this reason, if the user turns off the WIFI on their devices they will often get a notification by iOS or Google Android saying that location accuracy may decrease as a result.

Beacons

Another type of radio that many mobile devices support is Bluetooth. The original Bluetooth protocols (Bluetooth 1.0 – 3.0) known formally as the IEEE Standard 802.15.1–2005 ("IEEE Standard 802.15", 2015) were used to connect mobile devices, wireless earphones, speakers, and other devices. Communication between the devices required "pairing" of the devices, and its use was a serious drain on the battery of the mobile device. In recent years devices have added support for Bluetooth Low Energy ("Bluetooth Smart", 2016). This protocol, which is much less of a drain on the mobile device's battery, also has extensions such as beacon protocols.

Beacons are small devices that are installed in a venue and broadcast a signal using a protocol such as Apple's iBeacon protocol (Apple Inc, 2016) or the Eddystone open source protocol contributed by Google. The mobile device listens for such signals, noting that it is in the proximity of one or more specific beacons. The signal broadcast contains the following:

- **Beacon ID:** Unique ID associated with the particular beacon.
- **UUID:** The unique ID assigned to the owner of the beacon.
- **Major ID:** Typically used to indicate the venue ID (particular store, mall, hotel, etc.).
- **Minor ID:** A sub-area of the major location.

When the mobile app receives the data from the beacon, it also receives an estimate of the distance of the mobile device from the beacon in the form of a signal strength and/or distance estimate such as near, far, medium.

To be of use, the beacon information needs to be mapped to a location or function. This information may be stored directly in the mobile app, or it may be stored centrally by a server side solution. The mobile app must take the information received from the beacon and map it to the location, or send it to a server to be mapped to a location and returned to the mobile app.

The approaches mentioned so far are the most prevalent ones used today to identify a mobile device's location. However, there are others that are appearing in the market and identified in works such as GeoComputation by Abrahart and See (n.d.).

Table 2. Additional indoor tracking technologies

Technology	Description	Advantages	Disadvantages
Magnetic Field Based	Uses the mobile device's accelerometer, gyroscope, and compass to measure the magnetic field and compare it to a map of the magnetic fields in the building.	Claim accuracy of 1-2 meters. No special hardware installation in venue or on mobile device.	Requires mapping and fingerprinting of venue
Lighting Based	Uses the mobile device's camera to capture patterns of light flicker emitted by specially adapted fluorescent lighting.	Claim high level of accuracy	Camera must be on and in line of sight with lighting. Requires installation of special lighting, fingerprinting of venue, and an app on the mobile device.
Audio Beacons	Uses the mobile device's microphone to capture sounds emitted at different frequencies	Relatively inexpensive infrastructure	Microphone must be on. App required on mobile device.

Location Context

All of the technologies discussed provide raw data about the location of the mobile device. It might be in the form of an x,y,z coordinate or the proximity to a beacon or other transmitting device (WIFI, cellular, flickering light, audio signal, etc.). The raw data itself is not meaningful without context. Sometimes the application itself provides the context. For example, a navigation application containing maps (indoor or outdoor) of the surrounding area provides the context by showing the location of the mobile device in relation to the map. Other times the context is transmitted in place of, or together with, the raw location data. For example, a beacon name might indicate the context or a more sophisticated solution might provide context information and a system for associating context with pre-determined areas. IBM's Presence Insights, for example, enables the definition of zones, and the association of tags with those zones. Analytics and processing of location data can then be done based on the context, making solution development and analytics much easier and more intuitive.

Much research has been done to automatically understand the context of a mobile device user. Context can refer to many different things, such as the nature of the computing device being used, tasks being performed by a person, or the type of input device being used. However, the intent here is to understand the broader situational context of the person carrying the mobile device and whose location is being tracked. Location is just one parameter used in determining, for example, whether the person is working, doing a leisure activity, on vacation, etc. This area of research is often termed "pervasive computing". Claudio Bettini provides an overview of work done in this area in his paper, "A Survey of Context Modelling and Reasoning Techniques" (Bettini, Brdiczka, et al, 2010).

Vendor Approaches to Providing Location Information to Developers

Apple iOS

Location and Maps Programming Guide (Apple Inc., 2015)

In order for an iOS application to access the device's location, the user must approve the use of location services. Assuming such permission has been granted, there are multiple ways to obtain location information.

- **Standard Location Service:** Prior to using it, the developer indicates the level of accuracy desired. High levels of accuracy, such as those typically used for navigation, use the location tracking hardware (GPS, WIFI, accelerometer) of the mobile device, resulting in faster battery drainage.
- **Significant Change Location Service:** For applications with lower accuracy requirements, this API may be suitable. This is the least accurate of the location updates. Apple has not provided details of its accuracy, but developers have found that it can be between several hundred meters to several kilometers.
- **Region Monitoring:** Events are generated when a region is entered or exited. An application may be registered to a maximum of 20 regions
- **Geographical Region Monitoring:** An area defined by a circle of a specified radius around a known point on the earth's surface (i.e. longitude/latitude coordinates + radius). Uses GPS-based location tracking.

- **Beacon Based Region Monitoring:** An area defined by its proximity to be a Bluetooth low energy beacon.
- **Heading Information:** Leverages the magnetometer and thus can report the direction in which a device is pointing.
- **Course Information:** Leverages the GPS hardware to report the direction in which a device is moving. This is often used in navigation applications and is relevant only outdoors.

Google Android

Google has two different levels of location APIs. Android location APIs ("Google Location Services for Android", 2015) provide platform and sensor-specific APIs. They require the developer to have a deep understanding of the different sensing technologies, and when and how to use each of them. Google Location Services[9], which is part of Google Play Services, provides a higher level framework that hides the details of the sensing technologies, similar to Apple iOS's approach.

In the lower level Android Location APIs, the developer must indicate the type of sensing via predefined constants such as NETWORK_PROVIDER, which levels cell tower and WIFI sensing, or GPS_PROVIDER, which leverages the GPS capabilities of the mobile device. It is possible to get location updates from both of these sensing options by requesting location updates twice, each time indicating a different sensing technology. Location tracking may be requested at different levels of accuracy, such as fine-grained or course-grained location updates.

Google's example algorithm ("Android Developer Guide Location Strategies", 2015) for how to obtain the current best location when using android.location APIs is shown here. This highlights the complexity of obtaining the user's current location:

1. Listen for GPS and Network updates.
2. Retrieve cached network location.
3. Dismiss GPS location as it is now too old.
4. New cell-id fix is received.
5. WIFI based location is received.
6. New WIFI based location is dismissed due to larger error estimates.
7. GPS location replaces current best estimate.

When implementing more complex use cases, things get even more complicated. The developer has to access additional sensors, such as the accelerometer and possibly the gyroscope, to identify movement and proximity sensors to handle beacons.

For this reason Google created the higher level Google Location Services ("Google Location Services for Android", 2016), which provides the following higher level functionality:

- **Get Last Known Location:** The current location of the device.
- **Receive Location Updates:** Recurring updates regarding the device's location, enabling the developer to set update intervals as well as priorities between accuracy, power consumption.
- **Display a Location Address:** Transforms longitude and latitude into a location address.
- **Create and Monitor Geofences:** Sends notifications when a device enters, exits, or dwells for a certain amount of time in a predefined area as defined by latitude, longitude, and radius. Up to 100 geofences are supported per device.

Marketing/Contextual Information

Another very popular use case that leverages location information is contextual marketing. Companies that want to provide information or promotions can make them more effective by sending them at the appropriate time and place. Sending a promotion when a user is near a store is more likely to be effective than blindly sending a promotion. Once in a store, stadium, airport, or other such venue, companies can easily cross-sell and up-sell based on the departments/areas a user is currently visiting, has visited in the past, or related to visited areas.

Similarly, sending weather information upon arrival in a new city or providing information about a piece of artwork when a person is standing in front of it in a museum, are both examples of contextually relevant information. Granted, in these simple cases the context is defined solely by the location. If the person is an employee of the museum, for example, it may not make sense to send details about the art every time they walk past. Thus, location is but one, albeit important, piece of the puzzle.

Location tagging is another prevalent use case. For example, when a user posts information on social networking sites, the information may be augmented with its location. This technique could also be used to facilitate passive marketing. The location tag acts an additional way for a person to search for information about products or businesses. Personalized information can be provided on electronic displays that a shopper is standing in front of while in a venue. Less intrusive than push marketing, this allows a hybrid passive/active marketing approach.

Remarketing techniques popular on the web also become relevant for in-store shopping. On the web, if a customer abandons a basket, (i.e. puts products in a virtual basket but does not purchase them), they are often targeted by the same retailer with the same or similar products via banner ads or other mechanisms as they browse the internet. In a physical store, new location tracking technologies also provide the ability to identify where people have browsed and when people leave without purchasing. Thus, remarketing becomes relevant for physical store shoppers as well. Movement pattern analytics discussed later in this chapter describes some of the technical issues associated with implementing this type of use case.

For such use cases, there are several technical challenges. While most of these use cases are based on proximity to a static item or location, proximity alone is often not sufficient. Dwell time is also important to ensure that the person is not just passing by, and thus perhaps the information or promotion is not relevant. Therefore, such solutions often combine parameters such as dwell time and user type together with the location.

Asset/People Tracking and Optimization

Keeping track of people and assets is a challenge for any enterprise. Understanding where employees are is key to improving the efficiency of processes, as is keeping track of the tools, products, and other assets used by the business. While mobile location tracking is not necessarily relevant for tracking products, the mobile scanners (such as the Motorola MT2000 Series Mobile Computer) used by employees and other such equipment can be tracked in much the same way as a mobile phone. As tablets and other mobile devices become more prevalent in businesses, many employees can be tracked by the devices they use for their daily jobs. Understanding the location of both employees and customers enables enterprises to more accurately understand how much of employee time is spent interacting with customers versus doing other tasks. Different location tracking technologies may be used for the tracking of employees and enterprise assets, with the following questions used as guidelines:

- Can dedicated applications be installed on the devices being tracked, and if so, is there a way to prevent them from being turned off? If applications cannot be installed, then infrastructure-based tracking is the only option.
- What level of accuracy is required? For example, is it enough to know the general area / department of the venue, or is more detailed location required?
- Are geo-fenced based alerts needed to prevent assets from leaving the venue?

Whether to use infrastructure-based, device-based location tracking, or a combination of both, will be highly dependent on the use case and the types of devices being tracked. Another factor in the decision making is whether the enterprise allows employees to use their own devices on the job, known as BYOD (Bring Your Own Device). If so, the location tracking solution chosen must be one that supports many popular devices.

Movement Pattern Analytics

As mentioned previously, movement pattern analytics can be used to generate maps of previously unmapped indoor or outdoor venues. However, movement pattern analysis can also be useful for businesses trying to understand customer paths through their venue to make decisions about staffing locations, product positioning, infrastructure enhancement, or management. All of these use cases differ from the discussion on map generation because the mobile device users are usually anonymous visitors, meaning it cannot be assumed that they have downloaded an app to their mobile device and are running it. Thus, the location sensing has to be such that the venue is sensing the visitors (infrastructure-based) rather than the mobile device sensing its surroundings. Currently, the natural choices would be WIFI or small cell triangulation, both of which have accuracy limitations that are highly dependent on the number and positioning of the access points installed in the venue. Researchers in academia and industry have been working on algorithms to improve location accuracy in such situations. One such example is described in *Understanding Customer Behavior Using Indoor Location Analysis and Visualization* (Yaeli, Bak, et al, 2014).

Outdoors the issue is similar, yet somewhat simpler because of the existence of roads and paths. However, in third world countries or less developed areas the challenge is similar to indoor because of more open areas and fewer roads with which to correlate the location.

Dwell time analytics is in essence a subset of movement pattern analytics. When attempting to understand movement patterns, it is crucial to understand when the person is spending time in a specific area rather than passing through it. Since the radio signals used to track a person are noisy, it is often a non-trivial task to identify if the person is passing through or dwelling in a given location. Some movement within a given area is natural even when dwelling, thus algorithms for differentiating dwell from movement is another technological challenge. Tang and Song in their research, *Estimating the most likely space-time paths, dwell times and path uncertainties from vehicle trajectory data: A time geographic method* (Tang, Song, Miller & Zhou, 2015), addresses this challenge for both real-time and offline applications and "develop a method based on the potential path area for all feasible network-time paths."

Wang, Uzun, Bareth and Kupper also tackle this problem in their Tracommender project (2012), which uses location tracking and other contextual information to provide smart recommendations to smartphone users. They provide algorithms for dwell times and to compare paths of visitors, focused on improved recommendations.

Wellbeing Detection

Identifying the wellbeing of people is another area in which location tracking comes into play. Whether the person is a child whose parents are concerned about their safety, a dementia patient, or an elderly person living on their own, their movement patterns and location can be important to the people who care about them and care for them.

Geo-fencing capabilities are used to identify when a mobile device and its owner have entered or exited a predetermined area, either indoors or outdoors. Outdoors GPS is typically used for such functionality and requires that the device transmit its GPS location to a server, which then provides the information to approved people who have subscribed to the location. Apple's Find My Phone, Life360, and many other applications provide this type of functionality, as well as the ability to proactively send location information.

There are also companies such as Guidecare (Lin, 2015) doing indoor geo-fencing to identify when people who live alone at home may be in distress. Additional sensors may be leveraged, but location-based analytics are also key, as they provide an indication as to whether the person is moving about their house in a normal fashion (went to bed, got up in the morning, visited the kitchen to eat, etc.). These solutions are often based on beacons or other low cost, easy installation type infrastructure. However, they do require that the mobile device or a wearable device capable of receiving beacon signals be carried at all times while moving about the home. An alternative is the installation of more complex infrastructure in the home; such as infrared, video, WIFI, or other type of sensing. In uLocate (Chen, Harniss, et al, 2013), a project at the University of Washington, a solution combining WIFI tracking for indoor location and GPS for outdoors was created for tracking elderly people with disabilities. Alternatively, an RFID based approach is proposed in the paper, "RFID-based indoor location tracking to ensure the safety of the elderly in smart home environments" (Kim, Park & Jeong, 2013).

Similar type scenarios can be implemented for employee wellbeing and accident identification and prevention in mines, firefighting scenarios, warehouses, hospitals, etc.

ISSUES, CONTROVERSIES, PROBLEMS

As described so far, there are many benefits associated with location tracking. However, issues of privacy are not to be ignored. Any data collected that is associated with an identifiable human person is considered personally identifiable information (PII) and falls under privacy laws and guidelines. One can state that the location technologies discussed track not a human person, but rather the mobile device or wearable devices they are carrying or wearing. In fact, they track the unique identifier associated with one or more of the radios embedded in the device. For WIFI this is the MAC ID, for cellular this is the IMEI number, and for beacons it is the UUID of the application together with the major and minor IDs, which is transmitted from the mobile device to the service upon receipt of beacon information.

In recent years, companies such as Apple have attempted to address some of the privacy concerns associated with anonymously tracking via WIFI, for example, by randomly changing the MAC ID broadcast by a device prior to its connection to a given WIFI network. This prevents WIFI-based tracking technologies from identifying repeat visits of the same person.

The question arises whether these unique identifiers are to be considered personal information. If so, then tracking devices in this way constitutes the collection of personal information that falls under privacy laws. Privacy laws differ from country to country, but in more and more countries they are being considered as such, certainly when correlated with other information, such as a person's name, email, or address. Thus, obtaining consent is one way to address privacy.

Most companies today address consent by presenting long legal documents containing the terms and conditions of using their service or entering their facility. These may be accompanied by a small check box indicating that the terms and conditions are understood by the customer, or perhaps by placing a sign at the entrance of the venue (ex: informing of video surveillance or other tracking technologies.) Customers can either agree, or leave / not use the service at all. Other options are rarely provided.

In Europe it is mandatory, based on the European Data Protection Directive (Directive 95/46/EC), and the General Data Protection Regulation ratified by the European parliament in December 2015, to obtain a user's explicit consent to collect or process personal data unless the data is aggregated or obfuscated sufficiently. Although still considered in the grey zone in some places, more and more countries in Europe are considering tracking using IP Address, MAC ID, IMEI, and other such unique identifiers associated with mobile computing devices as the collection of personally identifiable information requiring explicit consent.

The United States does not have a similar data privacy law, but rather handles privacy on an industry basis. The US Health Insurance Portability and Accountability Act (HIPAA), for example, regulates the use and disclosure of personally identifiable health information. Criminal Justice Information (CJI) is governed by the Federal Information Security Management Act, providing the legal basis for handling the personal information collected by law enforcement agencies in the United States. In the area of commerce, payment card transactions are regulated via the Payment Card Industry Data Security Standard (PCI-DSS).

Companies wanting to use information such as a person's location have several options:

1. Do not store or process location information at an individual level. Aggregate the data to a summary level (ex: an average of 30 devices in the shoe department on Sunday mornings).
2. Obfuscate the data in such a way that it is not possible to reverse engineer it to obtain the person's identity.
3. Obtain the user's informed consent to capture and use his/her location data.

The last option is more of a challenge than it may seem. Making it clear to a person what data is being collected, for what purpose, with whom it will be shared, and how long it will be stored, is not a trivial undertaking. Even assuming this has been done, the enterprise that has obtained the data faces many challenges regarding how to abide by the laws governing who may access it, for what purpose, and how long it may be stored. The latter is driven by the fact that the data should not be stored for longer than is required to provide the service(s) for which there is consent, or according to local regulations and laws.

To abide by privacy laws and policies, once a person's location data is stored in an enterprise's data stores the enterprise has to do the following:

- Associate the purpose(s) for which the data was collected, with the data itself.
- Note the length of time it is necessary to store the data for the given purpose.
- Associate the privacy policies and consent terms and conditions that were relevant at the time of consent, as well as at the time of data collection, with the data itself.
- Provide proof that data was used only by appropriate people and for the approved purpose(s).
- Provide proof that the data was obfuscated or discarded once it was no longer needed for the approved purpose(s).

Each time the data is accessed, shared, or moved, all of these parameters have to be taken into account.

Unfortunately, today's storage solutions and IT infrastructure in general do not provide solutions for these requirements. European Union Research projects such as PrimeLife ("PrimeLife", 2011) have addressed some of these issues, and organizations such as IBM and Microsoft's Research departments are also actively working on solutions to ensure privacy. David Kotz from Dartmouth College has done research on people's privacy preferences as related to location. He and his team also created AnonySense (Shin, Cornelius, Kapadia, Triandopoulos & Kotz, 2015) to enable location-based crowd sourced insight without revealing the identity of the people or the mobile devices providing the location information. This was done by distributing what he called "sensing tasks" among participating anonymous mobile devices. Obviously, these privacy issues are not specific to location data, but addressing them is critical as more and more companies and organizations are collecting peoples' location data.

SOLUTIONS AND RECOMMENDATIONS

Developing location-based services requires knowledge of the underlying sensing technologies. Even though there are tools to hide these complexities from developers, some level of understanding is required to truly understand what level of accuracy and latency can be expected.

Determining the type of location sensing technology to leverage is largely driven by the use case and the main business goals as discussed in the previous section. Often there are multiple and sometimes conflicting goals. For example, a retailer or other enterprise may start out wanting to better understand the number and flow of customers in their venue. Subsequently, the natural progression is a desire for location-based interactions with customers. Especially for indoor venues, these require very different sensing capabilities. Infrastructure-based sensing is more appropriate for the former, since it will sense a larger percentage of the visiting population. The latter require customer opt-in. This can be done via infrastructure sensing by leveraging the WIFI captive portal. However, the preference is often for a mobile app-based approach, which tends to lend itself more to proximity-based sensing such as BLE beacons.

In short, when developing location-based mobile services the following should be taken into account:

- Business goals and focus use cases, examples of which have been described in this chapter.
- Sensing capabilities available on mobile devices carried by users – WIFI, BLE, accelerometer, etc.
- Willingness and/or feasibility of using infrastructure-based sensing.

In addition to determining the best location tracking method, all location-based solutions have to address the privacy implications. If the solution is to be used internationally it must be configurable to take into account different privacy laws and norms in different countries. A clear description of how location and other personal information is used should be presented to the user, and privacy settings should be more flexible than just opt-in or opt-out. One option could be to make the level of location granularity configurable based on the purpose for which it is being used. For example, the location granularity required to provide weather information is very different than what is required for a navigation application. Work such as at Carnegie Mellon University in the Caché (Amini, Linqvist, et al, 2010) project addresses this specific issue, caching detailed location information but allowing users to share only general geographical regions where they deem appropriate.

Ling Liu from the Georgia Institute of Technology provides a good overview of techniques that can be used to address location privacy in his presentation, "From Data Privacy to Location Privacy: Models & Algorithms" (2016). This paper provides a nice summary of many techniques and algorithms for anonymizing location information and when best to use each.

Mobile operating system providers could also enhance their offerings by providing information to users about which applications are using location information and how often it is used. Norman Sadeh's team at Carnegie Mellon took this approach and also provided "privacy nudges" (Alnuhimedi, Schaub, et al, 2015),, which included not only which applications were using the location information but also suggestions on how to alter settings to increase privacy. They found that significant numbers of study participants changed privacy settings as a result of such nudges.

FUTURE RESEARCH DIRECTIONS

Both the underlying sensing technologies as well as their uses will continue to change and develop. Key areas of research will be in algorithms for smoothing and improving the accuracy of the raw location data, as it tends to be quite noisy. In addition, advanced analytics to understand traffic patterns indoors and their relation to customer purchases and services consumption will continue to be of interest. The results of such research will enable enterprises to customize products and services offered to consumers.

In parallel, research to address how to communicate, manage, and enforce privacy policies will be of great importance as well. European Union projects such as Prime Life ("Prime Life", 2011) provide initial work in this space. However, browser-based internet interactions were their main area of focus. In addition, the solutions developed have not been adopted in the industry. This is at least partially due to the architectural complexity and performance overhead introduced. Thus, research focused on data privacy with location-based data and other sensor data being taken into account is of great importance, especially as the Internet of Things (IoT) gains in popularity. Location tracking can be seen almost as a subset of IoT, where IoT will be adding more and more information to it from additional types of sensors. Heart rate trackers, temperature sensors, sensors that track operations of appliances, cars, and homes are just a few examples. These sensors will be embedded in and on our bodies, in our cars and homes, and in the enterprises that serve us. That data on its own, and certainly when combined with location information, is extremely sensitive. Research for managing privacy settings in IoT will thus be of great importance, especially as all the new sensing data is added into the context inference engines that academic researchers such as Claudio Bettini (Bettini, Brdiczka, et al, 2010) have been working on for years.

CONCLUSION

Location technologies are changing the way people work and live. Both the use cases and the technologies are developing extremely rapidly. While telecommunications companies initially had the upper hand in identifying location, mobile phone manufacturers changed the playing field when they introduced GPS and other sensing technologies into mobile devices. Suddenly, location-based services could be provided by any app developer. In the indoor arena it remains to be seen who / what technologies will gain the most traction. Initially it appeared that infrastructure providers such as Motorola and Cisco would rule because they already had in place base infrastructure that could be used with some extensions and expansions. While they can provide solutions with coverage for both opted-in and anonymous tracking, to obtain the accuracy required for many indoor use cases significant infrastructure investments are required above what most customers already have. As a result, such companies are also adopting proximity sensing approaches via beacons and other less expensive options.

In short, it remains to be seen what sensing technologies will dominate in the coming years. Combinations will most likely be required to cover all but the most simple use cases.

As the technologies and use cases continue to develop, so will the need to address the privacy issues. Trust is key to any enterprise's business and brand, and ensuring the privacy of customer and employee data is key to building such trust.

REFERENCES

Abrahart, R. & See, L. (n.d.). *GeoComputation* (2nd ed.). Boca Raton, FL: CRC Press.

Almuhimedi, H., & Schaub, F. (2015). Your Location has been Shared 5,398 Times!: A Field Study on Mobile App Privacy Nudging. *Proceedings of the 33rd Annual ACM Conference on Human Factors in Computing Systems*. New York, NY: ACM.

Amini, S., & Linqvist, J. (2010). Caché: Caching location-enhanced content to improve user privacy. *Proceedings of the 9th International Conference on Mobile Systems, Applications and Services.*

Android Developer Guide Location APIs. Google Inc. (2015). Retrieved January 28, 2016 from http://developer.android.com/guide/topics/location/index.html

Android Developer Guide Location Strategies. Google Inc. (2015). Retrieved January 28, 2016 from http://developer.android.com/guide/topics/location/strategies.html

Bettini, C., & Brdiczka, O. (2010). A survey of context modelling and reasoning techniques. In Pervasive and Mobile Computing (pp. 161-180). Elsevier.

Bluetooth Smart (Low Energy) Technology. (n.d.). *Bluetooth SIG*. Retrieved January 28, 2016 from https://developer.bluetooth.org/TechnologyOverview/Pages/BLE.aspx

Chen, K.-Y., Harniss, M., Lim, J., Han, Y., Johnson, K., & Patel, S. (2013). uLocate: A Ubiquitous Location Tracking System for People Aging with Disabilities. In *8th International Conference on Body Area Networks, BODYNETS 2013*. doi:10.4108/icst.bodynets.2013.253584

Chen, T., Chiu, C., & Tu, T. (2003). *Mixing and Combining with OAO and TOA for the Enhanced Accuracy of Mobile Location*. Retrieved January 28, 20166, from http://ieeexplore.ieee.org/stamp/stamp.jsp?tp=&arnumber=1350199

Dupont, C., & Lookingbill, A. (2012). *System and Method for Managing Indoor Geolocation Conversions*. US Patent Office, Patent # US9147203 B1, USPTO.

Google Location Services for Android. Making Your App Location-Aware. (2015). Retrieved January 28, 2016 from http://developer.android.com/training/location/index.html

iBeacon for Developers, Apple Inc. (2016). Retrieved January 28, 2016 from https://developer.apple.com/ibeacon/

IEEE Standard 802.15-2005. (2011). *Part 15.1: Wireless Medium Access Control (MAC) and Physical Layer (PHY) Specifications for Wireless Personal Area Networks (WPANs)*. Author.

IEEE Standard for Information Technology. (2012). *Telecommunications and information exchange between systems Local and metropolitan area networks*. IEEE Standard 802.11.

Kim, S., Park, S., & Jeong, Y. (2013, December). RFID-based indoor location tracking to ensure the safety of the elderly in smart home environments. *Personal and Ubiquitous Computing*, *17*(8), 1699–1707. doi:10.1007/s00779-012-0604-4

Lin, W. (2015). *Guidercare Makes Smart Watch Ideal Solution for Elderly Care and Activity Tracking*. SMA100. Retrieved January 28, 2016 from http://www.mysmahome.com/COMPANY/4459/guidercare-makes-smart-watch-ideal-solution-for-elderly-care-and-activity-tracking.aspx

Liu, L. (2007). *From Data Privacy to Location Privacy: Models & Algorithms*. Retrieved March 6, 2016 from http://web.calstatela.edu/faculty/hpguo/Research/database/liu07.pdf

Location and Maps Programming Guide. Apple Inc. (2015). Retrieved January 28, 2016 from https://developer.apple.com/library/ios/documentation/UserExperience/Conceptual/LocationAwarenessPG/UsingGeocoders/UsingGeocoders.html#//apple_ref/doc/uid/TP40009497-CH4-SW1

PrimeLife. (2011). Retrieved January 28, 2016 from http://primelife.ercim.eu/

Shin, M., Cornelius, C., Kapadia, A., Triandopoulos, N., & Kotz, D. (2015, June). Location Privacy for Mobile Crowd Sensing through Population Mapping. *Sensors (Basel, Switzerland)*, *15*(7), 15285–15310. doi:10.3390/s150715285 PMID:26131676

Tang, J., Song, Y., Miller, H. J., & Zhou, X. (2015). Estimating the most likely space-time paths, dwell times and path uncertainties from vehicle trajectory data: A time geographic method. *Transportation Research Part C, Emerging Technologies*. doi:10.1016/j.trc.2015.08.014

Wang, Y., & Uzum, A. (2012). Tracommender – Exploiting Continuous Background Tracking Information on Smartphones for Location-Based Recommendations. In *Proceedings of the 5th International Conference, Mobile Wireless Middleware, Operation Systems, and Applications*. Berlin, Germany: Springer.

Yaeli, A., & Bak, P. (2014). Understanding Customer Behavior Using Indoor Location Analysis and Visualization. IBM Systems Journal, 58(5-6).

KEY TERMS AND DEFINITIONS

Accelerometer: A sensor found in mobile devices that measures acceleration. Used to note movement.

Bluetooth Low Energy: A wireless personal area network technology created by the Bluetooth Special Interest Group. Used for short range communication between devices.

Data Privacy: The aspect of information technology addressing how, when, by whom, and for what purpose data may legitimately be used based on local norms, laws, and enterprise policies.

GPS: Global Positioning System – Radio navigation system that allows land, sea, and airborne users to determine their location while in line of site of multiple positioning satellites orbiting the earth.

IMEI: A unique serial number that identifies a GSM or UMTS mobile device.

Location-Based Services: Products and services that leverage location to provide more personalization and/or accuracy.

MAC ID: Media access control address, the unique identifier of the component that interfaces with the WIFI network.

WIFI: The standard wireless local area network technology for connected computing devices to each other and the internet (IEEE Standard 802.11).

Chapter 10
Participatory Sensing for City-Scale Applications

Tridib Mukherjee
Xerox Research Center, India

Sharanya Eswaran
Xerox Research Center, India

Deepthi Chander
Xerox Research Center, India

Koustuv Dasgupta
Xerox Research Center, India

ABSTRACT

The rapid advancements in sensing, computation and communications have led to the proliferation of smart phones. People-centric sensing is a scientific paradigm which empowers citizens with sensor-embedded smartphones, to contribute to micro and macro-scale urban sensing applications – either implicitly (in an opportunistic manner) or explicitly (in a participatory manner). Community-based urban sensing applications, are typically participatory in nature. For instance, commuters reporting on a transit overload may explicitly need to provide an input through an app to report on the overload. This chapter will focus on the trends, challenges and applications of participatory sensing systems. Additionally, they will understand the solution requirements for effective deployments of such systems in real scenarios.

INTRODUCTION

In recent years, participatory sensing has evolved into a scientific paradigm that empowers citizens with sensor-embedded hand-held devices to contribute to micro and macro-scale urban sensing applications. Additionally, with the proliferation of social media and online blogs, city related issues are actively discussed by residents in open public forums on the web. It has thus become imperative for city agencies to properly analyze the information available through participatory sensing towards effective city planning. For example, a city agency may need to know which parts of the city has pollution issue because of garbage and how is it impacting the people. Similarly, a city transportation agency may want to get insights on whether the public transport services are commensurate with the demands in the city. The traffic department may want to know the spatio-temporal distribution of regular traffic problems and their potential causes. Insights on crime-infested areas may aid the police department to prioritize the most affected areas accordingly.

DOI: 10.4018/978-1-5225-0945-5.ch010

Many mobile crowdsourcing, crowdsensing, and human participatory sensing systems have explored the possibility of collecting implicit and explicit feedback from the residents on urban issues. Examples of such systems include: Moovit (Schwartz, 2015), Waze, Ushahidi (Ushahidi, n.d.), ParkNet (Mathur, 2010), Nericell (Mohan, 2008), PEIR (Mun, 2009) – to name a few. Many emerging cities are further experimenting strategies to engage with residents (citizens) to be the "catalysts" of change. A broad array of platforms, like dedicated Facebook pages, Twitter handles, and hashtags are being used to discuss local issues. Solutions like the Social Networking and Planning Project in Austin, and apps like SeeClickFix, allow citizens to express opinions about city planning, report problems like potholes or broken roads, or simply provide feedback to the agencies.

With the ready availability of Internet connections and smart phones, citizens are increasingly discussing their challenges in open public forums. Yet, while many of these initiatives give people a voice, and generate a lot of valuable data – there seems to be an inherent challenge in converting this data into actions. To worsen the situation, existing practices by many agencies are limited to manual surveys or call centers – that not only make the process slow and cumbersome, but are hardly scalable to growing scenarios. On one hand, it is important for the agencies to properly incentivize the residents to participate in providing meaningful feedback. From another perspective, it is also required for civic agencies to not only be aware of the problems, but also to possess the necessary capabilities to analyze the severity of the problems, often judge the reliability of the informants (reporters), and act upon the identified problems in a timely, accountable manner.

The basic challenge in such city-scale participatory sensing applications stems from the scale of the solution. This chapter will focus on a novel *urban sensing platform* (USP) that aggregates data from an eco-system of modern data sources (e.g., mobile sensing data, social media, web-based public forums, as well as the civic agencies' internal data) to derive valuable insights. Figure 1 depicts this vision. Challenges in terms of architecting the platform to gather and aggregate data in a scalable manner will be discussed. The aggregated data is then categorized to create meaningful summaries of reports gathered by the platform. In this context, this chapter will describe text-based categorization techniques which facilitate in determining events and to subsequently summarize them.

Figure 1. Urban Sensing Platform (USP) vision – a vehicle towards participatory sensing for city-scale applications

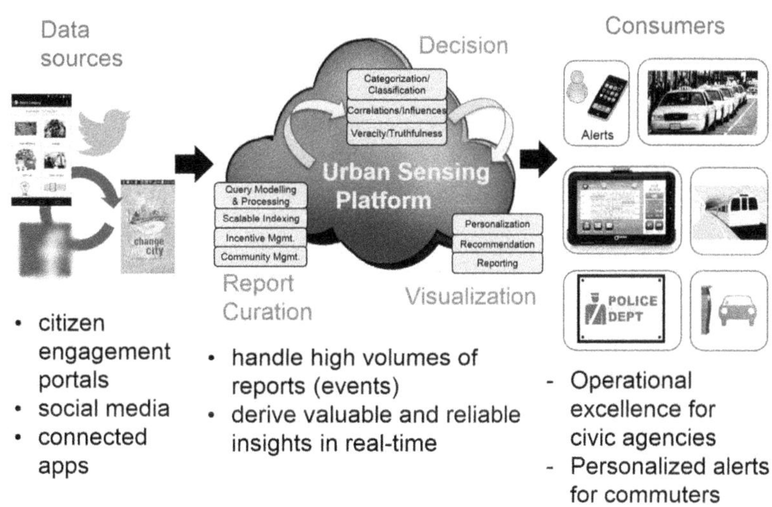

Another interesting aspect of participatory sensing is that of data veracity. Since the intent of reporters and ground truth of events are a-priori unknown, it becomes essential to simultaneously ascertain the occurrence likelihoods of events and the reliabilities of participants. This chapter will describe a data veracity framework that determines data veracity in participatory sensing systems. The effectiveness of personal sensing systems as well as community-based participatory sensing systems, such as the Favela Project, heavily rely on the engagement of reporters contributing with data sensed from their devices. Incentive strategies play a crucial role to promote engagement and to maintain a user pool. This chapter will also describe incentive strategies for participatory sensing systems and discuss the pros and cons of adopting various strategies. Finally, a case study of a participatory sensing deployment for developing regions will be discussed. In particular, challenges unique to geographically emerging markets, and solutions to address the same are highlighted.

DEFINITIONS

Before describing the elements of USP, first, some broad definitions are provided in this section

- **Curation:** The process of gathering reports from heterogeneous sources into a common standardized data repository.
- **Categorization:** The process of classifying the collected reports into pre-defined categories. Depending on the type of reports (e.g., text, images), this process may involve standard classification techniques (e.g., text classification from NLP, image classification).
- **Aggregation:** The process of combining related reports together to identify distinct issues/events. The relation of reports is mandated by the meta-parameters of the reports, i.e., time, location, and category. For any category, all the reports having timestamps within a time window as well as location within a radius (or generally within a spatial boundary) are considered related.
- **Verification:** The process of ascertaining truthfulness and veracity of the reports as well as the likelihoods of the events identified.
- **Incentivization:** The process of providing rewards or incentives by the civic agencies to the residents in order to receive desired reports on issues in the city.

URBAN SENSING PLATFORM

This section describes a manifestation of USP based only on text based reports. Figure 2 shows the major elements of an USP. Most of the elements (apart from categorization) is applicable for any other modalities (e.g., image, video) of reports from the residents. The report categorization may however need to employ specific classification techniques for different report modalities (e.g., image classification for reports having images). The classification itself is however not the focus of this chapter and the main objective is to provide insights on USP. After a report classification mechanism is identified, sets of reports can be processed concurrently by the *curation-categorization-aggregation-verification* pipeline in USP.

Figure 2. Major elements of an Urban Sensing Platform (USP)

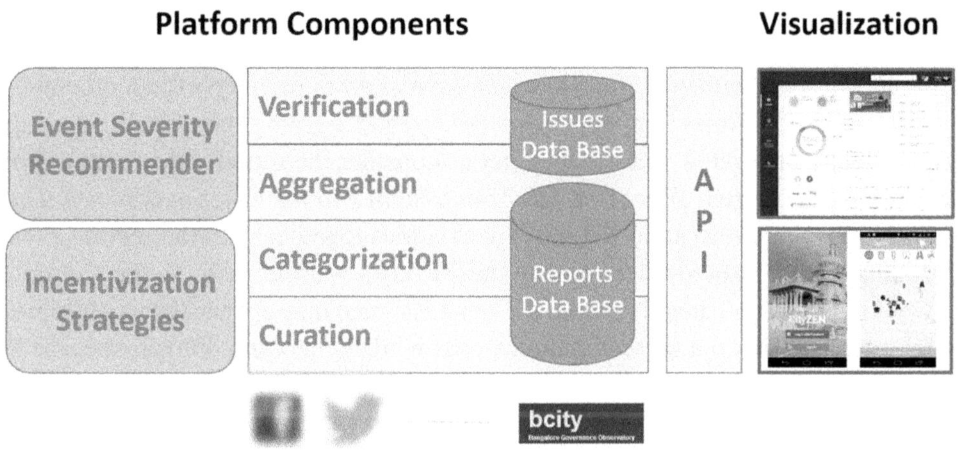

Data Curation

USP first collects the reports from heterogeneous sources and stores the reports in a reports database in a standardized form, in terms of the triple of < location; time; category > (irrespective of which source the report comes from). All the analysis can then be performed on the reports in the reports database. Sources can be either online, e.g., web blogs, social media reporting on city-related issues, or dedicated mobile apps. For the latter, data can be collected in the aforementioned standardized form – e.g., mobile app may need the user to input report category whereas time and location can be automatically inferred through time-stamping and GPS, respectively in the mobile devices (e.g., smartphones). One key aspect of getting reports from the mobile devices is to motivate the users (residents) to provide the required reports. Typically, in participatory sensing systems, this is done through incentivization, where appropriate rewards are provided to the residents for providing the reports. The form of the rewards can be monetary rewards, coupons or passes, value-added services by the agencies, as well as just reward points that translates to social recognition. Participatory incentive strategies to determine the reward points will be covered later in the chapter.

For other online sources, data collection needs to employ HTML parsing, and accessing JSONs or public APIs. For example, for an online complaints forum, called IChangeMyCity (IChangeMyCity, n.d.), in Bangalore, India, the JSON, containing complaints details, can be accessed directly. Similarly, Twitter has specific APIs for collecting public tweets. Most other online sources can be crawled using HTML parsing. Meta-data such as <location, time> and report-data such as text description need to be extracted from the gathered reports. In many cases, reports do not contain location information. To ascertain the location of such reports, a text-based approach needs to be employed to identify proper-nouns that prefix/suffix strong text-based location indicators (such as "at", "near", etc.). Google location APIs can then be queried to infer the location. Reports gathered from heterogeneous sources can thus be uniformly stored as <location, time, category> tuples in a report database. Although the location and time information of the reports can be extracted based on the aforementioned mechanisms, the category of reports need to go through a categorization module.

Data Categorization

The categorization process is described in this section for sample categories of interests to civic agencies (e.g., traffic, public transport, garbage, crime). For text-based reports, each report undergoes pre-processing involving – tokenizing, spell-correction, stop-word removal and n-gram generation. Extracted reports and comments are typically integrated. For instance, let us consider the report, "Borewell dried up in M.G road. Please look into the matter. *Your issue has been assigned to the water authorities. The status has been set to resolved*", where the (italicized) comment comes together with the report in the data base. Therefore, it becomes important to filter domain-specific keywords properly. In this example, Borewell may be a keyword in the Water category, while the word assigned may appear across multiple categories. Hence, it is important to filter out these common words while generating domain-specific keywords.

After the pre-processing step any well-known NLP techniques can be employed to perform classification. For example, a dictionary based rule engine can be used where a dictionary of unigrams, bigrams and trigrams are constructed using a training set. The document term matrix for the relevant terms of each report can be constructed. A relevance score for each word in each category can then be computed using TF-IDF (Ramos, 2003). For instance, Figure 3 illustrates keywords extracted for three sample categories. Words in the reports are then compared with this dictionary and categorized into one of the pre-defined categories based on the presence of some representative words.

Another option is to employ a Support Vector Machine (SVM) based classifier, which uses category-specific keywords (as shown in Figure 3) as feature vectors for each category. The keyword extraction is an off-line process and the keywords are taken as inputs to the classifier.

The categorization has been evaluated for the city of Bangalore, India. In this regard, data from around 25 sources across social media and public web-pages are being collected every 30 minutes. In our database, we have collected more than 61,669 reports across all the sources for around 8 months. The categorization had been performed for 10 categories, such as traffic, crime, roads, public transport, garbage, pollution, water, electricity, sewage, and illegal parking etc. Table 1 shows the accuracies of different classifiers on categorizing the reports from aforementioned online sources. The numbers are based on a labelled set of data, i.e. the set of data manually tagged to categories by multiple volunteers

Figure 3. Keywords, weighted based on term frequency, for three different categories

(a) Illegal Parking (b) Garbage (c) Garbage

to generate ground truth (and the outcomes of the classifiers are compared with the ground truth). Accuracy of a classifier is measured in terms of percentage of true positives w.r.t. the entire labelled dataset. From Table 1, it is clear that SVM has the best accuracy. Table 2 further shows the recall and precision of the SVM for each category. Recall denotes the number of correctly identified items out of the total number of items belonging to a category. Precision denotes the number of correctly identified items of the category, in the set of items predicted as belonging to this category.

The recall and precision get affected to some extent by the disambiguation problem, intrinsic to text analysis. For instance, "Poor lighting in the park" can be misclassified as an illegal parking problem because of the presence of the keyword park. When the discriminatory nature of the categories increase, the precision and recall tend to increase. For example, categories like, crime, pollution, garbage, electricity have discriminatory and un-ambiguous keywords. However, categories such as <illegal parking, roads, traffic, public transport> and <sewage, water> tend to have overlapping keywords. Thus, reports may inherently belong to multiple categories. For example, a report, "Garbage dumped on footpath is making it difficult for people to walk", can be tagged as belonging to both garbage and road categories. Therefore, multi-label classification may have to be incorporated in the categorizer. Improvements in categorization accuracies can further be achieved by using hybrid approaches such as topic modelling, to identify events that do not directly map into pre-defined categories (e.g., a report such as, "election rally near City Market at 4PM"), and yet relate to the pre-defined categories (such as traffic).

Table 1. Accuracy of different classifiers

Classification Techniques	Overall Accuracy
Dictionary Rule Based	72%
Multinominal Naïve Bayes	79%
Max Entropy Classifier	84%
Support Vector Machines (SVM)	87%

Table 2. Precision and recall of SVM classifiers for different categories

Categories	Recall	Precision
Traffic	70.3%	70%
Water	77%	76.5%
Illegal Parking	58.5%	69.5%
Pollution	82.5%	81.2%
Public Transport	63.2%	66.1%
Roads	79.6%	81.8%
Sewage	79.5%	69.2%
Garbage	85.2%	92%
Crime	94%	90%
Electricity	88.2%	81.1%

Data Aggregation

Once the reports are categorized, the aggregation module combines multiple related reports together to identify distinct events or issues. Specifically, the aggregation of reports into issues is to identify a $m: n$ mapping relationship, where m is the total number of reports, n is the total number of issues, and $n \leq m$ (i.e. one or more reports are attributed to each issue). To capture the ephemeral nature of issues, time can be divided into different slots or buckets. The granularity of the slots and the total time to be divided can vary across different categories. All reports of a category within a time slot and within a threshold radius of a landmark (or within a threshold of each other, in case nearest landmark is beyond the threshold) can thus be aggregated into a single issue. Algorithm 1 shows the process of aggregating the reports into issues. The process takes as input the sets ofl (new) reports R, existing issues I, landmarks L, categories C, and a threshold distance ρ. Based on these inputs, a new set of issues I' is returned. All the reports, $r \in R$, are aggregated into existing issues from I or into one or more new issues. All the new issues and the existing issues (updated with new reports) are part of the output issues set I'.

For each report, $r \in R$, it is checked if the category of the report, $C(r)$ is ephemeral, i.e. if a time based aggregation is required. In such a case, the existing issues of category $C(r)$, which are within the corresponding time slots, are extracted. Otherwise, all the existing issues of category $C(r)$ are extracted. If the new report is just another one of the existing issues, then the issues database is updated accordingly. Otherwise, a new issue is created in the issues database. If the Euclidean distance of the issue from the nearest landmark is less than ρ, then the new issue is pivoted to the nearest landmark. On the other hand, if the nearest landmark is farther than ρ, the latitude and longitude of the report are used as the location of the issue. For a new report of an existing issue, if the issue is not pivoted to a landmark, the location is updated to the mean latitude and longitude of all the reports (including the new report).

DATA VERIFICATION

A data veracity framework is now presented for verifying the truthfulness of citizen reports, by jointly ascertaining the reliability of sources and the likelihood of occurrence of the events reported. The framework is based on the maximum likelihood estimation model pioneered by (Wang, 2012). Most existing works assume (Wang, D., 2012; Wang, D., 2013; Wang, D., 2014; Wang, S., 2014; Wang, S., 2015) a binary report, i.e., a source reports either a 1 (event occurred) or 0 (did not occur). On top of this, how to incorporate any real-valued input for a report into the framework is further discussed in this section. When normalized to scale of 0 to 1, this can be interpreted in several ways, such as the probability of occurrence of event from the source's point of view, the confidence score of the source, a severity index for the event, frequency of sighting of event, etc. This additional degree of freedom helps capture and model data veracity in a better fashion.

The model is scalable and robust because it allows channel-level aggregation of reports, where a channel is defined as a common venue for reporting. For example, a channel may be a specific handle on Twitter or a Facebook group page, or a mobile app which allows multiple users to post their reports. In contrast, existing work is limited to individual source-level reporting. The robustness of the maximum-likelihood approach for estimation of source reliability lies in the availability of common sources who report across multiple events. However, in generic applications at city scale the data is typically sparse. Furthermore, as the geographic scale increases and the events are non-niche, the overlap of individual

Algorithm 1. Context-based report aggregation

```
Input: R, I, L, C, ρ.
Output: I'.
I' = I
for all r Є R do
  if C(r) is ephemeral then
    I" = {i Є I' | (C(i) = C(r)) and (LowTime(i) <= time(r) <= HighTime(i))}
    {extract existing issues of category C(r) in corresponding time bucket}
  else
    I" = {i Є I' | (C(i) = C(r))}
    {extract existing issues of category C(r)}
  end if
  if I" is not NULL then
    i = FindNearestIssue(r, I")
  end if
  if distance(i, r) <= ρ and I" is not NULL then
    R(i) = R(i) U {r}.  {add the report to the existing issue}
    if there is no l Є L such that location(i) = location(l) then
      update issue location to mean of all reports
    end if
  else
    create new issue i of the category C(r)
    set time of the issue i to that of the report r
    l = FindNearestLandmark(r, L)
    if distance(l, r) <= ρ then
      set the location of issue to landmark l
    else
      set the location of issue to report r
    end if
    I' = I' U {i}
  end if
end for
Return I'.
```

sources across issues is also hard to come by. The channel level aggregation of reports helps alleviate this problem. Figure 4 gives the system architecture of the Data Veracity framework.

- **Source-Claim Matrix (SI):** The aggregator module provides the events or issues identified (after report aggregation) as input to the Data Veracity framework. The issues are referred to as claims and the individual channels are the sources. This data is represented as a Source-Claim matrix. The element SI(i,j) represents the number of reports of the j-th claim (event) reported by source (channel) i. A claim j, can be "Traffic jam at Whitefield" and source i can be the channel, Traffline.

Figure 4. Data veracity framework: schematic

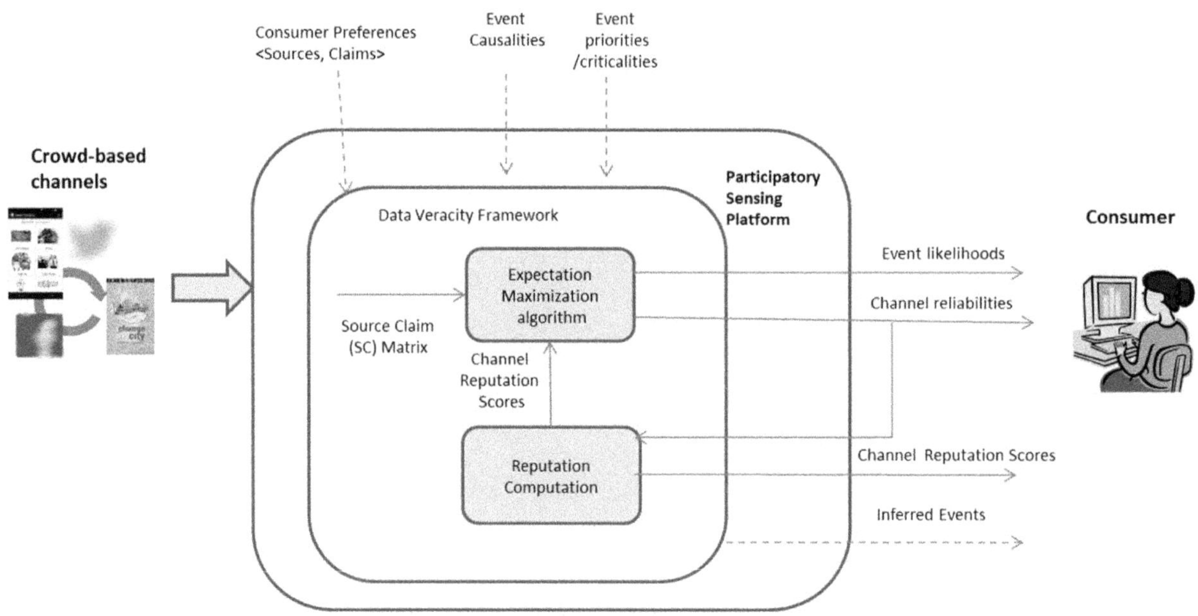

For instance, if Traffline submits 10 reports for the claim "Traffic Jam at Whitefield", SI(i,j)=10. On the other hand, if Traffline does not report on the claim j, SI(i,j)=0. The SI matrix is constructed for N claims over a period T for all sources reporting on the N claims. A key differentiation needed w.r.t the state of the art is to capture the weightages of different channels for different event types. The volume of reports from a source for a claim can be one manifestation of the weightage.

- **Expectation Maximization (EM) Algorithm:** This is a Maximum Likelihood based iterative algorithm that ascertains channel reliabilities and event likelihoods simultaneously using the aforementioned SI matrix as input. In this regard, a Binomial Veracity model is used. This model improves upon the aforementioned binary model. The binomial model captures the weightages of different channels for different types of events – a key requirement for urban sensing as mentioned above.
- **Consumer Preferences:** A consumer may be interested only in a subset of issues and subset of channels. For instance, a traffic agency maybe interested only on traffic related issues and sources such as Traffline that dedicatedly report on traffic. The SI matrix is trimmed based only on these chosen channels and claims (issues). The EM algorithm uses the trimmed SI matrix while computing source reliabilities and event likelihoods.
- **Event Causalities:** Some issues maybe intrinsically dependent. For instance, an accident will cause a traffic jam for a certain amount of time. Such dependencies are captured as a Bayesian network which feeds into the EM algorithm along with the observed SI matrix. These are then used to validate observations of the SI matrix by associating event likelihoods and source reliabilities with the causalities. Inferred events that surface based on the computed event likelihoods and reliabilities are also provided to the consumer with their associated likelihoods. For instance, low bus frequency can be used to infer 'bus bunching' at the bus depot with a certain probability.

- **Reputation Computation:** Historically accumulated source reliabilities, consumer category preferences, event criticalities and channel reliabilities computed based on the SI matrix are used to obtain the reputation score for each channel. The EM algorithm can be modified to incorporate the computed reputation scores, while weighing the event likelihood determined by each channel. For instance the reputation score of Traffline for traffic issues will be higher than that of a community page. Therefore, even if no reports on traffic may come from a community page, while Traffline has two reports on traffic, the traffic agency believes that there would be traffic. Similarly, a channel reporting on issues critical to a consumer is given a higher reputation score compared to a channel that misses out on reporting on critical issues. For instance a community page may report an illegal digging issue and will have a high reputation, compared to a newspaper channel that may miss out reporting on the same, with respect to the municipality tending to the location of the community. Note that the reputation scores can be used by the consumers internally, to design incentive mechanisms to elicit reports of high reliability from channels with high reputation.

The Data Veracity framework makes use of the aforementioned components to ascertain event likelihoods and source reliabilities simultaneously using an iterative approach as shown in Figure 5.

Binomial Veracity Model

Let S be the set of M sources (channels) $\{S_1, S_2, \ldots, S_M\}$; let I be the set of N claims (events) $\{I_1, I_2, \ldots, I_N\}$. SI is the observation matrix, where each entry S_iI_j is a value between 0 and 1. This can be taken as the probability or confidence with which source S_i thinks event I_j has occurred, or as the normalized number of reports that channel S_i has received for event I_j. Let Z be the vector of latent binary variables $[z_1, z_2, \ldots, z_N]$, where $z_j = 1$ when the event is true, and 0 otherwise. Let d_j be the prior probability of occurrence of event j. Let a_i be the probability that source S_i reports a true event, i.e., $P(S_iI_j = 1|I_j = true)$; let b_i be the probability of a false positive by source S_i, i.e., $P(S_iI_j = 1|I_j = false)$. Let $\theta = [a_1, a_2, \ldots, a_m; b_1, b_2, \ldots, b_m]$. The observation data SI is then treated as a binomial distribution, where for example, an entry of 0.8 is taken as 4 successes out of 5 trials in a binary event. The likelihood function for our model is then given as:

Figure 5. Expectation maximization algorithm: schematic

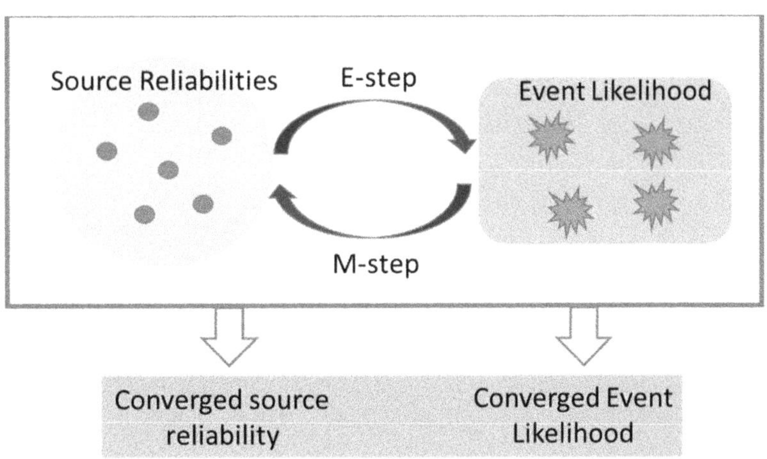

$$L(\theta;SC,Z) = p(SC,Z \mid \theta)$$
$$= \prod_{j=1}^{N}\left\{\prod_{i=1}^{M}\binom{n_{ij}}{k_{ij}} a_i^{k_{ij}}(1-a_i)^{(n_{ij}-k_{ij})} \times d_j \times z_j + \prod_{i=1}^{M}\binom{n_{ij}}{k_{ij}} b_i^{k_{ij}}(1-b_i)^{(n_{ij}-k_{ij})} \times (1-d_j) \times (1-z_j)\right\}$$

where, n_{ij} and k_{ij} are the indices of the binomial coefficient associated with $p(S_iI_j, z_j \mid a_i, b_i)$, which are obtained by the rational fraction approximation of S_iI_j. Expectation Maximization algorithm can be used for estimating the parameters of the equation above. The expectation is derived as follows:

$$E_{Z|SC,\theta}[\log L(\theta;SC,Z)]$$
$$= \sum_{j=1}^{N}\left\{p(z_j = 1 \mid SC_j,\theta) \times \left[\sum_{i=1}^{M} k_{ij}\log a_i + (n_{ij} - k_{ij})\log(1-a_i) + \log d_j\right]\right.$$
$$\left. + p(z_j = 0 \mid SC,\theta) \times \left[\sum_{i=1}^{M} k_{ij}\log b_i + (n_{ij} - k_{ij})\log(1-b_i) + \log(1-d_j)\right]\right\}$$

With further simplification, it can be derived that,

$$p(z_j = 1 \mid SC,\theta) = Z(j) = \frac{A(j) \times d_j}{A(j) \times d_j + B(j) \times (1-d_j)}$$

where

$$A(j) = p(SC_j,\theta \mid z_j = 1) = \prod_{i=1}^{M}\binom{n_{ij}}{k_{ij}} a_i^{k_{ij}}(1-a_i)^{n_{ij}-k_{ij}}$$
$$B(j) = p(SC_j,\theta \mid z_j = 0) = \prod_{i=1}^{M}\binom{n_{ij}}{k_{ij}} b_i^{k_{ij}}(1-b_i)^{n_{ij}-k_{ij}}$$

For the Maximization step, the optimal a_i and b_i in θ need to be derived that maximizes expectation value. By setting the first order derivatives of the expectation with respect to a_i and b_i, this can be derived to be:

$$a_i^* = \frac{\sum_{j \in SC_i} k_{ij} Z(j)}{\sum_{j=1}^{N} n_{ij} Z(j)}$$
$$b_i^* = \frac{\sum_{j \in SC_i} k_{ij}(1 - Z(j))}{\sum_{j=1}^{N} n_{ij}(1 - Z(j))}$$

The EM algorithm, as shown in Algorithm 2, iteratively estimates the source reliabilities θ and computes the event likelihoods *Z* based on the source reliabilities, until the values converge. The matrix *R* computed in step 13 gives the reliability of each source, i.e., the probability of the source reporting the correct state (true or false) of the event, with respect to each category of issue. If the issues have different levels of priorities or importance, the overall reliability of a source can be computed as *R* the weighted mean of reliabilities, i.e., reliability of source i, $R_i = \sum_{j=1}^{N} w_j R(i, j)$, where w_j is the priority of issue *j*. Figure 5 shows a schematic diagram of the algorithm.

Figure 6 shows the inferred likelihood (with at least 80% confidence) of various event categories, for the city of Bangalore, India (for a distribution of reports shown in Figure 7). The model computes the event likelihood based on the number of reports pertaining to that event across various channels of different reliabilities. The confidence is a measure of the estimation error associated with the event likelihoods deduced by the veracity model. It should be noted that the maximum event likelihood is around 65% because of the nature of the data available, and due to the fact that the likelihood heavily depends on the number of reports across different channels. As the number of reports corresponding to an event increases across different channels, the likelihood (as well as the corresponding confidence) also increases.

Incorporating Causality

The causality relationships between different events when exploited appropriately, provide better insights into the likelihood of occurrence of events. For instance, a road accident often causes traffic jam. When estimating the likelihood of occurrence of a traffic jam, it is therefore helpful to also consider whether or not a road accident has occurred. Similarly, the causality information also helps detect malicious collusional reports. For example, if there is a high number of reports on traffic jam at a certain loca-

Algorithm 2. Binomial EM algorithm

```
Step 1:  while θ does not converge do
Step 2:    for j = 1: N do
Step 3:      Compute Z(j)
Step 4:      d_j = Z(j) {To be used in the next iteration}
Step 5:    end for
Step 6:    for i = 1: M do
Step 7:      Compute a_i and b_i, using Z(j) from step 3
Step 8:      Update θ with a_i and b_i from step 7
Step 9:    end for
Step 10: end while
Step 11: for i = 1: M do
Step 12:   for j = 1: N do
Step 13:     R(i, j) = d_j a_i + (1 - d_j)(1 - b_i)
Step 14:   end for
Step 15: end for
```

Figure 6. Event likelihood for an 80% confidence across categories in the city of Bangalore, India

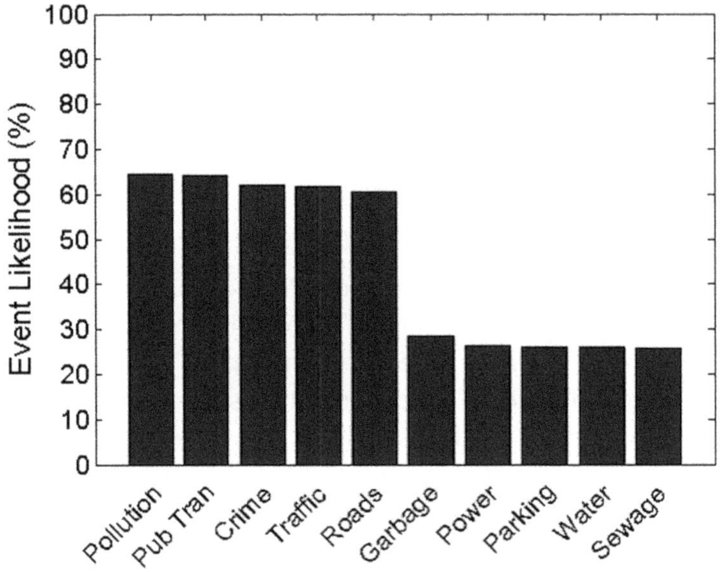

Figure 7. Source (channel) wise distribution of reports for different categories for Bangalore, India

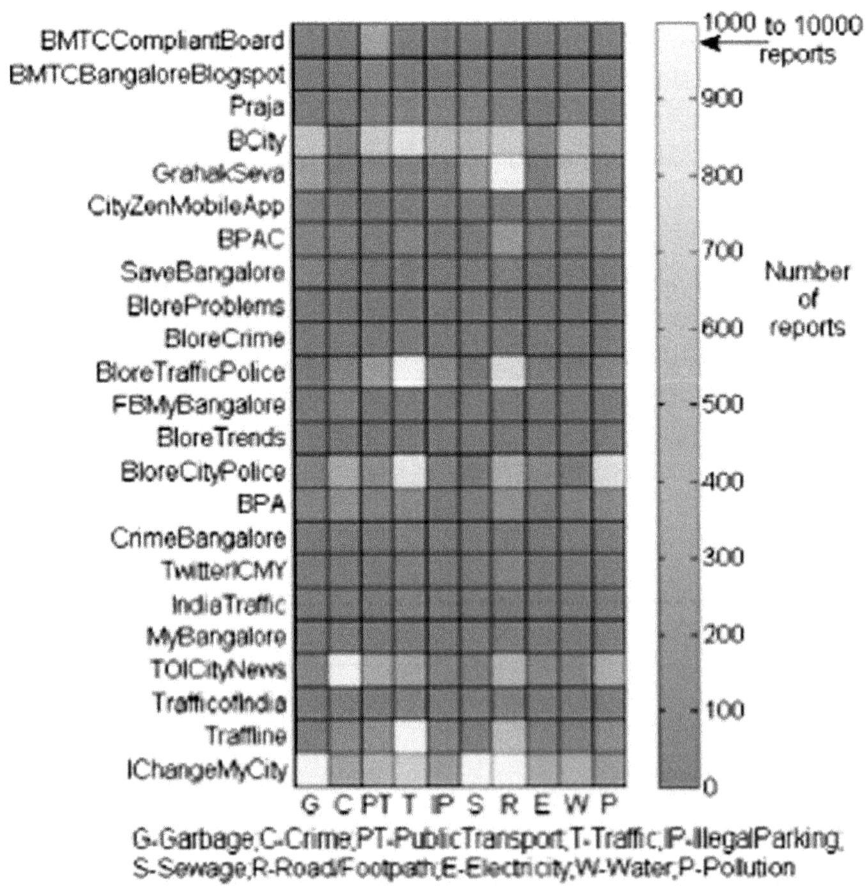

tion, however there are only a few reports about events that may potentially cause a traffic jam, such as accidents or road repair work, then this raises a question about the veracity of the traffic jam reports. Causality also helps with data sparsity, because some events can be inferred based on other events, which can be used to boost the existing observed data, i.e., they can be thought of as indirect, deduced reports.

The Binomial Veracity model is extended such that all of these benefits of causality information are captured. The causality information is structured as a Bayesian network, where the conditional probabilities of the child nodes with respect to the different states of the parent nodes are assumed to be given as input (which can be obtained from historical data). It should be noted here that (Wang, 2015) deals with the same problem. However, the Bayesian network in (Wang, 2015) consists of both issue and source observations as nodes. This means that the computational complexity of the Bayesian approach in (Wang, 2015) grows quadratically (O($MN^2(W+1)$ to be specific, where M is the number of sources, N is the number of events, and W is the number of nodes in the largest cluster of the Bayesian network). On the other hand, the complexity grows linearly in N, i.e., $O(N^2(W+1)$, when the Bayesian network consists of only the issues and the causal relationships among them. Accordingly, the likelihood function L is now defined as:

$$L(\theta;SC,Z)=\prod_{\forall j} p\left(SC_j, z_j \mid \theta, z_{k:\forall k\in 1:N}\right)$$

i.e., the likelihood of all events are considered as conditioning evidence when computing the likelihood of an event. It may be noted here that due to the conditional independence property of Bayesian networks, only node j's parents, children, and parents of children (i.e., the Markov blanket of node j denoted as $MB(j)$) will impact the likelihood of event j, and the other nodes (events) can be ignored. By following a similar derivation as described previously, the E-step in Eq. 2 now becomes:

$$Z(j)=p(z_j=1\mid SC,\theta,z_{k:\forall k\in MB(j)})=\frac{A(j)\times d'_j}{A(j)\times d'_j+B(j)\times(1-d'_j)}$$

where d'_j is the conditional probability $p\left(z_j=1 \middle| SI,\theta,z_{k;\forall k\in MB(j)}\right)$. Incorporating the probabilities of parents events conditionally helps leverage that information to boost the estimation of the likelihood of event; incorporating probabilities of child events conditionally helps mitigate the effects of collusion, and also helps with data sparsity. The algorithm remains the same as Algorithm 2, except for replacing $Z(j)$ computation in step 3.

Source Priorities with Reputation Score

Often, not all sources are treated equally; some sources, such as city officials and Government authorities are intrinsically more reputed than, say, a tabloid forum. In order to factor this in, each source is associated with a reputation score. This reputation score φ consists of a static component φ_{stat} and a dynamic component φ_{dyn}, given as $\varphi(i)=\delta\varphi_{stat}(i)+(1-\delta)\varphi_{dyn}(i)$ where $0\le\delta\le1$. φ_{stat} comes from prior knowledge, for example, the city officials will have a higher static score than a layman. φ_{dyn}

evolves over time from the reliability computed in the veracity framework from past reports, based on a beta reputation score function (Josang, 2002) computed using the positive feedback (a_i) and the negative feedback (b_i), i.e.,

$$\Psi_{dyn}(i) = \frac{\sum_{t=1}^{T}(a_i^t + 1)\lambda^{(T-t)}}{a_i^t + b_i^t + 2}$$

where $0 \leq \gamma \leq 1$ serves as a forgetting factor, and T is the time window of history. In order to capture this score in the veracity model, the optimum values of a_i^* and b_i^* are mildly modified as follows:

$$a_i^* = \frac{\sum_{j \in SC_i} \Psi(i,j) * k_{ij} * Z(j)}{\sum_{j=1}^{N} \Psi(i,j) n_{ij} * Z(j)}$$

$$b_i^* = \frac{\sum_{j \in SC_i} \Psi(i,j) k_{ij} * (1 - Z(j))}{\sum_{j=1}^{N} \Psi(i,j) * n_{ij} * (1 - Z(j))}$$

The same set of reports (for the city of Bangalore, India), used for evaluating the categorization, have also been used to show the efficacy of the veracity model. Figure 7 shows the distribution of the reports across the sources for different categories for the city of Bangalore, India. It can be observed that certain sources are more popular compared to others, across all categories. However, certain sources post large number of reports on specific categories (e.g., Traffline for the traffic category). Figure 8 shows the reliabilities of these sources for traffic and crime issues by applying the veracity framework. There is some impact of the volume of reports from the sources on their reliabilities for separate categories. For instance, the official reporting channels for the city, such as Bangalore Traffic Police is more trustworthy for traffic information; while Bangalore City Police and city tabloids are more trustworthy for crime reports. The reliabilities however are not directly proportional to the number of reports from the sources. Interestingly, Bangalore City Police has lesser reports on crime compared to traffic (Figure 7). However, the veracity framework can correctly estimate that the reliability of the Bangalore City Police for crime is higher than for traffic. Similarly, the veracity framework estimates higher reliability for Bangalore City Police compared to a tabloid, Times of India, for reports on crime, even though there are higher number of crime reports in Times of India than from Bangalore City Police.

INCENTIVIZATION FOR PARTICIPATORY SENSING

A key challenge in participatory sensing systems has been the design of incentive mechanisms that motivate individuals to contribute data to consuming applications. Incentive mechanisms for Participatory Sensing focuses on various aspects of incentive design with different assumptions about the underlying

Figure 8. Source reliability for traffic and crime issues

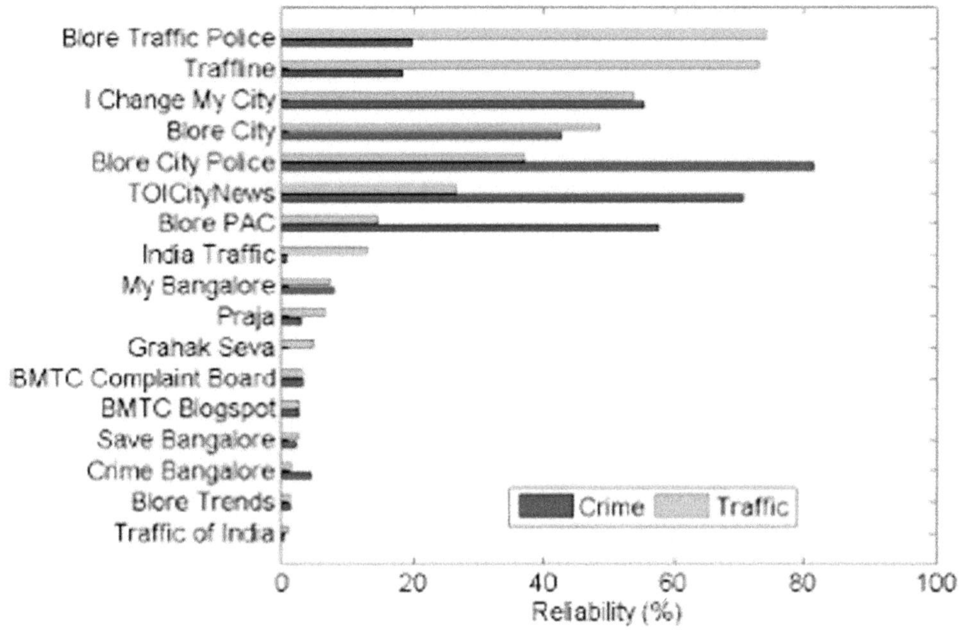

scenario. (Jaimes, 2012; Koutsopoulos, 2013; Krontiris, 2012) propose reverse-auction based incentive mechanisms, where reporters declare their expected costs as bids to the system and the incentive mechanism chooses a subset of the lowest bidding reporters. The reverse-auction based mechanism mandates the system to have prior knowledge of individual bidding profiles. Reverse-auction schemes also suffer from the handshaking overhead and lack of scalability of the reverse auction process.

(Thepvilojanapong, 2013) attributes a utility function for Points Of Interest (POIs). The nature of this function is that it increases if recent reports have not been obtained for the particular POI. The incentive is however not optimal and is determined purely based on the current utility of a POI. In (Zhao, 2014), a behavior based incentive mechanism with budget constraints is designed, which applies sequential all-pay auctions in mobile social networks (MSNs). Although (Zhao, 2014) recognizes the online arrival pattern of reporters, they most importantly require knowledge of bidding profiles of reporters and select only a certain subset of reporters depending on the budget.

For city-scale applications over USP, usually, consumers of citizen reports (e.g. city agencies) typically associate non-uniform utilities (or values) to different reports based on the spatio-temporal context of the reports. For example, higher number of traffic congestion reports may be warranted near an airport in early morning hours, than similar reports from a sparse residential area. In such cases, the design of an incentive mechanism must motivate participants, via appropriate rewards (or payments), to provide higher utility reports when compared to less valued ones. The main challenges in designing such incentive mechanisms for participatory sensing in city-scale applications are:

- **Unknown and Dynamic Report Supply:** The contributing reporters are mobile. The participatory sensing system maybe unaware of the profiles of the prospective reporters and hence their reporting likelihood and expected costs at any instant of time at any location. In a practical setting it is not scalable to manage and compute incentives on-the-fly using a highly dynamic reporter profile database. Therefore most of the existing reverse-auction based bidding solutions fail. The constraint of lack of reporter profile knowledge in a real-world scenario translates to a lack of knowledge of expected rewards. This in turn results in a poor estimate of participant reporting/engagement likelihood and consequently to an estimate of supply pool distribution. An interesting challenge is that of critical events which may warrant on-the-fly high demand for reports pertaining to a particular context (e.g., an accident in ITPL road). A reverse-auction based solution will be infeasible considering the scalability of the problem and the handshaking overhead involved in the bidding process.

- **Non-Uniform Consumer Demand for Reports with Different Contexts:** Different spatio-temporal scenarios have different utilities/values to the consumer. Hence the demand for the number of reports is also different for different contexts. Therefore the designed incentive mechanism must be sensitive to this non-uniform contextual report demand distribution. It should be able to motivate a report supply pool distribution that closely follows the demand distribution.
- **Budget Constraints:** In the absence of a guaranteed minimum reward desired by each reporter (which, without loss of generality, may not be the same for all reporters), reporters fail to engage with the participatory sensing system over longer time periods. This implies that the reward paid to each reporter must necessarily have a fixed incentive component, over and above which, a dynamic payment can be paid based factors such as report relevance. Further, an economically feasible incentive mechanism cannot adopt a greedy approach of promising high rewards to all reporters to ensure high participation.

Incentivization Framework

The incentive mechanism design is an optimization problem which aims at maximizing the utility of reports for the consumer, while minimizing the payment made by the system. Since an USP does not warrant any reporter profile knowledge, it is particularly suitable for unknown, dynamic reporter pools and is ideal for critical-event on-the-fly reporting. Report value is defined as the desired number of reports belonging to a particular spatio-temporal context. The optimization framework assumes a consumer-defined minimum reward r_{min} that the consumer would pay a reporter purely based on the contextual demand of the report and irrespective of the submitted report quality. This discourages reporters from dropping from the system as opposed to a case where incentivization is purely determined by the submitted report relevance/quality, where the reporter may obtain no reward if the report is deemed unfit, for instance, due to its poor quality. The framework further assumes a maximum reward, $r_{max,}$ which can be given by the system for any task. A closed-loop system can further allow the consumers to re-examine the reward range $[r_{min,}\ r_{max}]$, from which the rewards are picked, if the difference between the demand and supply distributions exceeds a certain consumer-defined threshold. Figure 5 presents the overview of the system, where the consumer submits requirements in terms of:

- **Consumer Utility:** The desired utility of reports pertaining to a specific context or spatio-temporal attributes is provided by consumers. In this regard, value is specifically defined as the desired number of reports belonging to a particular spatio-temporal context.
- **Set of Events:** The actual set of events for which reports are required (for e.g. Bus frequency, over-speeding)

- **Reporter Eligibility:** The desired spatio-temporal context of reporters (e.g., reporters travelling in 500D during 8-11AM on Route A). Reports bearing the desired context and reporters who receive a payment above their expected costs (rational reporters) are considered to be eligible reporters.
- **Range of Rewards:** The minimum reward and maximum reward the consumer is willing to pay for a single report. The total budget is not pre- defined in our solution, in order to discourage hard-limiting the number of reports or discarding reporter submissions.

Incentive Strategy

The incentive mechanism uses these inputs and the set of reporters already registered in the system to generate a set of rewards based on an optimization framework. The rewards generated are such that they maximize the receipt of reports of high utility to the consumer while minimizing the payment made by the system. These rewards are declared to the reporters and the supply distribution of reports received is studied by the system. An iterative algorithm in the Proposed Solution uses the received supply distribution in order to estimate the expected costs of reporters. The set of expected costs is then fed to the optimization framework to re-determine a new set of rewards based on the consumer-defined inputs, the set of reporters in the system and the estimated costs. This iterative Optimization process learns the set of expected costs of reporters in logarithmic number of iterations (trials). It is further assumed that if the supply distribution does not follow the Consumer Utility based demand distribution (within a consumer/system-defined threshold) of reports even after the convergence of the algorithm, the Consumer is requested to submit a new reward range to recompute the rewards. Such a feature facilitates the consumer in achieving a desired report demand distribution from the unknown pool of reporters.

Probably Approximately Correct-Multi-Armed Bandit (PAC-MAB) algorithms have been used to learn true participant behaviours in crowdsensing scenarios. (Jamieson, 2013) finds the near optimal arm and theoretically proves an upper bound on the number of samples or trials required. (Karnin, 2013) studies the problem of finding the best arm in minimum number of trials for a given confidence; and

Figure 9. Incentivization framework

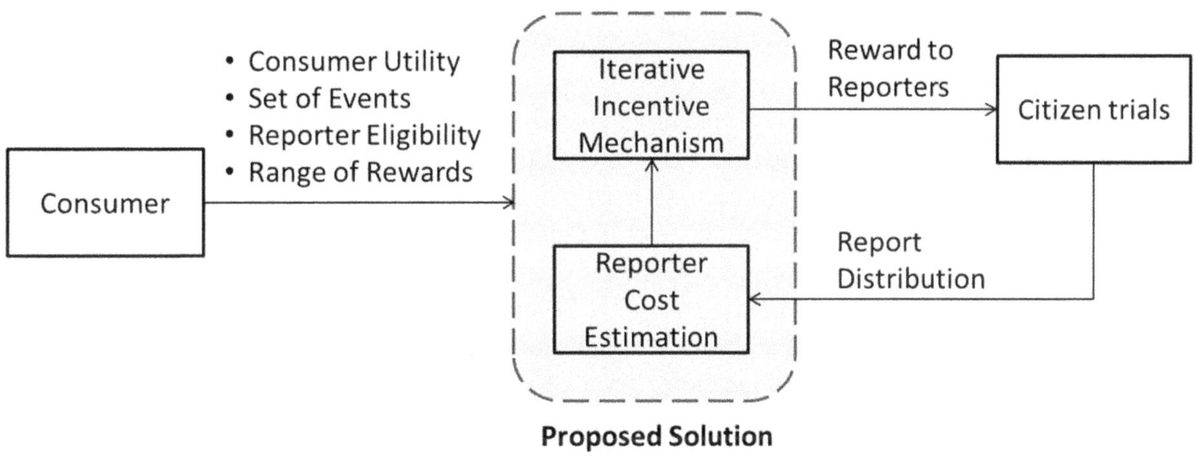

with a high confidence for a fixed number of trials. A MAB mechanism is also proposed in (Jain, 2014) for demand response, which makes monetary offers to strategic requesters who have unknown response characteristics. However, these techniques require a large number of trials to achieve performance bounds with a high confidence and ignore the cost associated with each trial. The number of trials run for every event (parallelly) is dependent on the desired confidence level and the extent of deviation from the desired number of reports that can be tolerated by the consumer. Based on the observations of the actual behaviour in the reward trials, the rewards can be updated. This update can follow a binary search technique within the reward range – thus reducing the number of trials (Biswas, 2015).

CONCLUSION AND FUTURE RESEARCH DIRECTIONS

This paper introduced Urban Sensing Platform (USP), which curates city-related issues being discussed in public forums, e.g., social media, online blogs, complaint boards etc., and generates actionable insights for the agencies to prioritize their operations accordingly. In this regard, USP relies on NLP techniques to identify categories of issues being discussed, report aggregation to identify distinct issues near city landmarks, and estimating likelihood of issues and source reliabilities to generate actionable insights in near real-time. Additionally, the chapter discussed about ways to incentivize the residents in providing specific feedback to the city agencies. The incentivization can ensure that desired number reports can be extracted from an unknown and dynamic population. Some of the key future directions in designing participatory sensing platforms for civic agencies include the following:

1. **Multi-Label Text Classification:** It is imperative to assign multiple categories to reports and correlate overlapping categories. For example, a report such as "big traffic due to bad road condition" can be tagged with both "traffic" and "road" categories.
2. **Un-Supervized Text Clustering and Topic Modelling:** One of the key challenges for participatory sensing at a city-scale is the ability to identify unknown problems, as opposed to identified problems of pre-defined categories (discussed in this chapter). Thus, in many real situations it is imperative to identify un-expected problem categories in an urban environment using un-supervised text clustering and topic modelling mechanisms.
3. **Processing Multi-Modal Data from Residents:** This chapter principally focused on text based reports from the residents. However, in reality, there can be images and videos shared by the residents. In fact, in many situations, it may be infeasible to type in information by the resident in real-time as opposed just clicking a picture and sharing. Hence, multimodal data processing is an important future direction for participatory sensing in city-scale applications.
4. **Complimenting with Smartphone and Other Device Sensing:** Modern smartphones are equipped with many different sensing capabilities, such as accelerometer. Enriching the information collected from resident reports with various sensor data can further generate more robust knowledge base. For example, if a resident reports "big traffic now at Bomanahalli and am stuck", and their phone GPS does verify the current location of the resident to be indeed Bomanahalli, but the accelerometer data suggests a fast movement at that location, then a USP can infer that the report is invalid. This may enhance the data veracity framework and impact the resident's reputations and consequently the incentives provided to the resident. Moreover, with the move towards IoT in general, and connected

vehicles to be specific, richer context knowledge generation about the environment is becoming increasingly feasible. Integrating with all these complementary information sources is imperative for generating more precise insights about the events.

5. **Predictive Analysis:** Learning from the event patterns and providing predictions about events can provide forecasts to city agencies as well as to the residents themselves. This can enable more proactive operations in the city.

REFERENCES

Biswas, A., Chander, D., Dasgupta, K., Mukherjee, K., Singh, M., & Mukherjee, T. (2015). PISCES: Participatory Incentive Strategies for Effective Community Engagement in Smart Cities. In AAAI HCOMP.

Huang, K. L., Kanhere, S. S., & Hu, W. (2010). Are You Contributing Trustworthy Data? The Case for a Reputation System in Participatory Sensing. In ACM Modeling, Analysis and Simulation of Wireless and Mobile Systems (MSWiM).

IChangeMyCity. (n.d.). Retrieved from the Janaagraha Wiki: http://www.ichangemycity.com/ichange-mystreet

Jaimes, L., Vergara-Laurens, I., & Labrador, M. (2012). A location-based incentive mechanism for participatory sensing systems with budget constraints. In IEEE PerCom. doi:10.1109/PerCom.2012.6199855

Jain, S., Narayanaswamy, B., & Narahari, Y. (2014). *A multiarmed bandit incentive mechanism for crowdsourcing demand response in smart grids*. AAAI.

Jamieson, K., Malloy, M., Nowak, R., & Bubeck, S. (2013). *UCB: An optimal exploration algorithm for multi-armed bandits*. arXiv preprint arXiv:1312.7308

Josang, A., & Ismail, R. (2002). The Beta Reputation System. In *15th Bled Electronic Commerce Conference*.

Karnin, Z., Koren, T., & Somekh, O. (2013). *Almost optimal exploration in multi-armed bandits*. ICML.

Koutsopoulos, I. (2013). *Optimal incentive-driven design of participatory sensing systems*. INFOCOM. doi:10.1109/INFCOM.2013.6566934

Krontiris, I., & Albers, A. (2012). Monetary incentives in participatory sensing using multi-attributive auctions. *International Journal of Parallel Emerg. Distrib. Syst.*, *27*(4), 317–336. doi:10.1080/17445760.2012.686170

Mathur, S., Jin, T., Kasturirangan, N., Chandrasekaran, J., Xue, W., Gruteser, M., & Trappe, W. (2010). ParkNet: drive-by sensing of road-side parking statistics. In *Proceedings of the 8th international conference on Mobile systems, applications, and services (MobiSys '10)*. doi:10.1145/1814433.1814448

Mohan, P., Padmanabhan, V. N., & Ramjee, R. (2008). Nericell: rich monitoring of road and traffic conditions using mobile smartphones. In *Proceedings of the 6th ACM conference on Embedded network sensor systems (SenSys '08)*. doi:10.1145/1460412.1460444

Mun, M., Reddy, S., Shilton, K., Yau, N., Burke, J., Estrin, D., Hansen, M.... & Boda, P. (2009). PEIR, the personal environmental impact report, as a platform for participatory sensing systems research. In *ACM MobiSys*.

Ramos, J. (2003). Using tf-idf to determine word relevance in document queries. In *Proceedings of the first instructional conference on machine learning*.

Schwartz, A. (2015). Moovit Crowdsources Public Transit Data, So You'll Never Get Stuck Waiting For The Bus Again. *Fast Company*. Retrieved from http://www.fastcoexist.com/3041915/moovit-crowdsources-public-transit-data-so-youll-never-get-stuck-waiting-for-the-bus-again

Thepvilojanapong, N., Zhang, K., Tsujimori, T., Ohta, Y., Zhao, Y., & Tobe, Y. (2013). Participation-Aware Incentive for Active Crowd Sensing. In *Proceedings of the 11th IEEE/IFIP International Conference on Embedded and Ubiquitous Computing (EUC)*.

Ushahidi. (n.d.). Retrieved from the Ushahidi Wiki: https://wiki.ushahidi.com/pages/viewpage.action?pageId=13598724

Wang, D., Kaplan, L., & Abdelzaher, T. (2014). On Truth Discovery in Social Sensing with Conflicting Observations: A Maximum Likelihood Estimation Approach. *ACM Transactions on Sensor Networks*, *10*(2), 30. doi:10.1145/2530289

Wang, D., Kaplan, L., Abdelzaher, T., & Aggarwal, C. (2013). On Credibility Tradeoffs in Assured Social Sensing. *JSAC, 31*(6), 1026 – 1037.

Wang, D., Kaplan, L., Le, H., & Abdelzaher, T. (2012). On Truth Discovery in Social Sensing: A Maximum Likelihood Estimation Approach. In *11th ACM/IEEE Conference on Information Processing in Sensor Networks (IPSN)*. doi:10.1145/2185677.2185737

Wang, S., Su, L., Li, S., & Hu, S. (2015). Scalable Social Sensing of Interdependent Phenomena. In *The 14th ACM/IEEE Conference on Information Processing in Sensor Networks (IPSN)*. doi:10.1145/2737095.2737114

Wang, S., Wang, D., Su, L., & Kaplan, L. (2014). Towards Cyber-physical Systems in Social Spaces: The Data Reliability Challenge. In *IEEE 35th Real-Time Systems Symposium (RTSS)*.

Zhao, D., Li, X. Y., & Ma, H. (2014). How to crowdsource tasks truthfully without sacrificing utility: online incentive mechanisms with budget constraint. In Proceedings of IEEE INFOCOM. doi:10.1109/INFOCOM.2014.6848053

Chapter 11
Mobile Application and User Analytics

Venkatraman Ramakrishna
IBM Research, India

Kuntal Dey
IBM Research, India

ABSTRACT

Mobile analytics is the systematic study of mobile device and application usage, and application performance, for the purpose of improving service quality. This chapter motivates the need for mobile analytics as an essential cog in the emerging economy built around devices, applications, and communication. A taxonomy of mobile analytics problems is presented, and technical details of a typical mobile analytics solution are discussed. Scale, heterogeneity, dynamically changing environments, and diverse privacy requirements pose challenges to collecting and processing data for such analysis. This chapter examines how analytics solutions handle these challenges. The core of the chapter consists of a technical section describing the general architecture of a mobile analytics solution, procedures to collect and process data, event monitoring infrastructure, system administration processes, and privacy management policies. Case studies of a number of analytics solutions available as commercial products or prototypes are presented.

INTRODUCTION

So-called "smart phones" and tablets are now multipurpose devices that have, for many people, replaced the traditional desktop or laptop computer. Voice communication is just one application ("app") among many. Mobile devices support not just personal applications like shopping, games, and Internet surfing, but also limited office use; increased productivity is incentive enough to overcome security concerns inherent in letting employees keep confidential email and documents on their personal devices. *App stores* centered on mobile operating systems are hubs of innovative software development today (Cuadrado, 2012). Since mobile devices are accessories, user behavior and habits can be understood by monitoring location and other context, helping app distributors offer more relevant services to users. Given the plethora of choices available to users, commercial success in the mobile software industry is dependent

DOI: 10.4018/978-1-5225-0945-5.ch011

upon understanding app usage and providing improved and more innovative features. Some examples will illustrate the kinds of insights that are sought after:

- The designer of an arcade game mobile app wants to know the demographics of users who download the free version from an app store versus those who pay.
- A restaurant-finder service-cum-mobile app designer wants to know why her service is popular among young professionals in New York but not in Los Angeles.
- The IT administrator of a corporation wants to understand why employees in Zurich saw their email apps crash intermittently yesterday while London employees did not. Could the predominance of iOS devices in the London office (as opposed to the predominance of Android in Zurich) explain this?

Such questions can be answered by systematically studying how mobile devices and applications are used, and by monitoring application performance. This process is termed *mobile analytics* and its purpose is to gain insights, discover patterns, and improve service quality. Mobile analytics requires:

1. A system architecture and mechanisms to generate analyzable data, and
2. Algorithms to process the data and gain desired insights.

Architectures are scenario-independent and have few variations across mobile analytics solutions, whereas data types and processing algorithms vary widely with scenario and a comprehensive coverage of these would be too large to fit in one chapter. Therefore, we focus mainly on architectural issues in this chapter.

Mobile analytics involves the collection, processing, and presentation of data. It relies on diverse disciplines like statistics, modeling, programming, and communications. A good analytics solution must separate the wheat from the chaff, and drill down into the essential factors that explain how users and applications behave. While fields like big data analytics, computer forensics, and IT operations research share these characteristics and attempt to solve similar problems, there are specific reasons, listed below, that motivates the treatment of mobile analytics as a special field of inquiry.

- **Business Reasons:** The ecosystem surrounding mobile apps is commercially lucrative and highly competitive. Understanding user demographics, gaining visibility into user experience, and avoiding service failures and downtimes, is vital. A holistic analytical framework specific to mobile apps and users enables developers, business managers, and IT operators, to do a better job of serving customers.
- **Technical Reasons:** Mobile computing environments have special characteristics that present unique analytics challenges, differentiating from traditional analytics. Some of the key differentiating challenges are given below:
    - Mobility results in dynamic context (especially location) changes, and there is a crucial need to collect contextual data.
    - Distributed client-server applications, and dependencies on multiple remote web services (Loreto2009), are common; e.g., email, maps, restaurant-finder, and gaming. This requires collecting and correlating data from dispersed and moving sources.

- There are scaling and fault tolerance challenges in collecting and filtering data in environments without centralized control over many application components.
- Mobile devices are constrained in resources (battery, bandwidth, memory, CPU, etc.), and collecting and processing data without negatively impact user experience is challenging.
- Collection of data from personal devices impacts user privacy. Yet, mobile computing analytics has the following in common with infrastructural or big data analytics, and computer forensics.
- Large data sets are generated at a rapid pace, presenting scalability and performance challenges.
- Much of the core analysis relies on long-standing statistical and modeling techniques.
- Well-known modes of data presentation and user interfaces exist, irrespective of the nature of the data produced and the target consumer. A mobile analytics solution designer can rely on past research in statistical and modeling techniques, and scaling, coordination, and fault tolerance in distributed computing. But architectures and interfaces for monitoring and analysis have special needs and common characteristics, as we will discover in this chapter when we examine the designs of mobile analytics products, whether developed by dedicated companies like Crittercism and Tealeaf or companies like Google that offer analytics services to bolster their existing businesses. Mobility and dispersed data generation sources pose not just challenges but also aid in devising solutions in the following ways:
- Each data production unit (e.g., mobile device) produces relatively little data at low rates. Filtering and processing may be distributed to these units to achieve better scale and accuracy.
- Since most data is produced on mobile devices with associated user identity and context information, a lot of metadata (for analytics) comes for free.

This chapter covers the challenges inherent in analyzing dynamic mobile user and application behavior, and describes a generic system architecture for mobile analytics inspired by state-of-the-art products available in the market. We believe this architecture will serve as a reference for solution designers as well as researchers. Readers are assumed to be familiar with statistics and data modeling techniques, and to understand the basics of distributed systems. In-depth coverage of machine learning or Big Data analytics are beyond the scope of this chapter.

APPLICATION RANGE AND SCENARIOS

Mobile analytics is indispensable to a broad range of application scenarios, and serves different roles and functions for each application.

Telecommunication Industry

Telecommunication network operations needs mobile analytics primarily to understand their users' needs and problems. A user may need better connectivity, more relevant voice and data plans, or value-added services. Dropped calls, low signal strength, and geographical regions with poor connectivity, are typical problems a user may face.

Identifying needs and problems are challenging because telecommunication networks are typically large scale operations covering millions of customers, and analyzable data is generated very rapidly. For example, using call detail records (CDRs) maintained by operators, location profiles can be generated for users and movement patterns modeled. Usage patterns of voice, data, and value-added services may provide other insights. Careful analysis may result in optimized infrastructure, improved connectivity, and more relevant service. SPSS (SPSS-Tool) and SAS (SAS-Tool) are examples of tools used to analyze telecom network usage.

Commerce

Advertisement and marketing are key to the success of e-commerce and m-commerce ventures. These companies need to analyze the effectiveness of their marketing and advertising campaigns by gauging user reaction to ads and promotions, and improve their commercial offerings by understanding mobile app usage. Such analysis can be app-centric or user-centric. Understanding the aggregate preference of users for a certain product sold by a specific app may help in redesigning the product to make it more attractive for users in general. Understanding a specific user's product preferences may help in creating personalized advertisements to serve that user's needs.

An example is app uptake analysis, which involves understanding where and how an app is downloaded, analyzing how customers interacts with those apps (buttons they click, forms they fill, and links they follow), and analyzing purchase transactions and payment modes. But this is challenging because it requires large-scale data collection infrastructure, and monitoring agents to be installed on customers' devices, which may pose security and privacy threats.

Third-Party Application Developers

Mobile analytics helps application developers identify design and performance issues in their apps. Some examples of what a developer may be interested to know are:

- Features of a given application that are used more frequently, as optimizing those parts are essential for application performance and a satisfactory user experience.
- Features of a given application that are under-utilized, and understand whether those sections are redundant or could have been designed differently.
- Buggy or poor-performing features.

This is useful for developers to identify points where a given application can be possibly improved in terms of quality, usability and performance.

Essential Service Providers

Essential service providers like utilities and emergency responders like hospitals and fire departments must allocate resources efficiently and be prepared for contingencies. Mobile analytics can help them do this better and more efficiently. For example, if a fire breaks out, alert notifications can be sent to the mobile devices of local residents and visitors. A list of residents can be generated through user profiling and demographics analysis, and a list of visitors through real-time location and movement analysis.

Healthcare and education services are also offered through mobile applications, whose effectiveness can be determined through monitoring and analysis of the user experience. Lastly, for governments and their agencies, which need to disseminate public services quickly and at high volumes, mobile analytics is an indispensable tool.

System Administration and Customer Service

For commercial and enterprise mobile applications that generate revenue, remote IT administrators are responsible for ensuring that these applications stay up and run smoothly, and providing customer service. Mobile analytics can help admins and helpdesk operators understand application performance and conduct diagnostics. Customers are also better served when problems are identified quickly and accurately.

To identify the causes of operational errors and performance bottlenecks, rapidly generated noisy data must be filtered, which is significantly challenging when an application does not run only on mobile clients but involves communication between clients and remote servers running within enterprise boundaries. An analytics system must be able to analyze not just client apps but also servers and network elements involved in running mobile applications, as a failure in one component (e.g., an authentication failure at a back-end server) could manifest itself in a different component (mobile user unable to make a purchase through his app). Identifying problems, collecting relevant evidence, and finding root causes, are challenging but possible through mobile analytics.

Analytics for such heterogeneous application scenarios face challenges not just in scale and rapid data processing, but are in designing good user interfaces. In subsequent sections, we will isolate these challenges and show how systems can be built to overcome them.

TAXONOMY OF MOBILE ANALYTICS CHALLENGES

Given such a wide and varied range of uses for mobile analytics, it is useful to classify analytics systems into mutually exclusive categories based on the goals they serve. Take, for instance, a music application backed by an online store (e.g., iTunes), which is a distributed system consisting of:

1. A music app running on the user's device,
2. An app store to serve up that app, and
3. Servers (deployed in a datacenter) maintaining user accounts and music inventories.

The music service providers want to attract more customers and gain more revenue. The third party running the app store (like Apple for iTunes) gets hosting fees from the music service provider, and therefore has an indirect stake in providing good customer service. This distributed music application must perform well end-to-end and be robust in the face of failure. These goals are shared by most commercial mobile apps, and can be broadly classified into three distinct categories.

Application Usage Analytics

Providing relevant service is the key to attracting customers and gaining revenue. Therefore, our music service provider needs to know its customers' music preferences, which can be analyzed along the lines of demographics (age, gender, language, ethnicity, etc.), customer location (geographic regions like cities or semantic locations like "work" or "home"), or music metadata (song name, artist, genre, period, etc.) Preferences could be correlated with social network memberships and device types (iOS, Android, etc.) used as well. Keeping users' privacy concerns in mind, application service providers need to understand the preferences of their users in aggregate to maximize their satisfaction.

Such app usage can be analyzed only by gathering data about users and their preferences. User demographics can be obtained from app stores to which these users subscribe and register personal information like age and gender. Agents deployed on servers can record service invocation statistics, while agents deployed on mobile devices can record device information (like device model and OS). Mobile devices can track user locations at desired granularities using inbuilt sensors, and with permission from users, their social network memberships from Facebook and other apps. Instrumented code and logs help to track usage activity of the target application.

The goal of this analysis is to answer specific queries about usage, and determine current and future trends. Questions like "what genre of music is most popular among teenagers in the US" and "what is the optimal price point for maximal sales" can be answered, and reports consisting of tables and graphs generated for executives and salespeople. Well-known statistical techniques like mean, median, standard deviation, distribution curves, correlation, and regression can be used to understand current usage as well as make reasonable projections about future usage. Standard database techniques can be used to collect, format and store data.

User Behavior Analytics

Our music service must provide personalized and relevant service to retain customers; e.g., music recommendations customized to customers' tastes, discount offers, targeted advertisements. This can only be done by understanding the customer's individual behavior and usage patterns. Since this is private data, it must be extracted with the users' full consent and using all reasonable safeguards. As in the application usage analytics case, a user's social network relationships, calling or texting patterns, and movement patterns, can be analyzed to profile the user, but the key difference between these two categories is that the focus here is on a single user and his/her devices as opposed to statistical aggregates of users.

Mobile devices, being user accessories, are the primary data sources for understanding user behavior and preferences. Agents deployed on these devices can track users' movement patterns and *presence zones* (Nadler, 2008), and more generally monitor user activity. Multiple mobile devices handled by a user are monitored to build accurate user profiles. For a given application, the most useful pieces of information include the contexts a user actively uses the application in, who the user frequently interacts with, what types of purchases the user makes, etc. Device logs can be analyzed to understand device usage patterns. Telecom networks' call detail records (CDRs) can be analyzed to understand a user's voice and data communication patterns as well as his relationships with others.

The goal of such analysis is to understand user preferences. The output is not in the form of reports, but more typically profiles or models that can be used to personalize the offered service, make it more relevant to a user, and generate promotional offers. Standard database techniques can be used to store and query data, and identity management techniques can be used to associate data across different devices handled by a user. Pattern detection and filtering is needed to weed out extraneous data that is irrelevant to building a user profile, for which statistical techniques can be used in conjunction with data mining and machine learning algorithms.

Forensics and Diagnostics

Failure or poor performance of the music service or its mobile app will negatively impact the service provider's bottom line as dissatisfied customers might turn to competitors' apps. A distributed system can fail in unanticipated ways and at different places (clients, servers, or network.) Forensics (or diagnostic analytics) is the art of detecting application failures and performance slowdowns due to bottlenecks, and identifying and fixing root causes. Unlike app or user analytics, it aims to provide better customer service, rather than expanding customer base and increasing revenue. Forensics is part of the IT helpdesk logistics pipeline, in which problems are typically diagnosed and fixed on a per-user basis, but diagnostic clues can be obtained from individual user data as well as statistically aggregated data.

For a standalone app, the mobile device is the sole data source for forensics, but for a distributed client-server app, data obtained from the servers and intermediate network components like gateways and routers will aid in identifying fault sources and fixing performance.

A forensic analyst is interested in application traces, rather than user behavior or app usage data. Both application-level data and system-level data are useful to diagnose an application crash, service failure, or poor performance. The former include logs and application traces recorded by instrumented application code. The latter includes vital health indicators, including CPU usage, memory usage, I/O activity, network activity, and availability of hardware resources. Forensic analytics systems typically offer dashboards with charts, logs, and traces, using which a system administrator or an IT operator can track system health and trace faults partly or wholly to their sources. Such systems may also send problem notifications in the forms of emails or text messages. Big data techniques may be necessary to filter relevant logs and health data. With fault and performance models unlikely to be adequately understood at the outset, advanced statistical techniques (like correlation or clustering) are useful, and so are machine learning and pattern detection techniques to detect performance anomalies. Program debugging techniques help developers parse crash dumps and application traces.

Table 1 summarizes the differences between the three categories as well as their common usage of database systems, statistical techniques, and access control frameworks to guard private user data.

Table 1. Summary of purposes of using different categories of mobile application analytics frameworks

	Application Usage	User Behavior	Diagnostics
Target Users	Analysts, executives, salespeople	Salespeople, advertisers, app developers	App developers, system administrators, IT helpdesk operators
Goals Served	Answer queries, generate reports, project trends	Targeted advertisements, forecasts, app enhancements	Fix app problems, app enhancements, improve usability
Nature of Data Collected	User registration info from app stores; service invocation data from servers; device, location, and social network data from mobile devices	Contextual, behavioral, social network data, and device logs from mobile devices; call records from telecom networks	Contextual data and application traces from mobile devices; logs and health info from mobile devices, servers, and network
Techniques and Concepts Used	Statistics, databases, predictive analytics	Pattern detection, statistics, databases, predictive analytics, data mining, log analytics, machine learning, identity management	Pattern detection, pattern matching, statistics (correlation), databases, predictive analytics, data mining, log analytics, big data analytics, machine learning, code debugging

CROSS-DOMAIN RESEARCH AND IMPLEMENTATION CHALLENGES

A complete mobile analytics solution for any combination of purposes discussed in the previous section involves building a multi-component distributed system. Challenges in building these components spans the areas of data warehousing, storing and indexing large amounts of data, application adaptation and debugging, security and privacy, user interface development performance, and scalability. These are not new research areas, and a number of products exist to serve these functions. Examples include data warehousing systems like Cognos (Cognos-Tool) and Informatica (Informatica-Tool), big data processing platforms like Hadoop (Hadoop-Tool), mobile user interface design tools such as Fluid (FluidUI-Tool) and InVision (InVision-Tool), and application scalability management and testing tools such as IBM Rational/MobileFirst (MobileFirst-Tool). But mobile analytics requires novel ways of system-building to put these components together and to handle challenges of scale, privacy concerns, and legacy applications. The challenges involved in building the various blocks of a mobile analytics solution and in putting them together are described here.

Data Collection, Warehousing, and Processing

To analyze a mobile application, data must be collected from clients as well as server components of that application, and filtered for further processing. Data source diversity, accurate data extraction, and scalability of the collection process, are three key challenges involved in data collection and warehousing. Mobile devices running the applications are heterogeneous, running different operating systems, software stacks, and application versions. They may cover a widely distributed geographical area and face varying network conditions.

ETL (*extract-transform-load*) is a well-known process for collecting, filtering, and storing data. In the *extract* phase, data is extracted from the diverse sources (like application log files.) In the *transform* phase, *adaptors* filter out relevant data and format it into a form amenable to analysis. Subsequently, the data is *load*ed into the data warehouse, and indexing, aggregation and mirroring (for fault tolerance) are carried out for fast and accurate processing. While the ETL process is well-defined in theory, adapting it to specific mobile analytics scenarios is often an architectural and implementation challenge requiring novel designs.

User and Data Privacy

Analytics systems need to extract data from the source mobile devices, typically by instrumenting the mobile applications or deploying agents. Since users' devices contain private and often confidential data, leakage and unauthorized access is a major concern, as high profile cases of enterprise servers getting hacked (e.g., eBay, Target, Home Depot, Sony Pictures) indicate (Weiss2015). Access to such data may also be regulated by company or government policy (e.g., HIPAA for medical data in the US, European Union's Data Protection Directive). Another problem with analytics systems is the lack of transparency about what the instrumented applications and agents are actually doing. By allowing a mobile analytics system to understand her behavior and habits, a user might inadvertently be exposing her private data to spammers, telemarketers, and identity thieves. Better service in exchange for privacy is typically not an acceptable tradeoff. Analytics solution providers also have a stake in certifying that their products enforce data security and abide by privacy policies, without which fewer users may choose to enable monitoring, resulting in inferior analytics.

A commercial analytics system must:

1. Offer mechanisms to let users manage their privacy easily and understand the tradeoffs involved in sharing data, and
2. Abide by mandatory privacy policies imposed by a government or corporate entity.

Monitoring agents on devices should be designed with privacy in mind, and collected data should be properly safeguarded in analytics server repositories.

User Interfaces

Good user interfaces help application owners, helpdesk operations, administrators, and app developers monitor and analyze different aspects of an app. Interfaces for report generation and presentation must be interactive. Users need to be able to query the analytics system using general parameters like date-time, geography, and mobile application components (e.g., clients, servers, and databases, for a client-server application), or audience-specific parameters (audience demographics, geo-spatial distribution of users, popular in-app browsing patterns etc.).

Interactive dashboards are typically used for user queries and report displays. Here, the challenge lies in creating a dashboard that lets different types of users (e.g., mobile app administrators, system administrators, operations teams, application developers, business analysts) view the analytics results they need and are privy to. For example, a business analyst may be more interested in the demographic breakup of an app usage where a system administrator is more interested in application performance. Dashboards must provide role-specific features and also ensure proper access control.

Developers should use responsive web themes that work well across device types (mobiles, tablets, laptops etc.) to design interactive dashboards. Notifying appropriate users who need to know about events as they happen is critical (e.g., notifying a system administrator immediately when a system goes down), as is notification using the appropriate modality. For example, a serious problem (like a server crash) might necessitate a page, text message, or automated phone call, while lower priority issues could trigger emails.

Application Adaptation and User Experience

Monitoring a mobile application requires the ability to track its runtime activity, which can be done through adaptation or instrumention. Instrumentation provides the ability to monitor certain features of target applications that are of particular interest (e.g., payment service in an e-commerce application).

The most frequently used instrumentation methodology is the injection of monitoring code in the relevant parts of the application source. Injection of wrappers around existing (already-compiled) apps is another aproach. Designing a good instrumentation library package is challenging for the following reasons:

- Applications must be instrumented with minimal impact on user experience. Users of instrumented apps should not need to take additional actions (such as, click additional buttons) or suffer degraded application performance (wait for a long time after clicking a button).

- The impact of an instrumented app on device resources, such as battery, CPU usage and network bandwidth, and other apps concurrently running on the mobile device, must be minimal.
- The instrumentation must respect abide by privacy and security policies set by all the parties (app developer, client, service providers, advertisers etc.).
- Performance and scalability of instrumented versions of apps must be comparable to their uninstrumented versions.

Performance and Scalability

Monitoring large numbers of application instances and analyzing high volumes of rapidly generated data accurately are challenging problems, and a mobile analytics system must scale with data volume and speed. It must also aggregate and filter large amounts of data, extract insights, and respond quickly to user queries. Near-real time monitoring of client applications and users may be necessary in some scenarios.

Mobile analytics systems performance challenges vary with domain, problem, and the types of insights sought. But the challenges of *summarizing*, *indexing*, and *aggregating* data to respond quickly to user queries is common to all scenarios. For instance, consider a system administrator who need to identify and analyze all application accesses from New York between 6:00 AM and 12:00 noon over the past 30 days. If date is the primary index key, then data for all 30 days will be fetched first, and location and time filters will be applied next. But if secondary indices based on location and time ranges had already been generated, query execution performance would improve at the cost of higher disk and memory usage.

Also, while retaining all historical data will enable all queries to be answered as precisely as possible, it is impractical to store ever-growing data perpetually. This necessitates a *data retirement* policy, whereby old data is archived or deleted. The challenge lies in determining the right policy whereby stored data is sufficient to respond to most common user queries without using too much disk space. Responding to user queries not just accurately but also quickly is another challenge. Fast algorithms for data aggregation need to be implemented to minimize response delays. Commercial mobile analytics products must be well-tested for performance and scalability before release.

Security

Most security challenges faced by mobile analytics systems are common to all large IT systems and can be handled by secure communication protocols (Dierks, 2008), encrypted storage mechanisms (Al-Sabri, 2013), access control systems (Ferraiolo, 1995), and mechanisms to thwart DoS attacks (Zargar, 2013). A unique security challenge in mobile analytics is the possibility of malicious entities taking over mobile devices and generating spurious data that is unrepresentative of real users and app usage patterns. If a large number of mobile devices generate bogus data, not only will the analytics system be spammed but the analyzed results will also be worthless. Hence, putting proper filters to detect bogus data and ensure the veracity of source data is an important security requirement in mobile analytics.

Big Data Analytics

Big data analytics may be applicable in certain scenarios to analyzing the massive bulk of data generated by large numbers of mobile application instances. This data can be centrally collected using ETL for further processing. But typically being bulky and unformatted, such data may need to be processed using big data platforms and algorithms, like Hadoop and MapReduce. Using big data analysis tools, it is possible to analyze mobile applications and devices generating large volumes of data at high speed. The key challenge lies in mapping analytics requirements to big data problems and generating automatic reports for user consumption.

STATE-OF-THE-ART SOLUTION ARCHITECTURE

Though mobile analytics systems vary in terms of goals, target users, and data analysis methods, their system architectures have much in common. A general architecture can be extrapolated from existing research and a study of available products. This section describes an end-to-end high-level architecture from data collection to presentation, delves into the details of each component of this architecture, and discusses privacy management. This architecture does not represent a currently implemented system but is rather the embodiment of an ideal, drawing from the best features of various analytics products available in the marketplace like Tealeaf (TeaLeaf-Tool), Crittercism (Crittercism-Tool), Aeternity (Aternity-Tool), Splunk (Splunk-Tool), and IBM's Mobile Infrastructure Analytics Service (MIAS) (Cherbakov2014). We will discuss the characteristics of these products in the next section.

End-to-End Architecture

A mobile analytics system is a pipeline that converts raw data into insights ranging from app usage statistics to root cause detection of faults (Figure 1). At one end lie data sources consisting of mobile devices, servers communicating with the applications on the devices, app stores, and networking elements. At the other end lie user interfaces through which processed data and insights can be presented to analysts, developers, and system administrators. In the middle of the pipeline lie functions and processes that convert raw data into useful results.

The first stage of the pipeline involves collecting data from, and monitoring events on, devices and applications, and uploading data to the analytics system. Mobile applications may be standalone, or distributed applications dependent on remote services. Therefore, data sources include not just mobile devices but also back-end servers and databases, and networking elements like routers and wireless access points that enable client-server communication. The architecture diagram illustrates the most general data collection system to monitors application flows, but few analytics products monitor the end-to-end application and all the hardware elements involved. Most pick and choose from a mix of devices, servers, and network session activity. For example, Tealeaf monitors clients as well as network activity, but not servers. Crittercism and Aternity monitor the mobile devices and apps, though the former can be combined with tools like Splunk for end-to-end monitoring. Splunk and MIAS monitor clients, servers, and network activity.

The last stage of the pipeline illustrates various presentation interfaces and delivery modalities. Target users are business analysts, IT system administrators, app developers, or helpdesk assistants. Outputs typically consist of charts and tables that can be viewed on demand on GUIs like web browsers and email clients, or notifications or emails to catch a user's attention.

In the middle lies the analytics system that converts raw data into useful output. Filtering out irrelevant data is the first step, and this task can be performed by a (domain-aware) filtering subsystem or be distributed among the monitored devices and the analytics servers that receive the observed data.

Figure 1. General architecture of a mobile analytics pipeline

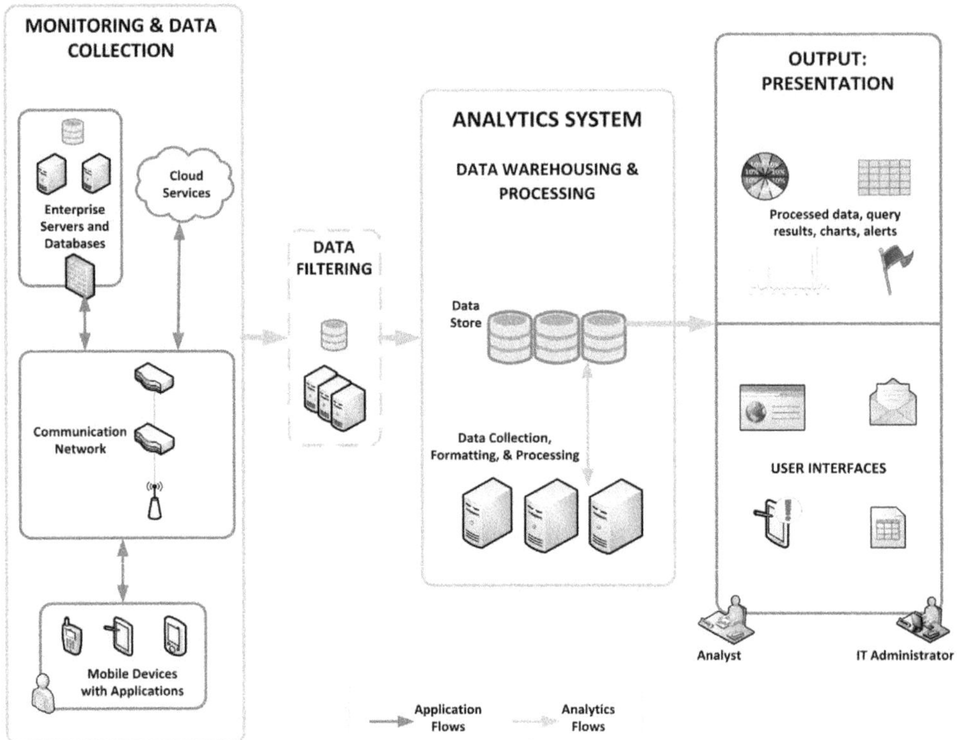

The analytics system itself consists of infrastructure to format, securely store, and process, data. Standard database or data warehousing systems like SQL, DB2, Oracle, Cognos or Informatica can be used to store and retrieve data. Analysis may involve statistical computations, machine learning, pattern detection, or log analytics, and is typically carried out on servers (in any configuration: centralized, distributed, or in clouds) having access to the data.

Event Monitoring

Analysis of application usage and performance requires continuous monitoring of state (of devices and applications) and activity (of applications and their users), or *events* occurring during an application run. The architecture for a generic monitoring system that can support both usage analytics and forensics is illustrated in Figure 2. Event and state monitoring agents are deployed throughout the system and application stacks on the devices participating in an end-to-end application flow. These agents communicate data to the analytics system either individually or in groups with centralized coordinators, using standard or custom network protocols (typically HTTP). Filtering modules may be configured on the data sources for sanity checking, formatting, and filtering out irrelevant data. The analytics system may perform additional filtering on the data it receives from the agents.

The state an analytics system needs to monitor refers to health indicators or vital statistics of devices, like CPU activity, memory usage, disk usage, disk I/O activity, and network traffic. Diagnostics is the main purpose for monitoring state, as these parameters indicate how optimally a device or app is

Figure 2. A mobile analytics architecture to support both analytics and forensics

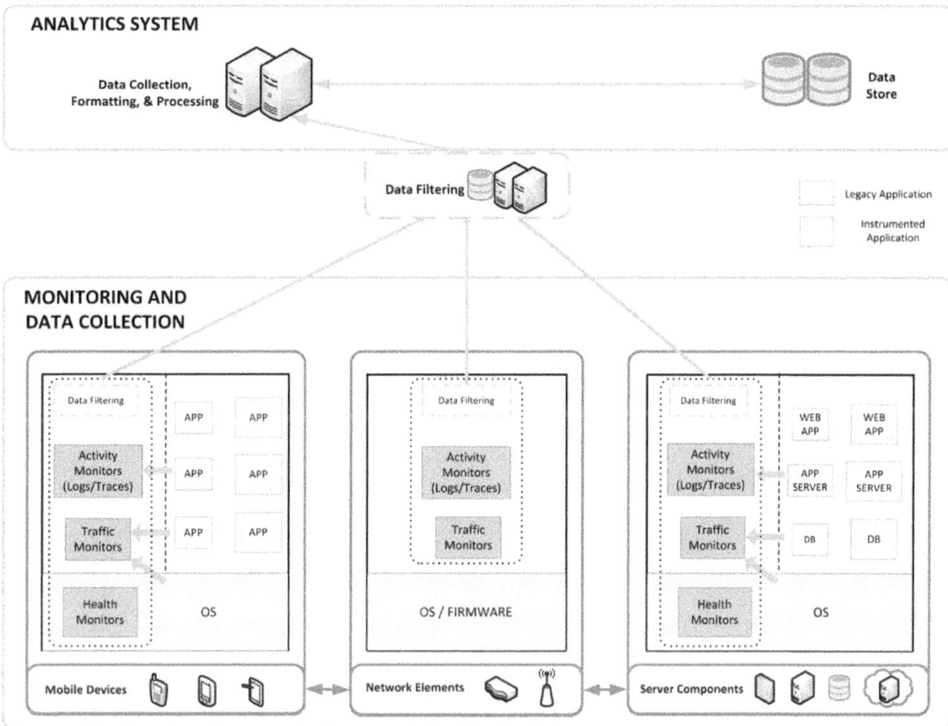

performing at a given instant, and also what the performance trends are. Standard OS mechanisms, like the */proc* file system on Linux and the *procmon* tool on Windows, can be used to monitor end device health. On networking elements like routers and switches running custom firmware, analytics agents may limit themselves to observing traffic.

Application activity and runtime events can be monitored *passively* or *actively*. In passive monitoring, applications are observed from the outside by agents that track runtime logs, traces, and network sessions (by monitoring interfaces using packet sniffers like Wireshark). In active, *synthetic*, or in-band monitoring, events are tracked by the applications themselves, done in one of two ways. In pre-compile wrapping (e.g., Tealeaf, Crittercism), the application source code is instrumented with library calls that track function calls, network sessions, and anything else a developer deems necessary for analytics. In post-compile wrapping (e.g., Splunk, Aternity), extra instructions are compiled into object code. Active monitoring mechanisms must be lightweight and not impact application performance or break application functionality.

Data Collection and Maintenance

Internal details of the analytics system are illustrated in Figure 3. The data collection procedure follows the *extract-transform-load* (ETL) paradigm used in data warehousing. Monitoring agents on application devices extract data and send it to one or more *storage servers* (typically web servers) using secure protocols like HTTPS. Transformation of the data into a format suitable for storage and analysis can happen both at sources and the analytics system. Finally this data is loaded into the data stores.

Figure 3. Data collection architecture for a mobile analytics system

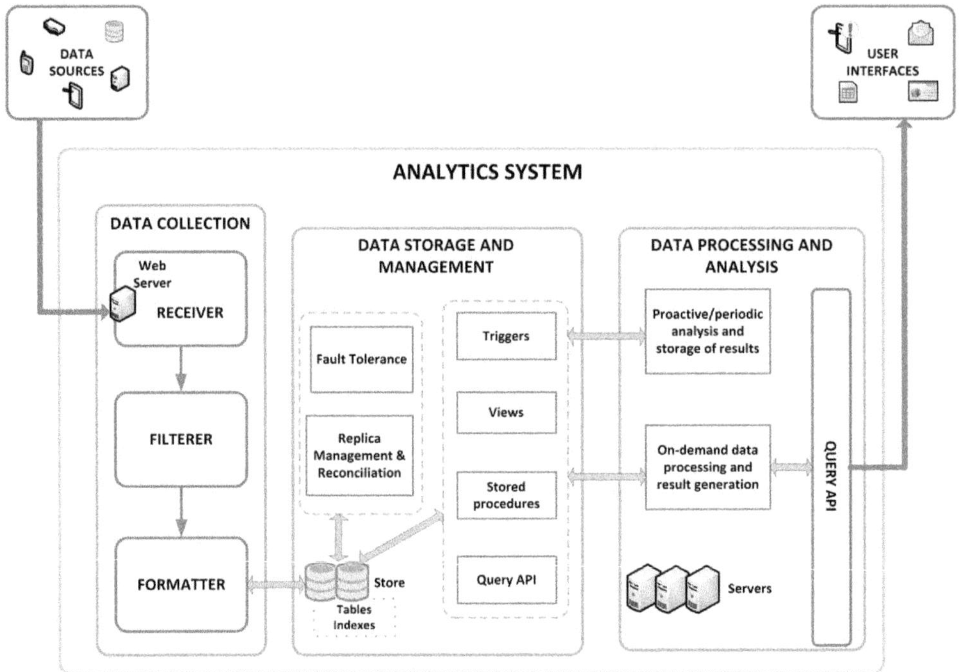

Schemas or data models are typically decided prior to agent deployment. Using monitoring agents and instrumented code, state and activity data can be extracted from the sources. Cleanup and formatting according to the schemas can be done by the agents, or sent in a raw form to the analytics system if the format is known; e.g., log entries written by well-known software (like Apache Web Server) in published formats. The analytics system may additionally annotate the data with source information. In some scenarios, it may not be possible to instrument code or deploy filters on the application devices. To handle large quantities of such data generated rapidly by multiple sources, curation techniques used in big data analytics (Stonebraker2013) can be inserted into the ETL pipeline to make the data amenable to analysis.

Standard database techniques can be used for data storage and maintenance in dedicated enterprise servers, data centers, or clouds. Protocols for replica reconciliation, fault tolerance, and recovery mechanisms, must be in place to guard against data loss. Transactional integrity must be a core feature of the database. Databases can be relational, document-oriented, object-oriented, or semantic, depending on the source data types. Similarly, indexing can be based on B-trees (as is done in relational databases), Coherence (used by Oracle with in-memory hash tables), Lucene indexes (Apache), etc. These databases must offer query APIs (like SQL) with wrappers for remote requests (like an HTTP RESTful request), and support stored procedures and triggers to modify data and generate useful output for analytics service customers.

Data Processing and User Interfaces

Analysis begins after application state and event data are collected and stored. The actual processes and algorithms vary with scenario and desired output, but certain characteristics are common to all mobile analytics systems. Broadly, analytics processes can produce results in two modes:

1. On-demand, and
2. Pro-actively (Figure 3).

On-demand analytics involves running computation procedures upon user request. Typically, an interface, like a dashboard on a web browser or some other GUI, lies between a user and the analytics system to convert a user's requests into database queries. These queries typically pertain to understanding how application usage varies along different dimensions, application preferences of users, or statistics about application function invocations. The output, generated through statistical computation, may consist of charts or tables on dashboards, or reports delivered as web pages, spreadsheets, or e-mail. For example, the music store discussed in the Taxonomy section could determine what genre of music is popular among users in a given demographic (or location) by extracting the right information through database queries and then running statistical averages and list sorting procedures. Preconfigured views and stored procedures may compute and store information often desired by users, for efficiency.

Pro-active, or continuous, processing is used to extract information, like behavioral patterns and impending or existing problems, from raw data. For example, collected data may reveal a user's movement and app usage patterns, but analytics is needed to correlate the two. Similarly, when an application crashes or performs poorly, the source data only indicates what the application and its user have been doing, without explicitly identifying a problem or its cause. Mining source data to identify patterns, link data across schemas, identify anomalous behavior and trace its root causes, just to name a few examples, require advanced statistical methods as well as machine learning techniques, such as correlation, clustering (like *k*-means), pattern detection and matching, and detecting outliers. Analysis for particular scenarios may draw on fields like log analytics, identity management, and code debugging. Developers or operators may configure seed values and default models, but analytics algorithms must learn over time to avoid missing useful insights. Examples of real-world insights include:

1. Detecting that a user's application tastes have changed, and
2. A device malfunction caused an app to fail consistently while invoking a remote service.

Learning techniques could be used to build and continuously refine models of users' tastes based on observed users' action data. Similarly, application malfunction patterns could be modeled and continuously refined, and application trace data could be matched against these patterns to determine if problems have occurred.

Continuous data mining produces results that are stored for on-demand viewing (through dashboards and reports) or alert notifications (through emails, text messages, or push notifications sent to users' devices) containing clues about problem root causes or information about impending faults predicted by analytics.

Privacy and Security Management

Protecting users' mobile devices and preventing data leakage to untrusted analytics systems against users' wishes is one function of mobile analytics privacy management. This is largely within user control as privacy protection mechanisms are built into most mobile device operating systems like Android or iOS. Monitoring agents and instrumented apps must explicitly request for permission to access core services (like GPS) (Au, 2011). Application isolation models prevent cross-app data sharing unless explicitly allowed by the user (Dwivedi, 2010). Containers or sandboxes add another layer of protection (Jaramillo, 2013).

Monitoring agents and code instrumentation can also be designed to be *polite* and privacy-conscious. Based on the principle of least privilege, no more access to resources ought to be asked for than absolutely necessary to avoid imposing a cognitive burden on the user, and the implication of allowing such access ought to be made as clear as possible. Furthermore, analytics systems can collect and store data (e.g., user location) in obfuscated form or at granularities that minimize identity theft risks (Bakken2004). User identity information collected for aggregate statistical analysis can be stored in anonymized or pseudonymized (a randomly created unique ID instead of a user's real ID is associated with data) (Duckham, 2005) forms.

Preventing unauthorized access to already-extracted data is another function of mobile analytics privacy management, where threat emanates from attackers targeting analytics services and data stores. Standard security mechanisms for communication (like TLS and HTTPS) and storage (like encrypted file systems) can be used to protect extracted data in transit to the analytics system. Different analytics consumers need access to different portions of the data, and the principle of least privilege can be applied. For example, analysts and advertisers need not be given person-specific information where views that return aggregate statistical information will suffice. System administrators and helpdesk operators only need traces of particular app runs for debugging purposes, but not the identity of the app users. Such privacy rules can be enforced using a variety of mechanisms ranging from ACLs and capabilities to trust management and policy management tools like KeyNote (Blaze, 1999), Ponder (Damianou2001), XACML, and IBM Tivoli Security Policy manager (Buecker, 2009).

CASE STUDIES: MOBILE APPLICATION ANALYTICS PLATFORMS

Having seen what an idealized generic mobile analytics architecture looks like, let us see how products available in the marketplace address the challenges we described earlier. As we have seen, a mobile application may have both client and server components communicating periodically with each other. But end-to-end monitoring requires extensive infrastructure, so products typically restrict themselves to monitoring either client apps or server infrastructure, though a select few monitor both. Some prominent and popular commercial products covering the range of analytics services and monitoring capabilities described earlier are discussed below. We selected Crittercism and Tealeaf as examples of primarily client-focused analytics tools, Google Analytics for its focus on application and user analytics, and Splunk and IBM MIAS for their end-to-end analytics capabilities spanning client and server.

Crittercism

Crittercism is a mobile analysis tool for performing client application analytics. It uses instrumented application code to infer device and application characteristics, and thereby generate reports. SDKs for iOS, Windows and Android are available. It is also compatible with platforms like HTML5, PhoneGap, Xamarin and Unity.

- **Capabilities:** Crittercism offers several analytics capabilities. Application runs are tracked and aggregated in daily and monthly active user accounts using uniquely generated identifiers. Crash reporting is provided by recording device state and stack trace information. Application crashes are grouped to make identification of frequent crash types easier, and to generate alerts. Crittercism can dump hexadecimal stack traces as human-readable function names, or *symbolication* in their terminology. The product supports user-inserted log messages (called *breadcrumbs*) of up to 140 characters of length, and can specifically monitor custom-defined transactions and generate metadata to analyze user experience during crashed sessions.
- **Installation:** Installing Crittercism and instrumenting mobile applications with it is straightforward, making it an attractive option for users. For instance, on the Android platform, one needs to download the Crittercism JAR (Java archive file) in the *libs* folder of an application, add an INTERNET permission request in the manifest file, and inject and initialize relevant Crittercism code into the mobile application source. An example code snippet to import the Crittercism library and initialize data collection procedures, after including the JAR in the *lib* and setting up appropriate permissions, is as follows:

```
import com.crittercism.app.Crittercism;
Crittercism.initialize(getApplicationContext(), "CRITTERCISM_APP_ID");
```

So-instrumented, the application will now record all device and application parameters that Crittercism is capable of monitoring, and send them to the Crittercism system for analysis and report generation.

Splunk

Splunk is a business intelligence toolkit for enterprises that provides analytics for servers, mobile devices and network activity, and generates dashboards, alerts, and reports. Specifically, Splunk MINT provides real-time performance metrics, real-time data analytics, network and carrier monitoring, and location-specific monitoring. It associates mobile application data with other enterprise data to provide end-to-end visibility into mobile application transactions and the impact of those transactions on the overall business process.

Splunk provides several dashboard views. The *real-time mobile insights* view tracks sessions, application users, total number of app crashes, fraction of users that have seen an application crash, session durations, and number of sessions per user on average (user retention). The *network monitoring* view shows the total number of requests, request latency (milliseconds) and HTTP transaction status. Splunk lets users define custom transactions, and monitors those transactions to collect data on number of attempts made, duration of each transaction, completion time, and failure rates. This helps a user easily identify transaction performance bottlenecks, such as frequently crashing or slow-executing modules.

Google Analytics

Google Analytics (GoogleAnalytics-Tool) is arguably the most popular web analytics platform in terms of the number of websites using it. As mobile applications have grown popular, Google Analytics has been adapted to work with the mobile application ecosystem. Google Analytics for Mobile Apps (GoogleAnalyticsMobile-Tool) works seamlessly with the web analytics platform, and offers a customizable interactive dashboard with similar look and feel.

Google Analytics for Mobile Apps requires the app developer to inject instrumentation code in his application. It generates reports on event tracking, flow visualization, app crash reporting, and exception reporting, and also provides a custom report creation platform. Reports can be generated post-facto or in real time. Several data analytics and reporting capabilities are supported, such as acquisition reporting, audience reporting, e-commerce reporting (in-app purchases), app-specific reporting, conversion reporting (effectiveness of sales with respect to stated goals), cohort reporting (identifying effective marketing strategies that have produced high user engagement), and other custom report generation capabilities. One can build custom visualizations and integrate Google Analytics data with external business data. Google Analytics also supports A/B split testing and alert generation.

IBM TeaLeaf

IBM TeaLeaf, like Crittercism, is a client-side application analytics platform that requires instrumentation of mobile application code to track device and application characteristics. TeaLeaf supports fine-grained custom instrumentation, using deep embedding of code inside mobile applications.

The TeaLeaf dashboard is capable of reporting application usage statistics and statistics of device characteristics like battery usage, network traffic, memory usage, operating system details, access demographics, client-server HTTP sessions, and several other parameters. It can generate crash reports, and supports visual analytics and client session replays. Tealeaf instrumentation code at a certain point in the application can trigger screen captures. TeaLeaf can capture all touch-screen gestures, such as tapping, swiping, scrolling, zooming and screen rotation. While debugging, these screenshots may help the developer to understand application runtime flows. Figures 4, 5, 6 and 7 show sample screenshots of TeaLeaf dashboard views.

Though primarily used to monitor mobile applications running on end-user devices, Tealeaf can also monitor HTTP sessions between the client and a remote website. TeaLeaf supports sites running HTML5 and based on responsive web design (RWD), and both native and hybrid Android and iOS applications. Session-tracking is done by monitoring HTTP requests and responses using a *packet forwarder* tool that runs on the HTTP server. HTTP session attributes like request and response times, round-trip latency, response codes, messages, and timeouts can be recorded using this packet forwarder, which is essentially a packet sniffer. This tool tracks incoming HTTP requests from mobile devices and the HTTP responses returned by the server. To enable packet forwarding, the server code need not be instrumented; only a standard *TealeafTarget.jsp* file needs to be exported by the web server.

Figure 4. A dashboard overview of TeaLeaf showing a distribution of device models, users by country, mobile operating system preference and session counts for each carrier, in a simulated test setting

Figure 5. A dashboard overview of TeaLeaf showing a distribution of session counts per user, average time on site per user and session count details per user per mobile carrier, in a simulated test setting

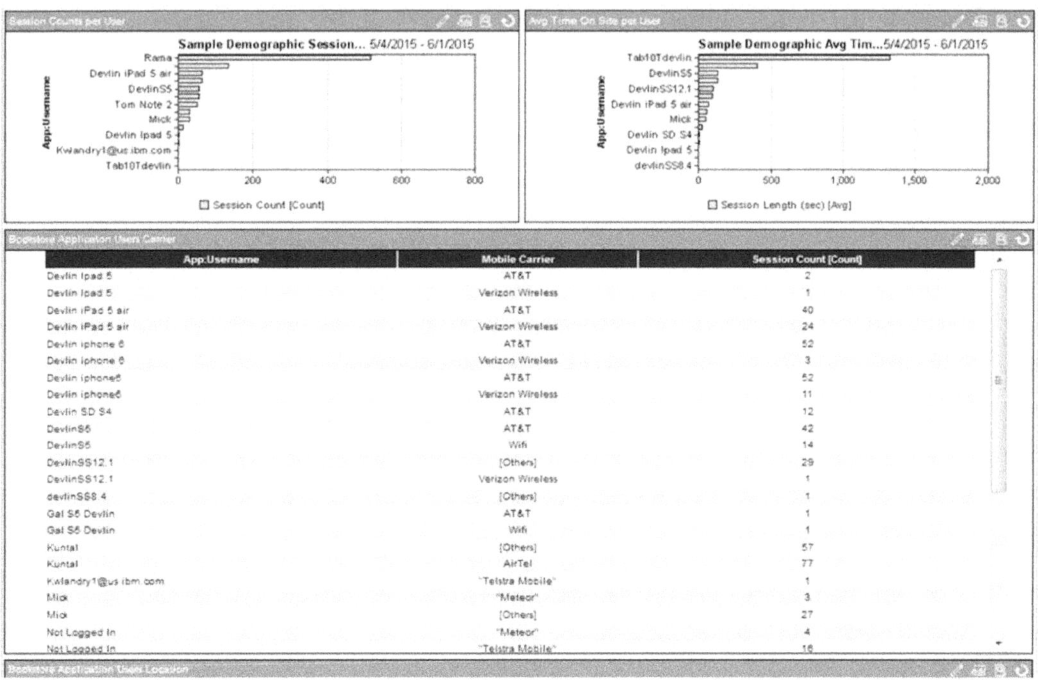

Figure 6. A list of completed TeaLeaf sessions in a simulated test environment; each of the icons lets a user (administrator, etc.) drill down deeper and investigate.

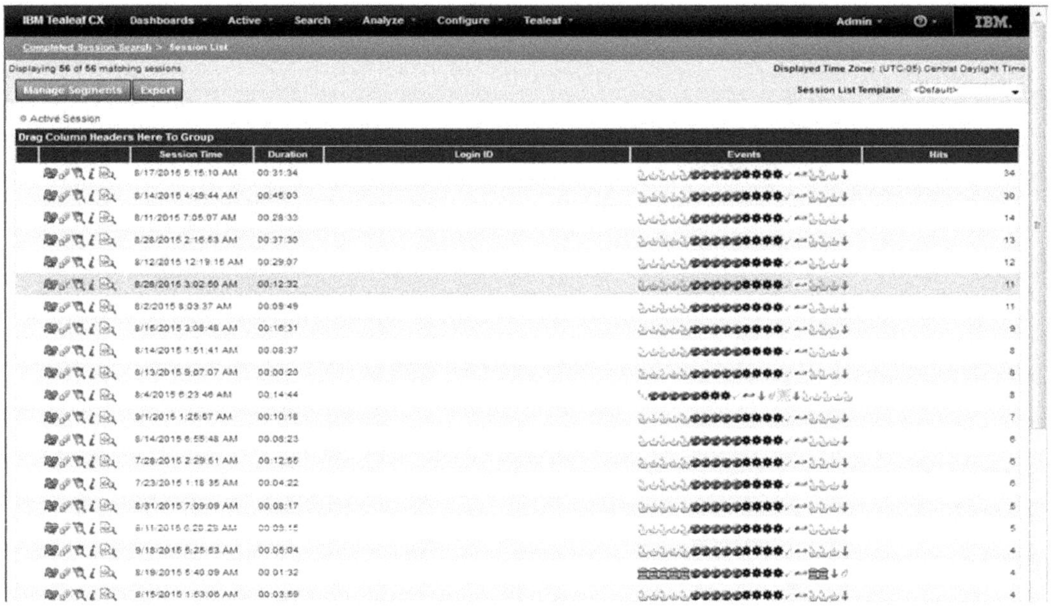

Figure 7. Browser-based session replay for TeaLeaf; one can also use TeaLeaf's RTV (RealTea Viewer) for a more app-specific visual look in a setting where visual recording is enabled.

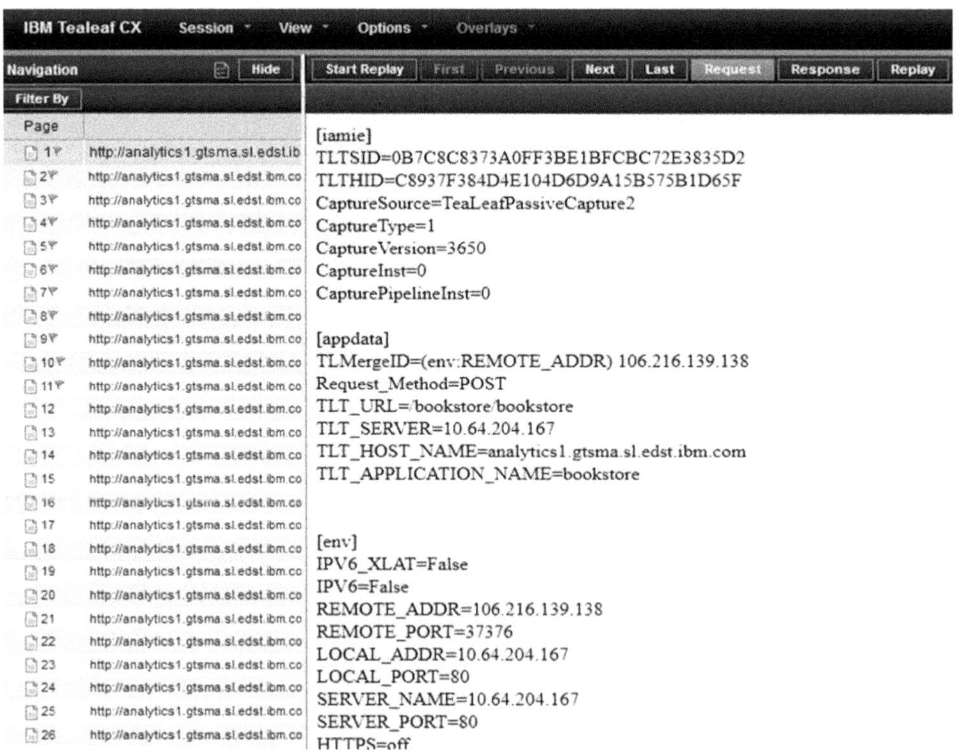

IBM Mobile Infrastructure Analytics Service

Unlike many mobile analytics tools that focus on client monitoring, the *IBM Mobile Infrastructure Analytics Service* (IBM MIAS) monitors end-to-end client-server network sessions in combination with active monitoring of back-end server components. This aids in detecting faults and anomalous performance, and filtering out relevant evidence from server activity for debugging purposes.

- **Architecture:** MIAS relies on the IBM WebSphere Analytics Platform (IWAP) (Lobo2013) for data warehousing. IWAP provides a platform for monitoring, collecting, storing, indexing, and querying data, and is publicly offered with certain versions of IBM MobileFirst. Data monitoring on mobile devices and servers is done using agents called *sensors*. The client-side data, including network session activity and device state, is collected using sensors in conjunction with third-party analytics software (like TeaLeaf, which is offered by default with MIAS, though other toolkits like Crittercism can be plugged in easily.) The application backend, typically comprising of an application server and a database server, is monitored using sensors that collect logs generated by these servers. By default, IBM MIAS is configured to monitor IBM WebSphere Application Server and IBM DB2. Collected data is stored in a Lucene-indexed database within IWAP. Stored procedures called *pipes* are used to process and analyze data. The MIAS architecture is illustrated in Figure 8.

Figure 8. Client-server architecture of MIAS including analytics components

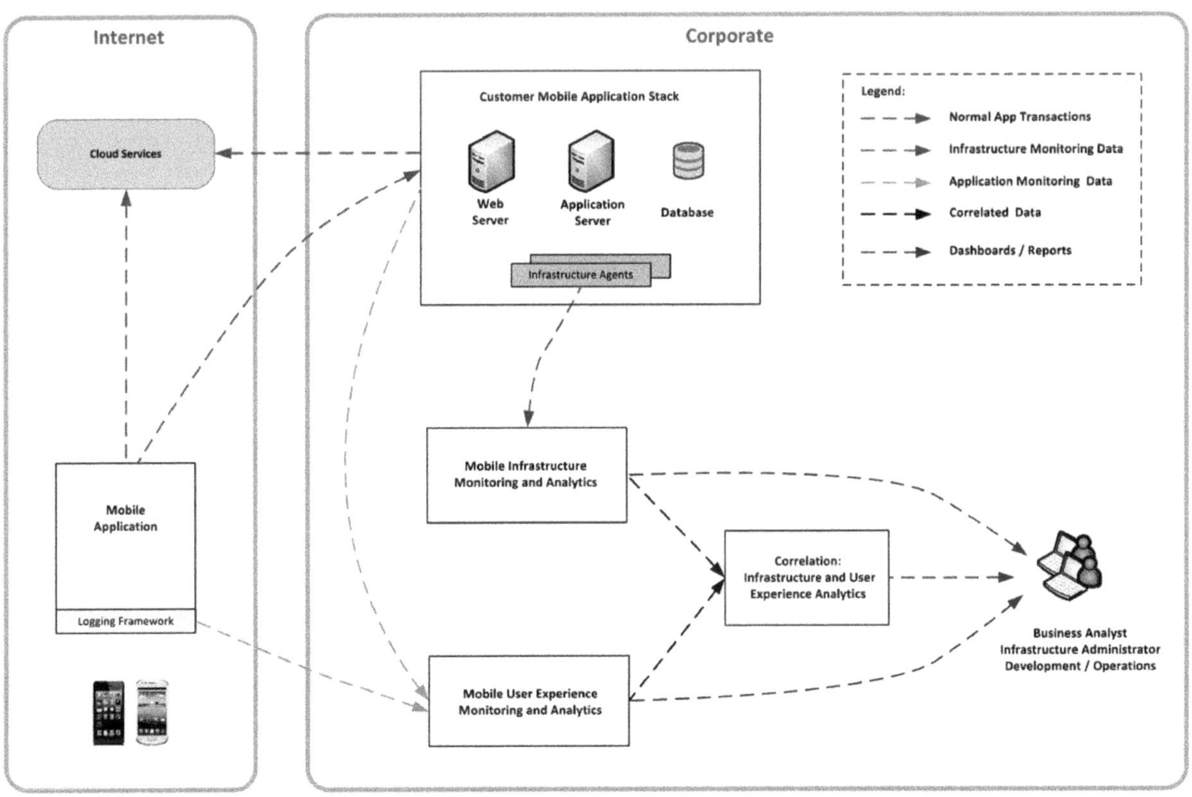

- **Features:** MIAS presents the user with analytics results through a one-stop browser-based dashboard containing information about mobile devices, application network sessions, application server and database activity, and server health. The user may view mobile application activity as analyzed by Tealeaf. Then he may proceed to view the networking activity of applications containing statistical information about round-trip latency and connection success rates along different dimensions (Figure 9). Then he can view logging activity of application servers (such as IBM WAS) and databases, along with server and database running status (Figure 10). Lastly, the user can view server health parameters such as CPU and memory usage, disk I/O and networking activity, and disk space availability (Figure 11). The user may select arbitrary time ranges on the dashboard, which results in the execution of one or more IWAP pipes to generate new analytics results, typically statistics and graphs.
- **Forensics:** MIAS helps administrators and debuggers identify how and where a client-server application is going wrong by tracking and displaying faults and performance dips in networking activity. HTTP connections between mobile app and application server are monitored using Tealeaf's *packet capture* feature, and all attributes of individual sessions are recorded in the IWAP database for analysis. An "Alerts" button in red (default is grey and disabled) catches the user's attention when one or more unattended problems are detected in the selected time period (Figure 8). Selecting the button results in the display of a list of faults and suspected performance issues. Every HTTP session that does not return a successful response code (between 200 and 300) is listed as a fault, and every session with unusually high round-trip latency is listed as a performance anomaly (Figure 12).

Figure 9. HTTP sessions analytics; note the "Alerts" button in red at the top right

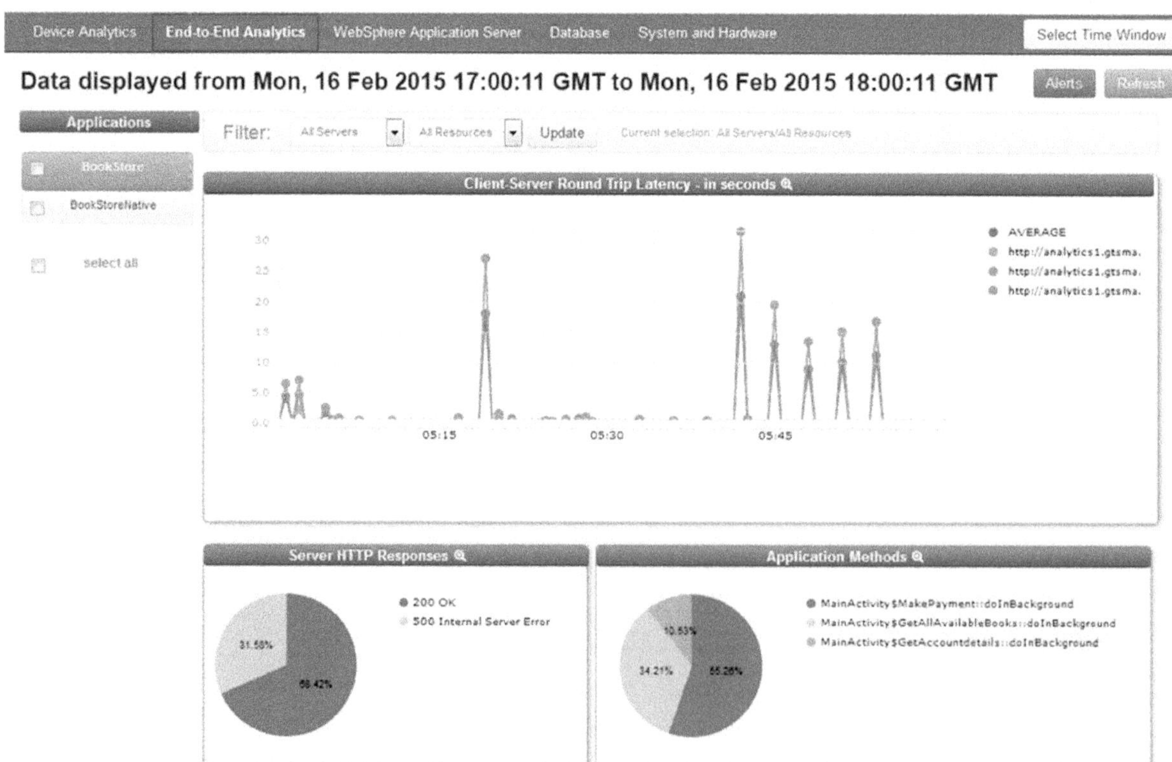

Figure 10. IBM WAS analytics; red/green indicate that an instance is down/up

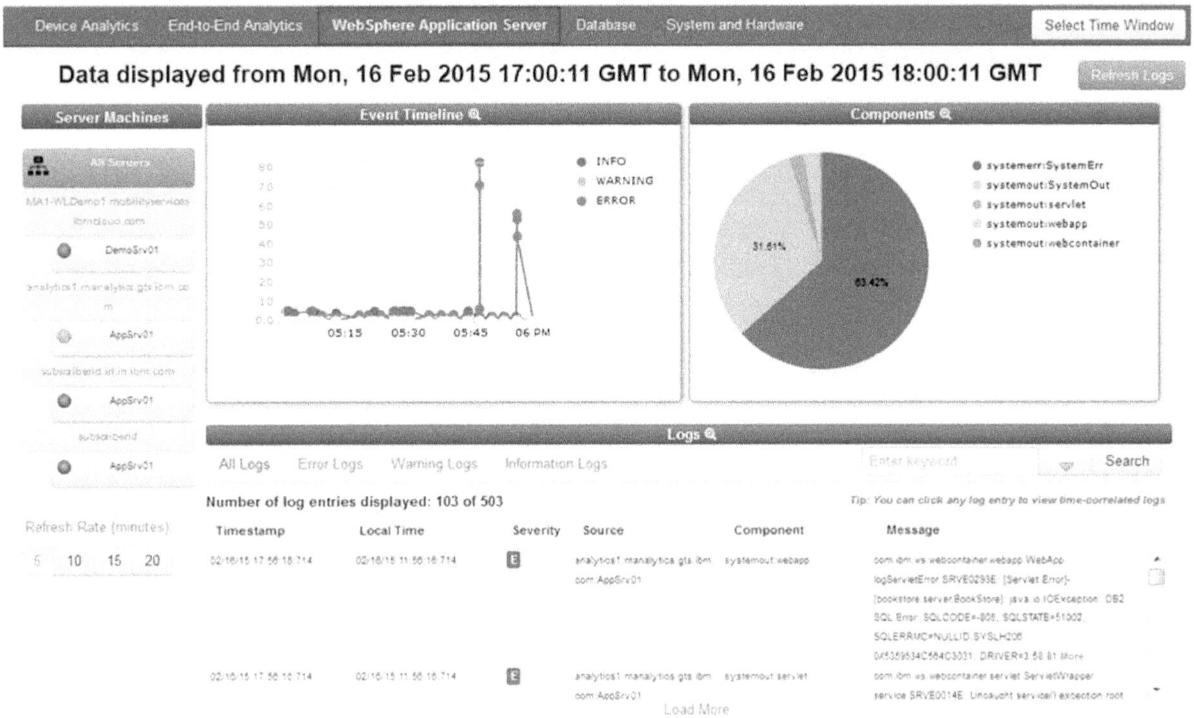

Figure 11. Server Health Indicators; Clockwise from top left: I/O, N/W, Disk Space, CPU/Memory; Note red and green buttons (indicating up or down) against server names in the column at the left

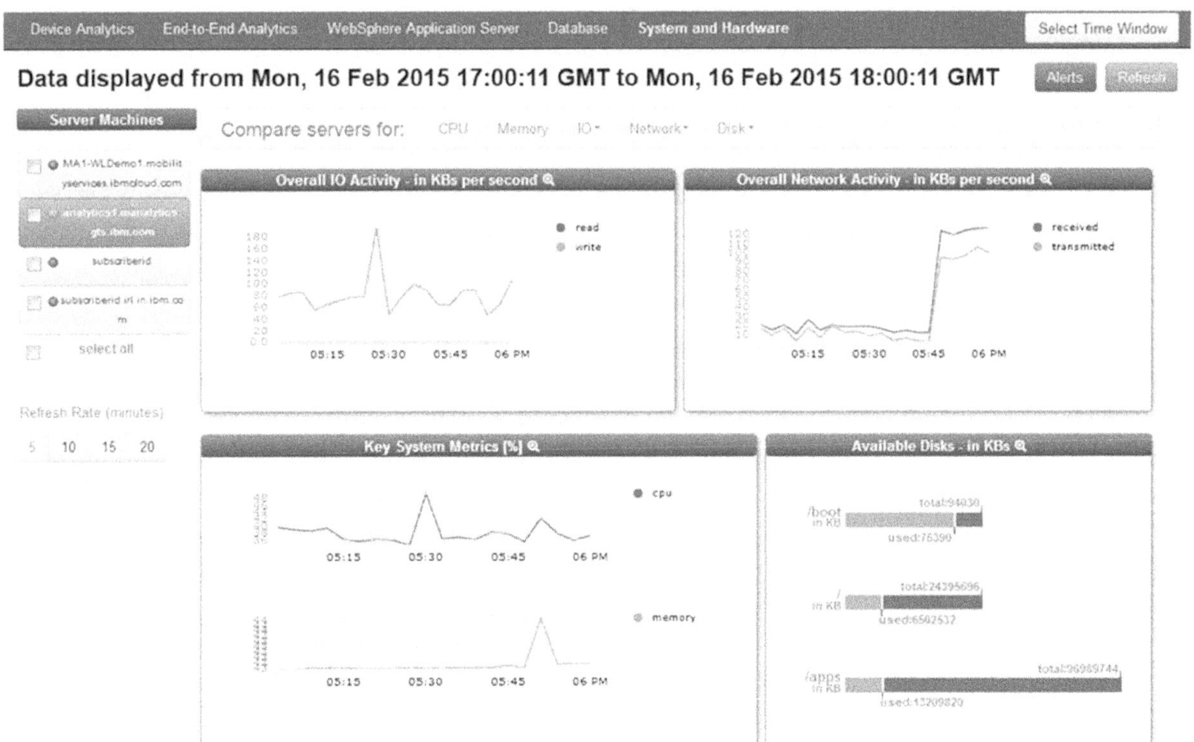

Figure 12. Two-month listing of faulty HTTP sessions, with a mix of failures and performance anomalies; Note various metadata for each listed connection, and the "Check Logs" button for debugging

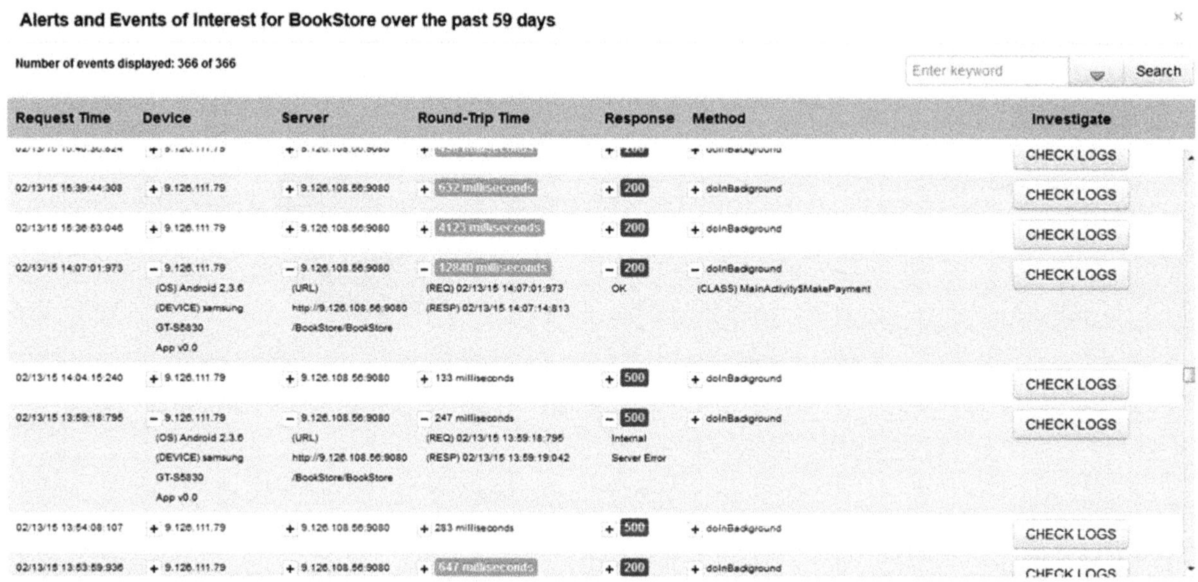

- **Performance Anomalies:** MIAS uses a combination of clustering and basic statistical methods to detect performance outliers, or HTTP sessions with unusually high round-trip times (RTTs). An IWAP pipe continuously tracks RTT values of recorded HTTP sessions, and puts them into two buckets (containing low and high values) using k-means clustering (choosing $k = 2$). After additional statistical computations, an *anomaly threshold* is determined, and connections with RTTs higher than that threshold are flagged for poor performance.
- **Debugging:** Faulty network sessions, identified by various attributes, can be examined by an analyst (Figure 12). Performance anomalies are indicated by entries with highlighted Round-Trip Times and 200 response codes (indicating success). To diagnose the root cause of a particular fault, one can drill down into the application activity by examining logs relevant to the network session. Using its knowledge of the session start and completion times, and synchronizing that with the server machines' clocks, MIAS runs temporal analysis procedures to extract a relevant set of server logs across components (application servers and databases) and displayed in reverse chronological order. Using text analysis, logs that are likely to be relevant to an application malfunction are highlighted (Figure 13). The dashboard also allows users to view log entry details, and highlights logging inactivity within application threads (Figure 14).

Figure 13. Note the highlighted logs indicating a database access control failure; also note application threads highlighted in different colors to aid in analysis.

Figure 14. Log entries classified by application threads, with thread inactivity highlighted.

Other Tools

While some of the most popular and well-equipped mobile analytics tools and platforms have been described in this section, a number of other products are also available in the market, like Localytics (Localytics-Tool), MixPanel (MixPanel-Tool), Flurry (Flurry-Tool), AppSee (AppSee-Tool), and Aternity. These tools have broadly similar system designs to the ones described in this section and offer similar capabilities, but some provide distinctive features for the application domains they are used in.

CONCLUSION: OPEN CHALLENGES AND FUTURE WORK

Though many commercial products for mobile analytics exist in the market, it is still an emerging and growing field. These products and toolkits are aggregations of components inspired by research in distributed systems, statistical and pattern detection algorithms, and data warehousing. With increasing need for comprehensive data capture and accurate insights, it is essential to go beyond mere business bottom line-enhancement objectives and study mobile analytics in a systematic way.

Although source code instrumentation generates the appropriate data points for analysis, it imposes additional cognitive burden on developers and increases the lengths of coding cycles. In practice, a learning curve of studying instrumentation manuals and experimentation cannot be avoided. Mechanisms to minimize or even eliminate the developer's role in instrumentation will help both developers and analysts. In addition, research should investigate the extent to which instrumentation can be automated or guided by high-level policies, and leave minimal footprint on device resources.

Forensics for distributed systems are still works-in-progress, though root cause determination (Nguyen, 2012) and log analytics (Fu, 2009) are well-researched subjects. Debugging applications whose clients-service interaction modes are not known beforehand is an area ripe for research. Log analytics research must address the problem of mining patterns from natural language log entries emitted by uninstrumented software.

Existing systems don't address the problem of spurious application traces and events generated by malicious sources. Research on spam detection (Blanzieri, 2008) and foiling phishing attacks (Khonji, 2013) in the context of email could be applied to identify invalid data and blacklist malicious data sources. To provide privacy assurances to users and give them incentives to trust a mobile analytics system, research in certifying application and code properties would be very useful. Prior work in cryptography, proof-carrying code (Necula, 2002), and the P3P standard (Cranor, 2003) could be leveraged.

Lastly, mobile analytics platforms today generate answers demanded by users or send event notifications, after which a human enters the loop. Future research should investigate how much of this user decision making can be automated, especially for forensics. Automating the IT helpdesk to determine and fix application problems would be a step toward true artificial intelligence.

REFERENCES

Al-Sabri, H. M., & Al-Saleem, S. M. (2013). Building a Cloud Storage Encryption (CSE) Architecture for Enhancing Cloud Security. *International Journal of Computer Science Issues*, *10*, 259–266.

AppSee. (n.d.). Retrieved April 18, 2016, from https://www.appsee.com

Aternity. (n.d.). Retrieved December 10, 2015, from http://www.aternity.com

Au, K. W. Y., Zhou, Y. F., Huang, Z., Gill, P., & Lie, D. (2011, October). Short paper: a look at smartphone permission models. In *Proceedings of the 1st ACM workshop on Security and privacy in smartphones and mobile devices* (pp. 63-68). ACM. doi:10.1145/2046614.2046626

Bakken, D. E., Parameswaran, R., Blough, D. M., Franz, A. A., & Palmer, T. J. (2004). Data obfuscation: Anonymity and desensitization of usable data sets. *IEEE Security and Privacy*, *2*(6), 34–41. doi:10.1109/MSP.2004.97

Blanzieri, E., & Bryl, A. (2008). A survey of learning-based techniques of email spam filtering. *Artificial Intelligence Review*, *29*(1), 63–92. doi:10.1007/s10462-009-9109-6

Blaze, M., Feigenbaum, J., Ioannidis, J., Keromytis, A. D. (1999). *The KeyNote Trust Management System Version 2*. RFC 2704.

Buecker, A., Forster, C., Muppidi, S., & Safabakhsh, B. (2009). *Flexible Policy Management for IT Security Services Using IBM Tivoli Security Policy Manager.* IBM Red Paper Publication REDP-451200. Retrieved December 10, 2015, from http://asmarterplanet.com/mobile-enterprise/blog/2014/12/mobile-infrastructure-analytics.html

Cognos. (n.d.). Retrieved December 10, 2015, from http://www.ibm.com/software/analytics/cognos

Cranor, L. F. (2003). P3P: Making privacy policies more useful. *IEEE Security and Privacy*, *1*(6), 50–55. doi:10.1109/MSECP.2003.1253568

Crittercism. (n.d.). Retrieved December 10, 2015, from http://www.crittercism.com

Cuadrado, F., & Dueñas, J. C. (2012). Mobile application stores: Success factors, existing approaches, and future developments. *Communications Magazine, IEEE*, *50*(11), 160–167. doi:10.1109/MCOM.2012.6353696

Damianou, N., Dulay, N., Lupu, E., & Sloman, M. (2001). The Ponder Policy Specification Language. In *Policies for Distributed Systems and Networks* (pp. 18–38). Springer Berlin Heidelberg. doi:10.1007/3-540-44569-2_2

Dierks, T., & Rescorla, E. (2008). *The Transport Layer Security (TLS) Protocol Version 1.2*. RFC 5246.

Duckham, M., & Kulik, L. (2005). A formal model of obfuscation and negotiation for location privacy. In *Pervasive Computing* (pp. 152–170). Springer Berlin Heidelberg. doi:10.1007/11428572_10

Dwivedi, H. (2010). *Mobile application security*. Tata McGraw-Hill Education.

Ferraiolo, D., Cugini, J., & Kuhn, D. R. (1995, December). Role-based access control (RBAC): Features and motivations. In *Proceedings of 11th annual computer security application conference* (pp. 241-48).

Fluid UI. (n.d.). Retrieved December 10, 2015, from https://www.fluidui.com

Flurry. (n.d.). Retrieved April 18, 2016, from http://www.flurry.com

Fu, Q., Lou, J. G., Wang, Y., & Li, J. (2009, December). Execution anomaly detection in distributed systems through unstructured log analysis. In *Data Mining, 2009. ICDM'09. Ninth IEEE International Conference on* (pp. 149-158). doi:10.1109/ICDM.2009.60

Google Analytics. (n.d.). Retrieved December 10, 2015, from https://www.google.com/analytics

Google Mobile Analytics. (n.d.). Retrieved December 10, 2015, from https://www.google.com/analytics/mobile

Hadoop. (n.d.). Retrieved December 10, 2015, from https://hadoop.apache.org

IBM MobileFirst. (n.d.). Retrieved December 10, 2015, from http://www.ibm.com/mobilefirst

IBM TeaLeaf. (n.d.). Retrieved December 10, 2015, from http:// www.ibm.com/software/info/tealeaf

Informatica. (n.d.). Retrieved December 10, 2015, from https://www.informatica.com

InVision. (n.d.). Retrieved December 10, 2015, from http://www.invisionapp.com

Jaramillo, D., Smart, R., Furht, B., & Agarwal, A. (2013, April). A secure extensible container for hybrid mobile applications. In Southeastcon, 2013 Proceedings of IEEE (pp. 1-5). doi:10.1109/SECON.2013.6567439

Khonji, M., Iraqi, Y., & Jones, A. (2013). Phishing detection: A literature survey. *IEEE Communications Surveys and Tutorials*, *15*(4), 2091–2121. doi:10.1109/SURV.2013.032213.00009

Localytics. (n.d.). Retrieved April 18, 2016, from https://www.localytics.com

Loreto, S., Mecklin, T., Opsenica, M., & Rissanen, H. M. (2009). Service broker architecture: Location business case and mashups. *Communications Magazine, IEEE*, *47*(4), 97–103. doi:10.1109/MCOM.2009.4907414

MixPanel. (n.d.). Retrieved April 18, 2016, from https://mixpanel.com

Mobile Analytics: Why You Should Care. (n.d.). Retrieved December 10, 2015, from http://asmarterplanet.com/mobile-enterprise/blog/2013/10/mobile-analytics-why-you-should-care.html

Nadler, S., Soroka, V., Fuchs, O., Korenshtein, R., & Sonsino, E. (2008). Presence Zones for Contextual Location Based Services. In Innovations in Clouds, Internet and Networks, 2008. ICIN.

Necula, G. C. (2002). *Proof-carrying code. Design and Implementation*. Springer Netherlands.

Nguyen, N., Kleinrock, L., & Reiher, P. (2012). Debugging Ubiquitous Computing Applications With the Interaction Analyzer. *International Journal on Advances in Software*, *5*(3 & 4), 2012.

SAS. (n.d.). Retrieved December 10, 2015, from https://www.sas.com

Splunk. (n.d.). Retrieved December 10, 2015, from http:// www.splunk.com

SPSS. (n.d.). Retrieved December 10, 2015, from http://www.ibm.com/software/analytics/spss

Stonebraker, M., Bruckner, D., Ilyas, I. F., Beskales, G., Cherniack, M., Zdonik, S. B., & Xu, S. (2013, January). Data Curation at Scale: The Data Tamer System. In *Proceedings of the Conference on Innovative Data Systems Research. CIDR 2013*.

Weiss, N. E., & Miller, R. S. (2015, February). The Target and Other Financial Data Breaches: Frequently Asked Questions. In *Congressional Research Service, Prepared forMembers and Committees of Congress February* (*Vol. 4*).

Zargar, S. T., Joshi, J., & Tipper, D. (2013). A survey of defense mechanisms against distributed denial of service (DDoS) flooding attacks. *IEEE Communications Surveys and Tutorials*, *15*(4), 2046–2069. doi:10.1109/SURV.2013.031413.00127

Chapter 12
Mobile + Cloud:
Opportunities and Challenges

Pushpendra Singh
Indraprastha Institute of Information Technology, India

ABSTRACT

A mobile phones provides portability and personalized computing with ubiquitous connectivity. This combination makes them an ideal choice to use for various applications of personal use. The portability of mobile devices is the most important and useful feature of mobile devices. However, portability is achieved at the high cost of limited power and computation ability of the mobile device. Cloud computing fulfills the need of providing more computation power to complete the tasks that cannot be done on a mobile platform. The cloud provides an always available platform and do not have typical limitations, e.g. limited battery and computation power, of mobile platforms. Therefore combining cloud computing with mobile provides us best of both worlds i.e. we have a computing platform available for us all the time which we move, and yet we can access services and perform tasks that require high-power computation.

INTRODUCTION

In the last few years, there has been a remarkable spread of mobile technologies in developed as well as in developing countries. The penetration of mobile technologies is now far more than that of the regular internet and land-line telephones. According to the 2014 report of Telecom Regulatory Authority of India (TRAI), there are 943.9 million wireless telephones with a teledensity of ~75% and share of ~97% of total telephones in India. The TRAI report also mentions that while wired internet covers around 10% of the population (mostly in metro cities), the mobile internet reaches deep in every demography and more importantly almost everyone has access to a mobile device either through their personal phone or shared phone of a family member. It makes mobile phone the most ubiquitous computing platform.

Mobile phone provides portability and personalized computing with ubiquitous connectivity. This combination makes them an ideal choice for various applications of personal use, e.g. to know about transportation medium, healthcare advice, education, or entertainment. The portability of mobile devices

DOI: 10.4018/978-1-5225-0945-5.ch012

is the most important and useful feature of mobile devices. However, portability is achieved at the high cost of limited power and computation ability of the mobile device.

For augmenting the computation ability of a mobile device, various solutions have been proposed which include the use of standard techniques for example Remote Procedure Calls (RPC). A survey by Satyanarayan (2010) provides a good overview of such techniques. An interesting approach, namely cyber-foraging, has been proposed by Balan et al. (2002, 2007). The cyber-foraging approach provides a novel insight of using existing nearby machines at one hop distance, called *surrogates*, for offloading the computation. Most importantly, they advocate that the surrogates need not be trusted or managed. They propose the system to implement cyber foraging and modify existing applications to make use of cyber foraging (Balan, Gergle, Satyanarayanan, & Herbsleb, 2007). The cyber foraging approach stands out from other proposed solutions in multiple aspects: it proposed the use of existing surrounding machines instead of deploying new infrastructure; the machines need not be managed or trustworthy; use of surrogates improves the experience of the user of the application, but, the absence of surrogates does not stop the execution of the application. The proposed solution exploits the fact that surrogates are only a single hop away, and a direct link can be established with thus reducing latency. In later work (Satyanarayanan, Bahl, Cáceres, & Davies, 2009), the authors propose the use of VM based *cloudlets* to enable cyber foraging. VM based cloudlets solve the problem of misconfiguration and allow a smooth transitioning of application code execution from the mobile device to cloudlet and vice-versa. The cyber-foraging approach has been used for augmenting mobile capabilities with fixed infrastructure in different resource-constrained environments (Flinn, 2012) (Lewis, Echeverría, Simanta, Bradshaw, & Root, 2014). Though, the cyber-foraging approach advocates and promotes the use of existing unmanaged infrastructure, however, the complexities associated with such a set-up cannot easily be overcome.

Therefore, Cloud computing has emerged as the most popular alternative to providing unlimited computing ability to a mobile device while leveraging the ubiquitous connectivity that a mobile device offers. Cloud computing fulfills the need of providing more computation power to complete the tasks that cannot be done on a mobile platform. The cloud provides an always available platform and does not have limitations, e.g. limited battery and computation power, typical of the mobile platform. Moreover, the cloud infrastructure is managed and trustworthy, thus, it frees the user of the mobile device from the task of managing trust and handle the uncertainty of interacting with an unmanaged device. Therefore combining cloud computing with mobile provides us best of both worlds i.e. we have a computing platform available for us while on the move, and yet we can access services and perform tasks that require high-power computation.

Mobile Cloud Computing platforms, which harness the power of cloud computing with mobile devices, are being extensively used in various domains such as healthcare, transportation, energy monitoring, education, etc. to provide novel solutions for existing research and social problems. Use of cloud platform has given rise to novel research areas such as participatory sensing, crowd-sourcing, etc. which explore how cloud systems can be used with mobile systems to create better systems. Moreover use of cloud for storage and computation frees the mobile device from these requirements and allows mobile to do other functions which a cloud cannot perform, for example, use of onboard sensors to detect vital parameters that can be used, by the cloud, in providing a personalized environment to the user or in detecting a pattern or a context necessary for the new generation of smart applications e.g. using accelerometers present on a phone to detect traffic and road conditions (Singh, Juneja, & Kapoor, 2013), (Garg, Singh, Ramanathan, & Sen, 2014), collecting cellular data to detect location in absence of a GPS hardware (Yadav et al. 2010), (Yadav, Naik, Singh, Singh, & Chandra, 2012), providing personalized mobile learning experience (Uther, Zipitria, Uther, & Singh, 2005), etc. Therefore, the cloud not only augments the computational ability of a mobile device but both the mobile device and the cloud complement each other by letting each component be used for what it is best capable for – mobile device for personalization and mobility and cloud for computation and storage.

Mobile Cloud Computing (MCC) can be defined as

... a combination of ubiquitous connectivity of mobile device and elastic resources of the cloud to enable a computing and storage platform for providing unrestricted mobility, personalization, storage, and computing on the go.

The combination of mobile and cloud has already given rise to a new generation of applications which provide a personalized experience using the data collected by the mobile device and then combining it with data from other users of the application and running intelligent algorithms, using cloud infrastructure. Waze (https://www.waze.com/), a community-based traffic and navigation app, is one such example application which has been possible only through the platform enabled by MCC. Waze uses mobile devices and cloud to let people provide real-time traffic information which is then made available to other users. It also allows its users to edit maps, connect with their friends, find routes with cheaper gas stations, etc.

In this chapter, we will be discussing the core techniques of mobile cloud computing, novel applications of mobile cloud computing, the challenges faced by mobile cloud systems, and the opportunities and future research areas for the field of mobile cloud computing.

CORE TECHNIQUES OF MOBILE CLOUD COMPUTING

The promise of cloud computing is to provide access to unlimited and uninterrupted computing, and mobile technology offers to compute anytime and anywhere. Combining the two provides us ubiquitous uninterrupted computing which leads to novel applications. The new generation of MCC applications primarily relies on two technologies that have emerged as the core for almost all novel applications. These two core technologies are participatory sensing and crowdsourcing.

Participatory Sensing

Participatory Sensing refers to the use of mobile devices as sensor nodes and location-aware data collection instruments (Burke et al., 2006; Estrin, 2010). The primary idea emerged from the wireless sensor networks which are a collection of sensor nodes and allow integration of sensing, computation, and connectivity in low-power devices which are then embedded in the physical world to collect data (Burke et al., 2006). Sensor networks have allowed the observation of previously unobservable phenomena and have been used extensively in sciences, industry and the military.

However, while traditional wireless sensors networks have been controlled and managed by research or industrial organizations, the mobile devices, to be used for sensing, are owned and controlled by everyday users. There is neither a central observer nor a controller for managing the devices. The devices are managed and controlled by their users all the time and therefore user participation is the key to the success of such a system.

Participatory sensing tasks include collection, analysis, and sharing of local data which may be of interest to other users. Nowadays, mobile devices come with a variety of sensors that capable of capturing images, audio, location and other data. The latest S series phones by Samsung can capture ambient temperature, luminosity, atmospheric pressure among various other parameters. The ubiquity of cellular and infrastructure networks, like Wi-Fi, and availability of multiple wireless radios on mobile devices makes it very easy to communicate with remote servers.

Different apps (http://urbancivics.com/soundcity_app.html; Lu, Pan, Lane, Choudhury, & Campbell, 2009) have been developed that allows us to sense environmental features, e.g., Noise, ambient temperature, or pollution levels such as the concentration of CO2 or similar gasses in the environment (Mun, Reddy, Shilton, & Yau, 2009). Once the data is sensed using mobile data, the data is usually uploaded to a Cloud Platform. The participatory sensing allows us to collect a large amount of data from diverse sources in a cost-effective and efficient manner. If the same data needs to be collected through some other means then it will require infrastructure to be deployed, for example, to measure noise levels, we will have to deploy sound sensors at multiple locations, systems to collect data, etc. Using a Mobile Cloud platform for the same only requires developing a mobile app that collects sound data on a mobile device and uploads it to cloud-based storage (Guo, Yu, Zhou, & Zhang, 2014).

The combination of mobile and cloud has resulted in creating efficient and cost-effective systems for collecting a large amount of data which directly adds to better analysis resulting in useful insights. Several such systems have been developed, and analysis of this user-collected data is throwing new insights (Mun, Reddy, Shilton, & Yau, 2009).

Crowdsourcing

The Merriam-Webster dictionary defines crowdsensing as the practice of obtaining needed services, ideas, or content by soliciting contributions from a large group of people and especially from the online community rather than from traditional employees or suppliers. In the context of mobile cloud computing, crowdsourcing is similar to participatory sensing in the manner that it also exploits that fact that a large number of users are carrying mobile devices which can be used to collect data. However, the crowdsourcing goes beyond participatory sensing in the way that user can also manually feed data which can then be used for analysis. Thus, the data collected goes beyond what can be automatically sensed using a mobile device. For example, while participatory sensing can allow collection of noise levels inside a bus or metro using a microphone present on a mobile phone, crowdsourcing allows participants to submit their perception of how crowded a metro/bus is, this data can then be used to provide more personalized routes (Bajaj et al., 2015).

Besides the sensors present on the mobile device, crowdsourcing also uses the data that a user feeds in social networking applications like Twitter and Facebook. This data complements the data obtained from sensors and can be used for further analysis. With a large amount of data being generated (primarily via mobile devices) on social networking platforms, crowdsourcing has emerged as the core technology for several new applications and has also been used successfully in various social movements e.g. Arab Spring in spreading latest information that was gathered via crowdsourcing.

Both participatory sensing or crowdsourcing could be pro-active or opportunistic in nature. Pro-active sensing refers to activities where sensing is one of the primary objectives, and a participating user is ready to take extra steps e.g. switching on some specific sensors or applications primarily for sensing only. Pro-active sensing requires dedicated users and necessitates having good incentive mechanisms for engaging a user in the system. Metrocognition is am an example of a mobile cloud application that requires a pro-active user to use the application and provide feedback.

A different paradigm is proposed in Opportunistic Sensing. Liang et al. (2011) describe Opportunistic sensing as, "a paradigm for signal and information processing in which a network of sensing systems can automatically discover and select sensor platforms based on an operational scenario, determine the appropriate set of features and optimal means for data collection based on these features, obtain missing information by querying resources available, and use appropriate methods to use the data, resulting in an adaptive network that automatically finds scenario-dependent, objective-driven opportunities with optimized performance."

Devices employing opportunistic sensing tend to optimize resources while fulfilling sensing requirements. Opportunistic sensing also usually do not require intervention on behalf of the user and can run in the background.

Frameworks for Mobile and Cloud Sensing

Given the importance of leveraging cloud platforms with mobile devices, several frameworks (Cornelius et al., 2008; Das, Mohan, Padmanabhan, Ramjee, & Sharma, 2010; Bajaj & Singh, 2015), have been proposed that simplify collecting data from mobile devices to the cloud. The basic role of a framework is to issue queries for data collection, collect the data from participating devices, and make this data available to other participants. The framework has to manage data about all the participants, e.g. their location and capabilities, as well as about the data being exchanged in the form of queries and their responses. A framework may support different types of applications, e.g. applications collecting data about road conditions, applications collecting health data for a report about individual health conditions or epidemic, applications for education, or new applications emerging in the domain of energy or societal development.

Given the nature of mobile cloud sensing applications, they only become successful when a large number of people participate in them. Participants of a common framework can then use the services offered by the framework to issue queries as per their requirements and ask other participants to submit data that satisfies the query requirements. These participants are then required to service queries to submit data to the cloud. Figure 1 shows a query, from Sahyog framework that requires accelerometer data to be submitted by at least one device for the time duration of five minutes from a device that is currently in the driving state. The device must be within the 500-meter radius of given location (Bajaj & Singh, 2015). This query can be serviced by any other participating device of the Sahyog framework.

The task of issuing queries is very complex and based on how queries are issued; the frameworks can be categorized as:

1. **Push-Based Frameworks:** In such frameworks, the framework pushes the queries to the participant devices.
2. **Pull-Based Frameworks:** In such frameworks, the participant devices pull the queries from the framework.

Both of the approaches have their pros and cons. The push-based approaches require the framework to know about participating devices all the times which then requires the participating devices to update

Figure 1. Query format of Sahyog framework
Bajaj & Singh, 2015.


```
{       "query":
        {
                "userID": 12345678910,
                "queryNo": 358782039353,
                "dataReqd": "Accelerometer",
                "fromTime": 1406809803000,
                "toTime": 1406810103000,
                "expiryTime": 1406811603000,
                "count":{
                        "min":1,
                        "max":1
                        }
                "frequency": 20000,
                "activity": "driving",
                "latitude": 28.04547,
                "longitude": 77.25483,
                "radius":500
        }
}
```


their information, e.g. location, to the framework all the time. Constantly updating information is resource consuming as well as privacy intrusive. Once a query is issued to a device, a device may choose to serve it or reject it, however receiving a query just to reject also consumes resources which are a drawback of push-based systems. So far not much work has been done on how to select participating devices in a push-based framework and most of the available frameworks simple broadcast the queries to all available devices or randomly select some of the devices and thus leaving the decision to serve or reject the queries to the devices itself. However, having information about available devices and capabilities may allow the framework to develop intelligent schemes for query allocation (Bajaj & Singh, 2015) attempts to do the same in the Sahyog framework (Bajaj & Singh, 2015). Sahyog aims to dispatch queries only to the devices that have higher resource levels than the other and thus tries to optimize resource consumption of all the participating devices.

The pull-based approach requires a participating device to pull the queries whenever it wants. In a pull-based approach, a participating device has better control over the information that it shares with the framework and can also optimize its resources by choosing to pull queries only when it feels that serving such queries will not adversely affect its resources. Since the devices are pulling a query, almost every query that is pulled is serviced and not rejected. However, the framework does not have any control over participating devices which means that the framework cannot guarantee if a query can be serviced or not or how long it will take to serve a query. The framework can also not develop any intelligent schemes for query allocation and thus may not optimize at a global level. This may result in starvation of some queries.

Because of the pros and cons of both of the schemes, currently available frameworks use a combination of query allocations (Cornelius et al., 2008; Bajaj & Singh, 2015; "ASHA Status of Selection and Training" (September 2014); Ra, Liu, La Porta, & Govindan, 2012; Haderer, Rouvoy, & Seinturier, 2013; Eugster, Garbinato, & Holzer, 2005). The Anonysense framework (Cornelius et al., 2008) uses a pull-based approach and allows the participating devices to fetch queries from the framework. Anonysense aims to provide higher privacy to participants by using pull-based queries because then the participating device need not to update their information with the framework. The authors believe that this will instill more confidence among the participants leading to better participation. For submitting queries, Anonysense proposes its own domain-specific language –AnonyTL – to specify contextual parameters.

Another pull-based sensing framework is BubbleSensing "Handheld Tele-ECG Instrument for Rural Health Care" (September 2014). The authors propose binding the task of servicing queries with physical locations. They designate some of the mobile nodes as "anchors"; the anchor nodes are then used to broadcast sensing queries to other mobile nodes present in the area. Any number of nodes can participate in a given task. Such an approach is usually not very good for resource conservation at the participating devices.

(Brouwers & Langendoen, 2012) have proposed "Pogo" which is a push-based framework. In Pogo, queries are issued by the server to the participating devices. Pogo also proposes some data optimization techniques, e.g. combining it with other cellular data exchange that a participating device may use while servicing a query. In Pogo, queries are submitted using JavaScript instead of a domain-specific language. APISense (Haderer et al., 2013) also supports push-based sensing and allow applications to request data from participants by using a set of contextual parameters, like the location of time intervals. Their approach is similar to Sahyog (Bajaj & Singh, 2015). In Medusa (Ra et al., 2012), applications can submit sensing queries using a domain-specific high-level programming language called MedScript. Medusa proposes the use of Amazon's Turk framework – the framework is commercially available and has a front-end where providers pick up tasks in exchange for a small payment - for query allocation to providers i.e. they rely on human intelligence to allocate queries, however how this approach can be sustainable in finance terms is a challenge.

Query allocation is still one of the biggest challenges in the available and upcoming frameworks and several approaches have been proposed to handle it. (Reddy, Estrin, & Srivastava, 2010) propose automated allocation of queries based on the geographical and temporal availability of the mobile device. They also propose to include the performance of a mobile device (in previous such tasks). Evolutionary algorithms have been proposed by (Pham, Sim, & Youn, 2011) for query allocation in a participatory sensing platform. They propose to use algorithms based on multi-objective Knapsack problem to ensure good quality data at low costs by selected devices. Use of mobility patterns to issue queries and to also predict the amount of sensor data collection by a given user has been proposed by "CommCare Accredited Social Health Activist (ASHA)" (September 2014), they also focus on incentivising users for data collection. However, their approach requires mobility data to be shared with the framework which may be privacy-intrusive. While they propose other techniques, they do not take into account the resource levels, e.g. available battery, at the participating device and cost that a participant device may incur regarding resource consumption.

Receiving and servicing queries, e.g. collecting GPS traces, consume resources like battery and may also cost financially for using the data network for communicating with the framework. Among the resources available at a device, battery consumption is still the primary concern for mobile device users. Therefore, optimizing the resource consumption on participating mobile devices is very important for the frameworks.

(Sheng, Tang, & Zhang, 2012) have proposed the use of scheduling for sensing tasks and aims to find the best schedule such that the sensing cost is minimized. For calculating schedules, they require mobility and location data of the participating devices. This data is uploaded to a cloud and then used for calculating schedules for a mobile device. Consistently uploading location data requires GPS or other locating mechanisms like cellular or Wi-Fi based to be run on participating device most of the time.

GPS, in particular, is very resource consuming and can drain the battery of a mobile device quickly. Additionally, the location data also needs to be sent to the server periodically which incurs battery as well as financial costs. (Baier, Durr, & Rothermel, 2013) aims to tackle this problem by proposing a method to optimize location updates and reducing numbers of such updates by using probabilistic locations instead on actual locations.(Lu, Li, Xu, Chen, & Ding, 2013) focuses on minimizing the total participation cost involved in a sensing task by selecting a subset of participants based on their predicted trajectories. Intelligent uploading strategies have been proposed by (Musolesi, Piraccini, Fodor, Corradi, & Campbell, 2010) and (Brouwers & Langendoen, 2012). (Musolesi et al., 2010) analyzes streams of data and then trade off the accuracy of sensing with energy consumptions, while Brouwers et al. (2012) proposes piggybacking the sensing data and combining the transfer of sensing data with other cellular data to avoid the long-tail effect of cellular data and thus conserving energy.

APPLICATIONS OF MOBILE AND CLOUD SENSING

Using Mobile and Cloud enabled sensing, several new applications have been developed that take advantage of unlimited storage and uninterrupted computation ability of the crowd combined with all-time availability and personalization provided by mobile devices. In this section, we will be discussing few such areas which have benefitted tremendously by Mobile Cloud Computing.

Healthcare

Providing affordable and accessible healthcare has emerged as a major challenge across the globe and more specifically for developing countries. There is a dearth of doctors in developing countries for example, according to a 2013 study by ICMR, the doctor to population ratio was 1:1800 with a predicted shortage of 600,000 doctors (Deo, 2013). Most of the public healthcare policies require a large amount of data to be collected and analyzed before being implemented in the field. With recent advances in sensors to detect health parameters, e.g. B.P. monitor or pulse oximeter, we can leverage the Mobile-Cloud platform to not only collect data for public healthcare but also providing basic health services. Already public health workers such as ASHAs (Accredited Social Health Activist) "ASHA Status of Selection and Training" (September 2014), in India have started taking advantage of tools like CommCare "CommCare Accredited Social Health Activist (ASHA)" (September 2014), ODK-Sensors (Hartung et al., 2010; Brunette et al., 2012) and handheld Tele-ECG instruments "Handheld Tele-ECG Instrument for Rural Health Care" (September 2014), These tools use a combination of mobile and cloud to enable easy collection of healthcare data. The same data can then be either used for policy formulation or assessing health situation.

Several cloud-based solutions have also emerged to allow healthcare data to be safely stored in the cloud. Sana "Sana Technology Platform" (September 2014) is one of the most popular systems that provides end-to-end telemedicine platform and is based on Mobile-Cloud technology. Sana has following key features [sana2]:

1. Multi-user mobile clients that can provide decision support, direct remote workers in the collection of data, establish a connection between workers and physicians and integrate with point of care applications and devices.

2. Middleware that provides secure communication layer for secure, optimized transmission for regions with poor network connectivity, e.g. developing countries or remote areas. The middleware also allows connection to a broad range of other web services.

Solutions like Sana could be very easily connected to medical databases such as OpenMRS "About OpenMRS" (September 2014), which is another free and open-source solution for storing medical health records. Several other solutions in the area of the mobile-cloud system have been proposed that enable easy data collection (Asthana & Singh, 2013; Srinivasan et al., 2013; Gupta et al., 2013).

With Health data, one of the major challenges is of data integrity. Many system issues such as mobility of the user, network connectivity, environmental context, etc. are inherent to the mobile cloud system and are likely to contribute to noise in the collected data. Human input, which forms an integral component of collected data, through crowdsourcing, will likely result in accidental and maliciously introduced data errors that should be effectively filtered out.

Systemic redundancies need to be developed to address for commonly prevalent issues such as theft, human fallibilities in data collection together with sensing errors. While solutions are available for supporting application failure (Kiehn, Raj, & Singh, 2014) in mobile computing scenarios, the challenges related to data integrity are still open research problems.

Transportation

It is expected that by 2050, around 70% of the world's entire population will be living in cities and thus public transportation has emerged as a major problem for municipalities worldwide.

Despite the growth in public multi-modal transit systems, which are necessary to manage better mobility, it is widely understood that it will not be sufficient in convincing people to use public transport instead of private vehicles. One way to tackle this problem is to offer personalized travel information to citizens to make their journeys more efficient and enjoyable. The travel information should not only be objective (e.g., bus timetable, live bus tracking), but personalized – since every passenger has different preferences and interests (e.g., crowdedness of trains, the heat of tube platforms, the sociability of the coaches).

However, to provide this personalized information, a lot of information needs to be collected about users and transportation. Mobile-cloud systems have emerged as an effective measure to collect this information. Mobile phones are used to collect information which is then transferred to the cloud for analysis. Various sensors present on a mobile phone can collect information that may be helpful in determining factors related to transportation. GPS information from a mobile device can easily be used to detect speed which can then tell about traffic flow. Accelerometers from the phones have been used to detect road conditions such as potholes, speed-breakers etc. Microphones have also been used to detect traffic conditions by using noise collected from the ambient environment. Different types of mobile-cloud systems have been proposed that use combination of these sensors to solve various problems related to transportation including driving pattern, traffic density, speed, flux, traffic state classification, and providing personalized route recommendation etc. (Baier, Durr, & Rothermel, 2013; Krause, Horvitz, Kansal, & Zhao, 2008; Campbell, Eisenman, Lane, Miluzzo, & Peterson, 2006; Cornelius et al., 2008; Das et al., 2010; Ganti, Ye, & Lei, 2011; Lane et al., 2010; Mohan, Padmanabhan, & Ramjee, 2008; Thiagarajan et al., 2009; Rana, Chou, Kanhere, Bulusu, & Hu, 2010; Lu, Pan, Lane, Choudhury, & Campbell, 2009; Xiong, Zhang, Wang, Gibson, & Zhu, 2015).

Apart from the mentioned uses, mobile-cloud systems are used in many other systems e.g. education (Uther, Singh, & Uther, (2005); Uther, Uther, Athanasopoulos, Singh, & Akahane-Yamada, (2007)), evaluating evacuation drills using mobility data (Bajaj, & Singh, (2015)), energy consumption analysis (Balaji, Xu, Nwokafor, Gupta, & Agarwal, 2013), etc. Novel usage of mobile-cloud systems is emerging every day and in future, it will be one of the prominent technology platforms to solve real-world problems.

CHALLENGES OF MOBILE CLOUD COMPUTING SYSTEMS

Mobile cloud computing faces many of the same challenges that are faced by mobile computing and cloud computing and gives rise to some unique challenges that need to be overcome for enabling applications that can make use of such a system. This section provides an overview of some of the prominent challenges.

Connectivity

Mobile cloud computing applications require ubiquitous connectivity to function. Irrespective of the latest advances in the networking field, the wireless and mobile networks are still susceptible to frequent disconnections, low bandwidth, etc. In developing countries, though the voice network is now prevalent, the data connection is not always present in the remote areas. Non-availability of data network results in inadequate use, or at worst failure, of the mobile cloud computing system. Satyanarayan et al. (1990) propose Coda to provide resiliency to server and network failures and provide operations in the disconnected state.

Another line of work to handle disconnections is being done in the area of opportunistic networks. The opportunistic Network (Pelusi, Passarella, & Conti (2006)) is a new line of research in the field of networking which aims to provide connectivity whenever an opportunity is present. Opportunistic networks recognize that, at times, mobile applications will suffer disconnections. However, these disconnections should not affect their ability to function correctly. Therefore, an application should be equipped to handle the loss of connectivity. The applications take disconnections as an environmental characteristic rather than an error. Various techniques have been suggested to take advantage of opportunistic networks for handling disconnections. An interesting line of work in this area is to use social relations and the presence of other physical objects, e.g. kiosks, to enable connectivity (Karamshuk, Boldrini, Conti, & Passarella, 2011; Boldrini, Conti, Iacopini, & Passarella, 2007; Boldrini, Conti, & Passarella, 2008). Such approaches have been proved very successful in developing countries where data connectivity is still not prevalent.

Asthana, Singh, & Jain, (2015) propose a method to encode DTMF tones so that data can be transferred over voice networks. They are able to achieve transfer rates of 1.2 Kbps in real settings and have argued that such a system is needed for emergency applications, example an emergency rescue or woman safety application where only GPS location needs to be sent to the authorities. Their work aims to provide data connectivity where voice network is present.

A notable mobile cloud application that uses opportunistic network techniques at its core is Open Data Kit (https://opendatakit.org/) which provides a set of tools that help create applications for mobile data collection very easily. The ODK framework consists of three parts: a method to create data collection forms or surveys; methods to send the data to a server while taking advantage of opportunistic networks techniques; an aggregation framework to collect data on the server and show and extract it into useful formats.

Use of opportunistic network techniques has been instrumental in making ODK the most used mobile cloud platform for collecting data using mobile devices and used across the world, especially in developing countries. The ODK framework has also been extended to collect health data in the area of mobile healthcare (Wadhwa, Singh, Singh, Kumar, 2015; Wadhwa, Mehra, Singh, Singh, Kumar, 2015).

With advances in networking technologies, 4G, 5G, Wi-Max, Software Defined Networks (SDN), cognitive radio, etc., the hope is that seamless connectivity will indeed be achieved in near future, and it will enable mobile cloud computing systems to be deployed in the wild.

Data Collection Challenges

The data collection has emerged as the killer application for mobile cloud systems. However, a collection of data brings several challenges with it.

Battery

Despite advances in physical resources, e.g. processing power, memory, the battery of mobile devices is still a limiting factor in using a mobile device for data collection. Collecting sensor data consumes battery of mobile devices, and this puts limitations on the act of collecting data itself. For example, continuously using GPS usually drains a phone's battery in 3-4 hours. So far, GPS is the only outdoor localization technique which provides high accuracy in outdoor environments. The high consumption by GPS limits its use for longer durations in any application. While many other localisations have been developed that avoid the use of GPS (Yadav, Naik, Singh, Singh, & Chandra, 2012), they are not very accurate can only be used for applications where the only approximation of a location is needed. Battery consumption of GPS is a major bottleneck in determining the location of the user on a continuous basis; location remains the most important factor in determining the context of a user, and high battery consumption is the limiting factor.

Similarly, collection by other sensors, e.g. Microphone, also consumes a high amount of battery. Recently many mobile cloud applications have surfaced which use audio collected by Microphone to determine context or provide services. Some recent examples are Google Now, Sir by Apple, and Cortana by Microsoft. All the mentioned application take input in speech and provide an answer. The activation of these applications is enabled by user uttering a pre-defined phrase, e.g. Google Now required users to utter "Hello Google" to activate the application. While such a functionality provides a better user interface and easy access to the application, enabling such a feature requires continuous use of Microphone which is energy consuming. For the same reason, Apple allows voice invocation of Siri only when the device is connected to a power supply.

Applications that require high computation e.g. public cryptography applications, image processing applications also result in high consumption and are avoided on mobile devices.

Privacy and Security

Another big challenge with the data collection is that of privacy and security. The data collected by mobile devices can reveal a lot about the individual. Today's mobile applications can track different physical states of interactions (Eagle & Pentland, 2006), her social circle, her workplace, etc. The collected data and its intelligent inference help create intelligent applications which help in everyday life, for example, giving real-time traffic information and suggesting navigational help, or tracking exercise routine to achieve a personal goal or the variety of applications that improve our social interactions. Some of the popular applications, e.g. Endomondo, Waze, Facebook, Twitter, have benefitted a lot from data collected from a mobile device and intelligent interferences drawn from the data. At the same time, revealing so much data may also cause a privacy breach. For example, an application that provides traffic information can easily detect the places that a user visits or application that uses voice command to activate a feature may silently record all the conversation and upload it to third-party servers. The work on Taintdroid (Enck et al. 2010, 2014) and others (Gibler, Crussell, Erickson, & Chen, 2012). has already shown that mobile applications often leak data to the third party. Different Software Engineering techniques are being used to detect mobile applications that leak data. The leaking of data by applications is of serious concern for open platforms like Android. From the OS, the support is provided in terms of selecting what type of data a user wants to share with the app or what permissions does the app have. However, this support is very limited, because, often the user has no other alternative than providing the app all the permissions that it requires or not install the app at all. The Apple iOS provides better control than Android for managing the permission of the app. However, many apps stop giving their best experience once they are denied permissions.

Mobile cloud applications are also vulnerable to the attacks on the cloud infrastructure. Recently reports have come out about the stealing of private photographs of celebrities from their cloud account (http://www.reuters.com/article/us-scarlettjohansson-idUSTRE78D5RW20110914). The attack on web services etc. are a known threat, and mobile clouds systems are also vulnerable to it. However, given that now mobile phones are used to collect information that is very private in nature – photos/videos, location traces – and of financial value – using of mobile for banking services, mobile wallet – the threat is more severe in causing damage to an individual. Identity theft is another rising challenge for mobile cloud systems. Identity theft refers to wrongful access to personal data such as name, employer, date of birth, credit card numbers, etc. Since most of the mobile cloud systems already store such data for different applications, identity theft can be done by getting unauthorized access to mobile cloud systems. Limitation of computation power and battery and human-computer interaction issues restrict the use of strong encryption mechanisms on mobile devices, and there is a need to develop encryption mechanisms that are low-cost in terms of computation and battery power and can be used on mobile devices without affecting the experience of the user.

Data Interoperability

While fetching data, via participatory sensing or crowd-sourcing, from mobile cloud systems, the data may come in different formats, and it is a challenge to understand and make sense of the data. For example, one temperature sensor may send the data in Fahrenheit while other may send the same data in Celsius. Moreover, how to make this data useful for an application i.e. how much information, also called meta-data, should be attached to the data so that an application can understand what the data is without human intervention. Semantic technologies are being used to tackle some of these challenges, and new ontologies have been developed to make data interoperable. However, adding meta-data to the data brings many additional challenges, for example, it increases the storage requirement and processing requirements for the data; while many ontologies have been proposed, there is still a lack of standard ontologies which postpone the problem of data interoperability to another level of working with different ontologies. With new systems coming up, interoperability has become very important, and new mechanisms are needed to handle it at data and application levels.

Inaccuracy in Data Collection

Another important challenge in data collection is to ensure the accuracy of the data collected through mobile cloud system. The Inaccuracy may come involuntarily or maliciously. The sensors that are present on the mobile device are not very accurate for reasons of size, cost, and battery needs, for example, GPS present on mobile devices is known to provide accuracy within few meters only, even when the perfect conditions are present, and therefore, it is still challenging to tell on which side of a narrow road a user is standing, or it is not possible to detect lane change using only GPS data. Same is the case with other sensors, e.g. ambient temperature or barometer, present on mobile devices. The inaccuracy is involuntary and occurs due to inherent nature of the sensors and other mechanisms present on a mobile device.

At the same time, malicious alterations to the data may also occur when such alterations benefit the end-user. Suppose a mobile health application uploads health data of a user to the cloud which is then used to determine the premium on insurance that the user will have to pay. In such cases, it is beneficial for the user to alter the data. Other scenarios could be to manipulate the data of a transport application for avoiding a speeding fine. Considering the growing use of mobile cloud systems in healthcare, Kotz et al. (2015) have proposed different methods for providing support for privacy and security issues (Shin, Cornelius, Kapadia, Triandopoulos, & Kotz, 2015; Mare, Sorber, Shin, Cornelius, & Kotz 2014).

There is a need to develop mechanisms that can ensure that data manipulation has not taken place, or there is enough provenance information attached to identify the breach.

CONCLUSION AND FUTURE DIRECTIONS

Mobile cloud systems are an emerging area of research and offer many opportunities. While considerable progress has been done, there are still many challenges that need to be overcome. The challenges mentioned above offer the opportunities to work in newly emerging areas of mobile cloud systems.

There is another emerging direction of research which is using the cloud to offload computation from a mobile device without affecting the functionality of the mobile application. The code offloading aims to fill the gap of resources between the mobile device and cloud by enabling dynamic resource allocation – including computing and storage resources – and parallel execution. Code offloading is promising to enable computing-intensive applications, e.g. face recognition, natural language processing, optical character recognition, on mobile devices and thus opening up new opportunities for application development. The emerging area of mobile code offloading needs to address the challenges of bandwidth and energy costs, when or when not to offload, or dynamic resource allocation.

In this chapter, we have covered the current research work in the domain of mobile-cloud systems. We have discussed the core techniques the mobile cloud system and various frameworks that have been proposed to enable mobile cloud systems. We have also discussed various applications that mobile cloud systems are offering. We discussed healthcare and transportation in detail where such systems are being used extensively. Other areas where mobile cloud systems have huge potential and use are Energy and

Education among others. More importantly, we have discussed the major challenges faced by mobile-cloud systems.

Mobile-cloud systems are an emerging area that will grow in importance over time. The challenges give us the opportunity to work on new problems, and despite the advancement in technology, there is still a gap in the expectations from mobile cloud systems and what they currently offer.

REFERENCES

About OpenMRS. (2014). Retrieved from http://openmrs.org/about/

ASHA Status of Selection and Training. (2014). Retrieved from http://nrhm.gov.in/communitisation/asha/asha-data.html

Asthana, S., & Singh, P. MVoice: a mobile based generic ICT tool. In *Proceedings of the 6th International Conference on Information and Communications Technologies and Development*. doi:10.1145/2517899.2517940

Asthana, S., Singh, P., & Jain, S. (n.d.). Adaptive Framework for Data Transmission over GSM Voice Channel for Developing Regions. In *New Technologies, Mobility and Security (NTMS)*. 7th International Conference IEEE.

Baier, P., Durr, F., & Rothermel, K. (2013). Efficient distribution of sensing queries in public sensing systems: In Mobile Ad-Hoc and Sensor Systems(MASS). *IEEE 10th International Conference,* (pp. 272-280).

Bajaj, G., Bouloukakis, G., Pathak, A., Singh, P., Georgantas, N., & Issarny, V. (2015). Toward Enabling Convenient Urban Transit through Mobile Crowdsensing. *Intelligent Transportation Systems (ITSC), IEEE 18th International Conference.*

Bajaj, G., & Singh, P. (2015). Sahyog: A middleware for mobile collaborative applications. In *New Technologies, Mobility and Security (NTMS). 7th International Conference IEEE*. doi:10.1109/NTMS.2015.7266518

Bajaj, G., & Singh, P. (2015). Sensing human activity for assessing participation in evacuation drills. In *Adjunct Proceedings of the ACM International Joint Conference on Pervasive and Ubiquitous Computing and Proceedings of the 2015 ACM International Symposium on Wearable Computers (UbiComp/ISWC'15 Adjunct)*. ACM. DOI: doi:10.1145/2800835.2801613

Balaji, B., Xu, J., Nwokafor, A., Gupta, R., & Agarwal, Y. (2013). Sentinel: occupancy based HVAC actuation using existing WiFi infrastructure within commercial buildings. In *Proceedings of the 11th ACM Conference on Embedded Networked Sensor Systems*. doi:10.1145/2517351.2517370

Balan, R., Flinn, J., Satyanarayanan, M., Sinnamohideen, S., & Yang, H.-I. (2002). The case for cyber foraging. *Proceedings of the 10th workshop on ACM SIGOPS European workshop: beyond the PC - EW10*. doi:10.1145/1133373.1133390

Balan, R. K., Gergle, D., Satyanarayanan, M., & Herbsleb, J. (2007). Simplifying cyber foraging for mobile devices. *Proceedings of the 5th International conference on Mobile systems, applications and services - MobiSys '07*. doi:10.1145/1247660.1247692

Biagioni, J., Gerlich, T., Merrifield, T., & Eriksson, J. (2011). EasyTracker: automatic transit tracking, mapping, and arrival time prediction using smart phones. In *Proceedings of the 9th ACM Conference on Embedded Networked Sensor Systems*. doi:10.1145/2070942.2070950

Boldrini, C., Conti, M., Iacopini, I., & Passarella, A. (2007). HiBOp: A history based routing protocol for opportunistic networks. *2007 IEEE International Symposium on a World of Wireless, Mobile and Multimedia Networks, WOWMOM*.

Boldrini, C., Conti, M., & Passarella, A. (2008). Exploiting users' social relations to forward data in opportunistic networks: The HiBOp solution. *Pervasive and Mobile Computing*, *4*(5), 633–657. doi:10.1016/j.pmcj.2008.04.003

Brouwers, N., & Langendoen, K. (2012). Pogo: a middleware for mobile phone sensing. In *Proceedings of the 13th International Middleware Conference*. doi:10.1007/978-3-642-35170-9_2

Brunette, W., Sodt, R., Chaudhri, R., Goel, M., Falcone, M., Van Orden, J., & Borriello, G. (2012). Open data kit sensors: a sensor integration framework for android at the application-level. In *Proceedings of the 10th International Conference on Mobile Systems, Applications, and Services*. ACM.

Burke, J., Estrin, D., Hansen, M., Ramanathan, N., Reddy, S., & Srivastava, M. B. (2006). Participatory sensing. In *Workshop on World-Sensor-Web (WSW'06): Mobile Device Centric Sensor Networks and Applications*, (pp. 117-134).

Campbell, A. T., Eisenman, S. B., Lane, N. D., Miluzzo, E., & Peterson, R. A. (2006). People-centric urban sensing. In *Proceedings of 2nd Annual International Workshop on Wireless Internet*. doi:10.1145/1234161.1234179

CommCare Accredited Social Health Activist (ASHA). (2014). Retrieved from http://www.commcarehq.org/users/commcareasha/

Cornelius, C., Kapadia, A., Kotz, D., Peebles, D., Shin, M., & Triandopoulos, N. (2008). Anonysense: Privacy-aware people-centric sensing. In *Proceedings of the 6th International Conference on Mobile Systems, Applications, and Services, MobiSys '08*.

Das, T., Mohan, P., Padmanabhan, V. N., Ramjee, R., & Sharma, A. (2010). Prism: platform for remote sensing using smart phones. In *Proceedings of the 8th International Conference on Mobile Systems, Applications, and Services*. doi:10.1145/1814433.1814442

Deo, M. G. (2013). Doctor population ratio for india-the reality. *The Indian Journal of Medical Research*, *137*(4), 632. PMID:23703329

Eagle, N., & Pentland, A. (2006). Reality mining: Sensing complex social systems. *Personal and Ubiquitous Computing*, *10*(4), 255–268. doi:10.1007/s00779-005-0046-3

Enck, W., Gilbert, P., Chun, B.-G., Cox, L. P., Jung, J., McDaniel, P., & Sheth, A. N. (2010). TaintDroid: An Information-Flow Tracking System for Realtime Privacy Monitoring on Smartphones. *Osdi*, *10*(49), 1–6.

Enck, W., Gilbert, P., Han, S., Tendulkar, V., Chun, B.-G., Cox, L. P., & Jung, J. et al. (2014). TaintDroid. *ACM Transactions on Computer Systems*, *32*(2), 1–29.

Eriksson, J., Girod, L., Hull, B., Newton, R., Madden, A., & Balakrishnan, H. (2008). The pothole patrol: using a mobile sensor network for road surface monitoring. In *Proceedings of the 6th International Conference on Mobile Systems, Applications, and Services*. Doi: doi:10.1145/1378600.1378605

Estrin, D. (2010). Participatory sensing: Applications and architecture. *IEEE Internet Computing*, *14*(1), 12–42. doi:10.1109/MIC.2010.12

Eugster, P. T., Garbinato, B., & Holzer, A. (2005). Location-based publish/subscribe: Network Computing and Applications.*4th IEEE International Symposium*. doi:10.1109/NCA.2005.29

Flinn, J. (2012). Cyber Foraging: Bridging Mobile and Cloud Computing. *Synthesis Lectures on Mobile and Pervasive Computing*, *7*(2), 1–103. doi:10.2200/S00447ED1V01Y201209MPC010

Ganti, R. K., Ye, F., & Lei, H. (2011). Mobile crowd sensing: Current state and future challenges. *Communications Magazine, IEEE*, *49*(11), 32–39. doi:10.1109/MCOM.2011.6069707

Garg, S., Singh, P., Ramanathan, P., & Sen, R. (2014). VividhaVahana: smartphone based vehicle classification and its applications in developing region.*Proceedings of the 11th International Conference on Mobile and Ubiquitous Systems: Computing, Networking and Services.* doi:10.4108/icst.mobiquitous.2014.257982

Gibler, C., Crussell, J., Erickson, J., & Chen, H. (2012). AndroidLeaks: Automatically detecting potential privacy leaks in Android applications on a large scale. Lecture Notes in Computer Science, 7344, 291-307.

Guo, B., Yu, Z., Zhou, X., & Zhang, D. (2014). From participatory sensing to Mobile Crowd Sensing. *2014 IEEE International Conference on Pervasive Computing and Communication Workshops, PERCOM WORKSHOPS 2014* (pp. 593-598).

Gupta, A., Thapar, J., Singh, A., Singh, P., Srinivasan, V., & Vardhan, V. (2013). Simplifying and Improving Mobile Based Data Collection. In *Proceedings of the Sixth International Conference on Information and Communications Technologies and Development Notes*. ACM Press. doi:10.1145/2517899.2517929

Gupta, A., Thapar, J., Singh, A., Singh, P., Srinivasan, V., & Vardhan, V. (n.d.). Simplifying and improving mobile based data collection. In *Proceedings of the 6th International Conference on Information and Communications Technologies and Development.*

Hachem, S., Pathak, A., & Issarny, V. (2013). Probabilistic registration for large-scale mobile participatory sensing. In *Pervasive Computing and Communications (PerCom), IEEE International Conference*, (pp. 132-140).

Haderer, N., Rouvoy, R., & Seinturier, L. (2013). A preliminary investigation of user incentives to leverage crowd sensing activities. *Pervasive Computing and Communications Workshops (PERCOM Workshops),2013 IEEE International Conference.*

Handheld Tele-ECG Instrument for Rural Health Care. (2014). Retrieved from http://www.barc.gov.in/technologies/ecg/ecg br.html

Hartung, C., Lerer, A., Anokwa, Y., Tseng, C., Brunette, W., & Borriello, G. (2010). Open data kit: Tools to build information services for developing regions. In *Proceedings of the 4th ACM/IEEE International Conference on Information and Communication Technologies and Development*. doi:10.1145/2369220.2369236

Huang, C. M., Lan, K. C., & Tsai, C. Z. (2008). A survey of opportunistic networks. *Proceedings - International Conference on Advanced Information Networking and Applications, AINA* (pp. 1672-1677). doi:10.1109/WAINA.2008.292

Jain, V., Sharma, A., & Subramanian, L. (2012). Road traffic congestion in the developing world. In *Proceedings of the 2nd ACM Symposium on Computing for Development.* doi:10.1145/2160601.2160616

Joseph, S. R. H., & Uther, M. (2009). Mobile devices for language learning: Multimedia approaches. *Research and Practice in Technology Enhanced Learning, 4*(1), 1–26. doi:10.1142/S179320680900060X

Karamshuk, D., Boldrini, C., Conti, M., & Passarella, A. (2011). Human mobility models for opportunistic networks. *IEEE Communications Magazine, 49*(12), 157–165. doi:10.1109/MCOM.2011.6094021

Kiehn, A., Raj, P., & Singh, P. (2014). A Causal Checkpointing Algorithm for Mobile Computing Environments. *LNCS, 8314*, 134–148.

Kotz, D., Fu, K., Gunter, C., & Rubin, A. (2015). Privacy and Security for Mobile and Cloud Frontiers in Healthcare. *Communications of the ACM, 58*(8), 21–23. doi:10.1145/2790830

Krause, A., Horvitz, E., Kansal, A., & Zhao, F. (2008). Toward Community Sensing. In *Proceedings of the 7th International Conference on Information processing in sensor networks*, (pp. 481-492).

Lane, N. D., Chon, Y., Zhou, L., Zhang, Y., Li, F., & Kim, D., … Cha, H. (2013). \Piggyback crowd sensing (pcs): energy efficient crowd sourcing of mobile sensor data by exploiting smartphone app opportunities. In *Proceedings of the 11th ACM Conference on Embedded Networked Sensor Systems.* doi:10.1145/2517351.2517372

Lane, N. D., Miluzzo, E., Lu, H., Peebles, D., Choudhury, T., & Campbell, A. T. (2010). A survey of mobile phone sensing. *Communications Magazine, IEEE, 48*(9), 140–150. doi:10.1109/MCOM.2010.5560598

Lewis, G. A., Echeverría, S., Simanta, S., Bradshaw, B., & Root, J. (2014). Cloudlet-Based Cyber-Foraging for Mobile Systems in Resource-Constrained Edge Environments. *ICSE, 14*, 412–415.

Liang, Q., Cheng, X., & Chen, D. (2011). Opportunistic Sensing in Wireless Sensor Networks: Theory and Application. *2011 IEEE Global Telecommunications Conference - GLOBECOM 2011, 63*(8), 1-5.

Lu, H., Pan, W., Lane, N. D., Choudhury, T., & Campbell, A. T. (2009). Soundsense: scalable sound sensing for people-centric applications on mobile phones. In *Proceedings of the 7th International Conference on Mobile Systems, Applications, and Services.* doi:10.1145/1555816.1555834

Lu, X., Li, D., Xu, B., Chen, W., & Ding, Z. (2013). Minimum cost collaborative sensing network with mobile phones. In Communications (ICC), 2013 IEEE International Conference.

Mare, S., Sorber, J., Shin, M., Cornelius, C., & Kotz, D. (2014). Hide-n-sense: Preserving privacy efficiently in wireless mhealth. *Mobile Networks and Applications, 19*(3), 331–344.

Mohan, P., Padmanabhan, V. N., & Ramjee, R. (2008). Nericell: rich monitoring of road and traffic conditions using mobile smartphones. In *Proceedings of the 6th ACM Conference on Embedded Network Sensor Systems*, (pp. 323-336). doi:10.1145/1460412.1460444

Mohan, P., Padmanabhan, V. N., & Ramjee, R. (2008). Nericell: rich monitoring of road and traffic conditions using mobile smartphones. In *Proceedings of the 6th ACM Conference on Embedded Network Sensor Systems*. doi:10.1145/1460412.1460444

Mun, M., Reddy, S., Shilton, K., & Yau, N. (2009). PEIR, the personal environmental impact report, as a platform for participatory sensing systems research. *MobiSys*, (pp. 55-68).

Musolesi, M., Piraccini, M., Fodor, K., Corradi, A., & Campbell, A. T. (2010). Supporting energy-efficient uploading strategies for continuous sensing applications on mobile phones. In *Pervasive Computing* (pp. 355–372). Springer. doi:10.1007/978-3-642-12654-3_21

Pelusi, L., Passarella, A., & Conti, M. (2006). Opportunistic networking: Data forwarding in disconnected mobile ad hoc networks. *IEEE Communications Magazine*, *44*(11), 134–141. doi:10.1109/MCOM.2006.248176

Pham, H. N., Sim, B. S., & Youn, H. Y. (2011). A novel approach for selecting the participants to collect data in participatory sensing: in Applications and the Internet (SAINT).*11th International Symposium IEEE/IPSJ*. doi:10.1109/SAINT.2011.17

Ra, M. R., Liu, B., La Porta, T. F., & Govindan, R. (2012). Medusa: A programming framework for crowd-sensing applications. In *Proceedings of the 10th International Conferenceon Mobile Systems, Applications, and Services, MobiSys '12*. doi:10.1145/2307636.2307668

Rana, R. K., Chou, C. T., Kanhere, S. S., Bulusu, N., & Hu, W. (2010). Ear-phone: an end-to-end Participatory urban noise mapping system. In *Proceedings of the 9th ACM/IEEE International Conference on Information Processing in Sensor Networks*. doi:10.1145/1791212.1791226

Reddy, S., Estrin, D., & Srivastava, M. (2010). Recruitment framework for participatory sensing data collections. *Pervasive Computing,* 138-155.

Sana Technology Platform. (2014). Retrieved from http://sana.mit.edu/platform/

Satyanarayanan, M. (2010). Mobile Computing: the Next Decade. *Scenario, 15*, 1-6. Retrieved from http://dl.acm.org/citation.cfm?id=1810936

Satyanarayanan, M., Bahl, P., Cáceres, R., & Davies, N. (2009). The case for VM-based cloudlets in mobile computing. *IEEE Pervasive Computing / IEEE Computer Society [and] IEEE Communications Society*, *8*(4), 14–23. doi:10.1109/MPRV.2009.82

Satyanarayanan, M., Kistler, J. J., Kumar, P., Okasaki, M. E., Siegel, E. H., & Steere, D. C. (1990). Coda: A Highly Available File System for a Distributed Workstation Environment. *IEEE Transactions on Computers*, *39*(4), 447–459. doi:10.1109/12.54838

Sen, R., Cross, A., Vashistha, A., Padmanabhan, V. N., Cutrell, E., & Thies, W. (2013). Accurate speed and density measurement for road traffic in India. In *Proceedings of the 3rd ACM Symposium on Computing for Development*. doi:10.1145/2442882.2442901

Sen, R., Maurya, A., Raman, B., Mehta, R., Kalyanaraman, R., Vankadhara, N., … Sharma, P. (2012). Kyun queue: a sensor network system to monitor road traffic queues. In *Proceedings of the 10th ACM Conference on Embedded Network Sensor Systems*. doi:10.1145/2426656.2426670

Sen, R., Raman, B., & Sharma, P. (2010). Horn-ok-please. In *Proceedings of the 8th International Conference on Mobile Systems, Applications, and Services*. Doi: doi:10.1145/1814433.1814449

Sheng, X., Tang, J., & Zhang, W. (2012). Energy-efficient collaborative sensing with mobile Phones: in INFOCOM. Proceedings IEEE, 1916-1924.

Shin, M., Cornelius, C., Kapadia, A., Triandopoulos, N., & Kotz, D. (2015). Location Privacy for Mobile Crowd Sensing through Population Mapping. *Sensors (Basel, Switzerland)*, *15*(7), 15285–15310. doi:10.3390/s150715285 PMID:26131676

Singh, A., Naik, V., Lal, S., Sengupta, R., Saxena, D., Singh, P., & Puri, A. (2011). Improving the efficiency of healthcare delivery system in underdeveloped rural areas. *2011 3rd International Conference on Communication Systems and Networks, COMSNETS 2011*.

Singh, P., Juneja, N., & Kapoor, S. (2013). Using mobile phone sensors to detect driving behavior. *Proceedings of the 3rd ACM Symposium on Computing for Development - ACM DEV '13*. doi:10.1145/2442882.2442941

Singh, P., Juneja, N., & Kapoor, S. (2013). Using mobile phone sensors to detect driving behavior. In *Proceedings of the 3rd ACM Symposium on Computing for Development*. doi:10.1145/2442882.2442941

Srinivasan, V., Vardhan, V., Kar, S., Asthana, S., Narayanan, R., & Singh, P. (2013). Airavat: An Automated System to Increase Transparency and Accountability in Social Welfare Schemes in India. In *Proceedings of the Sixth International Conference on Information and Communications Technologies and Development Notes*. ACM Press.

Srinivasan, V., Vardhan, V., Kar, S., Asthana, S., Narayanan, R., Singh, P., … Seth, A. (n.d.). Airavat: An automated system to increase transparency and accountability in social welfare schemes in India. In *Proceedings of the 6th International Conference on Information and Communications Technologies and Development*.

Thiagarajan, A., Ravindranath, L., Balakrishnan, H., Madden, S., & Girod, L. (2011). Accurate, low-energy trajectory mapping for mobile devices. In *Proceedings of the 8th USENIX Conference on Networked Systems Design and Implementation*.

Thiagarajan, A., Ravindranath, L., LaCurts, K., Madden, S., Balakrishnan, H., Toledo, S., & Eriksson, J. (2009). Vtrack: Accurate, energy-aware road traffic delay estimation using mobile phones. In *Proceedings of the 7th ACM Conference on Embedded Networked Sensor Systems, SenSys '09*. doi:10.1145/1644038.1644048

Thiagarajan, A., Ravindranath, L., LaCurts, K., Madden, S., Balakrishnan, H., Toledo, S., & Eriksson, J. (2009). VTrack: accurate, energy-aware road traffic delay estimation using mobile phones. In *Proceedings of the 7th ACM Conference on Embedded Networked Sensor Systems*. doi:10.1145/1644038.1644048

Torres, M. H. C., Haesevoets, R., & Holvoet, T. (2013). Coos: coordination support for mobile collaborative applications. In *Mobile and Ubiquitous Systems: Computing, Networking, and Services* (pp. 152–163). Springer. doi:10.1007/978-3-642-40238-8_13

Uther, M., Singh, P., & Uther, J. (2005). Mobile Adaptive CALL (MAC): An adaptive s/w for computer assisted language learning. *Proceedings - IEEE International Conference on Pervasive Services, ICPS 2005* (Vol. 2005, pp. 413-416).

Uther, M., Uther, J., Athanasopoulos, P., Singh, P., & Akahane-Yamada, R. (2007). Mobile adaptive CALL (MAC). A lightweight speech-based intervention for mobile language learners. *Proceedings of the Annual Conference of the International Speech Communication Association, INTERSPEECH* (Vol. 2, pp. 1445-1448).

Uther, M., Uther, J., Athanasopoulos, P., Singh, P., & Reiko, A. Y. (2007). *Mobile Adaptive CALL (MAC): A lightweight speech-based intervention for mobile language learners*. Academic Press.

Uther, M., Zipitria, I., Uther, J., & Singh, P. (2005). Mobile Adaptive CALL (MAC): A case-study in developing a mobile learning application for speech/audio language training. *Proceedings - IEEE International Workshop on Wireless and Mobile Technologies in Education, WMTE 2005* (Vol. 2005, pp. 187-191).

Wadhwa, R., Mehra, A., Singh, P., & Singh, M. (2015). A pub/sub based architecture to support public healthcare data exchange. *Communication Systems and Networks (COMSNETS).7th International Conference*. doi:10.1109/COMSNETS.2015.7098706

Wadhwa, R., Singh, P., Singh, M., & Kumar, S. (2015). An EMR-enabled medical sensor data collection framework. *Communication Systems and Networks (COMSNETS),7th International Conference.*

Xiong, H., Zhang, D., Wang, L., Gibson, J. P., & Zhu, J. (2015). \Eemc: Enabling energy-e_cient mobile crowd sensing with anonymous participants. *ACM Transactions on Intelligent Systems and Technology, 6*(3), 39. doi:10.1145/2644827

Yadav, K., Naik, V., Singh, A., Singh, P., & Chandra, U. (2012). Low energy and sufficiently accurate localization for non-smartphones. *Proceedings - 2012 IEEE 13th International Conference on Mobile Data Management, MDM 2012* (pp. 212-221). doi:10.1109/MDM.2012.32

Yadav, K., Naik, V., Singh, A., Singh, P., Kumaraguru, P., & Chandra, U. (2010). *Challenges and Novelties While Using Mobile Phones As ICT Devices for Indian Masses: Short Paper*. Academic Press.

Yoon, J., Noble, B., & Liu, M. (2007). Surface street traffic estimation. In *Proceedings of the 5th International Conference on Mobile Systems, Applications and Services*. Doi: doi:10.1145/1247660.1247686

You, C. W., Lane, N. D., Chen, F., Wang, R., Chen, Z., Bao, T. J., & Campbell, A. T. (2013). CarSafe app: alerting drowsy and distracted drivers using dual cameras on smartphones. In *Proceeding of the 11th Annual International Conference on Mobile Systems, Applications, and Services*. Doi: doi:10.1145/2462456.2465428

Compilation of References

Abdunabi, R., Sun, W., & Ray, I. (2014). Enforcing spatio-temporal access control in mobile applications. *Computing*, *96*(4), 313–353. doi:10.1007/s00607-013-0340-2

Aberer, K., Catasta, M., Radu, H., Ranvier, J. E., Vasirani, M., & Yan, Z. (2014, March). Memorysense: Reconstructing and ranking user memories on mobile devices. In *Pervasive Computing and Communications Workshops (PERCOM Workshops), 2014 IEEE International Conference on* (pp. 195-198). IEEE.

About OpenMRS. (2014). Retrieved from http://openmrs.org/about/

Abrahart, R. & See, L. (n.d.). *GeoComputation* (2nd ed.). Boca Raton, FL: CRC Press.

AHIMA. (2011, March). Security Audits of Electronic Health Information (Updated). *Journal of American Health Information Management Association*, *82*(3), 45–50.

Aitken, M. (2013, October). Patient Apps for Improved Healthcare: from Novelty to Mainstream. *imshealth*. Retrieved from http://www.imshealth.com/en/thought-leadership/ims-institute/reports/patient-apps-for-improved-healthcare#ims-form

Aitken, M. (n.d.). *Patient Apps for Improved Healthcare: From Novelty to Mainstream*. Retrieved from http://www.imshealth.com/portal/site/imshealth/menuitem.762a961826aad98f53c753c71ad8c22a/?vgnextoid=e0f913850c8b1410VgnVCM10000076192ca2RCRD

Almuhimedi, H., & Schaub, F. (2015). Your Location has been Shared 5,398 Times!: A Field Study on Mobile App Privacy Nudging. *Proceedings of the 33rd Annual ACM Conference on Human Factors in Computing Systems*. New York, NY: ACM.

Al-Sabri, H. M., & Al-Saleem, S. M. (2013). Building a Cloud Storage Encryption (CSE) Architecture for Enhancing Cloud Security. *International Journal of Computer Science Issues*, *10*, 259–266.

Alshehri, S., & Raj, R. (2013). Secure Access Control for Health Information Sharing Systems.*2013 IEEE International Conference on Healthcare Informatics, ICHI 2013*. doi:10.1109/ICHI.2013.40

Amalfitano, D., Fasolino, A. R., & Tramontana, P. (2011, March). A gui crawling-based technique for android mobile application testing. In *Software Testing, Verification and Validation Workshops (ICSTW), 2011 IEEE Fourth International Conference on* (pp. 252-261). IEEE. doi:10.1109/ICSTW.2011.77

Amalfitano, D., Fasolino, A. R., Tramontana, P., De Carmine, S., & Memon, A. M. (2012, September). Using GUI ripping for automated testing of Android applications. In *Proceedings of the 27th IEEE/ACM International Conference on Automated Software Engineering* (pp. 258-261). ACM. doi:10.1145/2351676.2351717

Amini, S., & Linqvist, J. (2010). Caché: Caching location-enhanced content to improve user privacy. *Proceedings of the 9th International Conference on Mobile Systems, Applications and Services*.

Android Developer Guide Location APIs. Google Inc. (2015). Retrieved January 28, 2016 from http://developer.android.com/guide/topics/location/index.html

Android Developer Guide Location Strategies. Google Inc. (2015). Retrieved January 28, 2016 from http://developer.android.com/guide/topics/location/strategies.html

Apktool. (n.d.). Retrieved May 31, 2016, from https://ibotpeaches.github.io/Apktool/

AppBrain. (2015). *Number of android applications.* Retrieved from http://www.appbrain.com/stats/number-of-android-apps

Apple Inc. (2013, April 23). *Debug Accessibility in iOS Simulator with the Accessibility Inspector.* Retrieved from https://developer.apple.com/library/ios/technotes/TestingAccessibilityOfiOSApps/TestAccessibilityiniOSSimulatorwithAccessibilityInspector/TestAccessibilityiniOSSimulatorwithAccessibilityInspector.html

Apple Inc. (2016a). *VoiceOver for iOS.* Retrieved from http://www.apple.com/accessibility/ios/voiceover/

Apple Inc. (2016b). *UIKit Function Reference.* Retrieved from https://developer.apple.com/library/ios/documentation/UIKit/Reference/UIKitFunctionReference/#//apple_ref/doc/uid/TP40006894-CH3-SW39

Apple Inc. (2016c) *UIAccessibility Protocol Reference.* Retrieved from https://developer.apple.com/library/ios/documentation/UIKit/Reference/UIAccessibility_Protocol

Apple. (2015). *ResearchKit and CareKit.* Retrieved from http://www.apple.com/researchkit/

Apple. (n.d.). *Healthkit.* Retrieved from: http://developer.apple.com/healthkit

AppSee. (n.d.). Retrieved April 18, 2016, from https://www.appsee.com

Aronszajn, N. (1950). Theory of Reproducing Kernels. *Transactions of the American Mathematical Society*, ▪▪▪, 68.

Artz, D., & Gil, Y. (2007, June). A Survey of Trust in Computer Science and the Semantic Web. *Journal of Web Semantics*, *5*(2), 58–71. doi:10.1016/j.websem.2007.03.002

ASHA Status of Selection and Training. (2014). Retrieved from http://nrhm.gov.in/communitisation/asha/asha-data.html

Ashcraft & Engler. (2002). Using Programmer-Written Compiler Extensions to Catch Security Holes. *S&P.*

ASM-Guide. (n.d.). Retrieved May 31, 2016, from http://download.forge.objectweb.org/asm/asm4-guide.pdf

Asthana, S., Singh, P., & Jain, S. (n.d.). Adaptive Framework for Data Transmission over GSM Voice Channel for Developing Regions. In *New Technologies, Mobility and Security (NTMS).* 7th International Conference IEEE.

Asthana, S., & Singh, P. MVoice: a mobile based generic ICT tool. In *Proceedings of the 6th International Conference on Information and Communications Technologies and Development.* doi:10.1145/2517899.2517940

Aternity. (n.d.). Retrieved December 10, 2015, from http://www.aternity.com

Au, K. W. Y., Zhou, Y. F., Huang, Z., Gill, P., & Lie, D. (2011, October). Short paper: a look at smartphone permission models. In *Proceedings of the 1st ACM workshop on Security and privacy in smartphones and mobile devices* (pp. 63-68). ACM. doi:10.1145/2046614.2046626

Ayewah, N., Pugh, W., Morgenthaler, J. D., Penix, J., & Zhou, Y. (2007). Using Findbugs on Production Software. In OOPSLA Companion. doi:10.1145/1297846.1297897

Azim, T., & Neamtiu, I. (2013, October). Targeted and depth-first exploration for systematic testing of android apps. In ACM SIGPLAN Notices (Vol. 48, No. 10, pp. 641-660). ACM. doi:10.1145/2509136.2509549

Baader, F., & Sattler, U. (2003, December). Description logics with aggregates and concrete domains. *Inf. Syst., 28*(8), 979–1004. Retrieved from doi:10.1016/S0306-4379(03)00003-6

Bacon, D. F., & Sweeney, P. F. (1996). *Fast Static Analysis of C++ Virtual Function Calls.* OOPSLA. doi:10.1145/236337.236371

Baier, P., Durr, F., & Rothermel, K. (2013). Efficient distribution of sensing queries in public sensing systems: In Mobile Ad-Hoc and Sensor Systems(MASS). *IEEE 10th International Conference,* (pp. 272-280).

Bajaj, G., & Singh, P. (2015). Sahyog: A middleware for mobile collaborative applications. In *New Technologies, Mobility and Security (NTMS).7th International Conference IEEE.* doi:10.1109/NTMS.2015.7266518

Bajaj, G., & Singh, P. (2015). Sensing human activity for assessing participation in evacuation drills. In *Adjunct Proceedings of the ACM International Joint Conference on Pervasive and Ubiquitous Computing and Proceedings of the 2015 ACM International Symposium on Wearable Computers (UbiComp/ISWC'15 Adjunct).* ACM. Doi:10.1145/2800835.2801613

Bajaj, G., Bouloukakis, G., Pathak, A., Singh, P., Georgantas, N., & Issarny, V. (2015). Toward Enabling Convenient Urban Transit through Mobile Crowdsensing. *Intelligent Transportation Systems (ITSC), IEEE 18th International Conference.*

Bakken, D. E., Parameswaran, R., Blough, D. M., Franz, A. A., & Palmer, T. J. (2004). Data obfuscation: Anonymity and desensitization of usable data sets. *IEEE Security and Privacy, 2*(6), 34–41. doi:10.1109/MSP.2004.97

Balaji, B., Xu, J., Nwokafor, A., Gupta, R., & Agarwal, Y. (2013). Sentinel: occupancy based HVAC actuation using existing WiFi infrastructure within commercial buildings. In *Proceedings of the 11th ACM Conference on Embedded Networked Sensor Systems.* doi:10.1145/2517351.2517370

Balan, R. K., Gergle, D., Satyanarayanan, M., & Herbsleb, J. (2007). Simplifying cyber foraging for mobile devices. *Proceedings of the 5th International conference on Mobile systems, applications and services - MobiSys '07.* doi:10.1145/1247660.1247692

Balan, R., Flinn, J., Satyanarayanan, M., Sinnamohideen, S., & Yang, H.-I. (2002). The case for cyber foraging. *Proceedings of the 10th workshop on ACM SIGOPS European workshop: beyond the PC - EW10.* doi:10.1145/1133373.1133390

Baloul, M., Cherrier, E., & Rosenberger, C. (2012). Challenge-based speaker recognition for mobile authentication. *2012 BIOSIG - Proceedings of the International Conference of the Biometrics Special Interest Group (BIOSIG)* (pp. 1-7). Darmstadt: IEEE.

Baltrunas, L., Ludwig, B., Peer, S., & Ricci, F. (2011). Context-aware places of interest recommendations for mobile users. In Design, User Experience, and Usability. Theory, Methods, Tools and Practice (pp. 531-540). Springer Berlin Heidelberg. doi:10.1007/978-3-642-21675-6_61

Baracaldo, N., Palanisamy, B., & Joshi, J. (2014). Geo-Social-RBAC: A Location-Based Socially Aware Access Control Framework. In Network and System Security (pp. 501-509). Springer International Publishing. doi:10.1007/978-3-319-11698-3_39

Baride, S., & Dutta, K. (2011). A cloud based software testing paradigm for mobile applications. *Software Engineering Notes, 36*(3), 1–4. doi:10.1145/1968587.1968601

Bechhofer, S., Van Harmelen, F., Hendler, J., Horrocks, I., McGuinness, D. L., Patel-Schneider, P. F., (2004). Owl web ontology language reference. *W3C recommendation, 10.*

Beimel, D., & Peleg, M. (2011). Using OWL and SWRL to represent and reason with situation-based access control policies. *Data & Knowledge Engineering, 70*(6), 596–615. doi:10.1016/j.datak.2011.03.006

Bertino, E., Catania, B., & Damiani, M. (2005). GEO-RBAC: A spatially aware RBAC. In *SACMAT '05 Proceedings of the tenth ACM symposium on Access control models and technologies* (pp. 29-37). Stockholm, Sweden: ACM.

Bettini, C., & Brdiczka, O. (2010). A survey of context modelling and reasoning techniques. In Pervasive and Mobile Computing (pp. 161-180). Elsevier.

Biagioni, J., Gerlich, T., Merrifield, T., & Eriksson, J. (2011). EasyTracker: automatic transit tracking, mapping, and arrival time prediction using smart phones. In *Proceedings of the 9th ACM Conference on Embedded Networked Sensor Systems*. doi:10.1145/2070942.2070950

Biometrics. (n.d.). *Biometrics*. Retrieved from http://dictionary.reference.com/browse/biometrics

Bishop, C. M. (2006). *Pattern Recognition and Machine Learning* (Vol. 1). Springer.

Biswas, A., Chander, D., Dasgupta, K., Mukherjee, K., Singh, M., & Mukherjee, T. (2015). PISCES: Participatory Incentive Strategies for Effective Community Engagement in Smart Cities. In AAAI HCOMP.

Bitbucket. (n.d.). Retrieved May 31, 2016, from https://bitbucket.org/pxb1988/dex2jar

Blanzieri, E., & Bryl, A. (2008). A survey of learning-based techniques of email spam filtering. *Artificial Intelligence Review*, *29*(1), 63–92. doi:10.1007/s10462-009-9109-6

Blaze, M., Feigenbaum, J., Ioannidis, J., Keromytis, A. D. (1999). *The KeyNote Trust Management System Version 2.* RFC 2704.

Bluetooth Smart (Low Energy) Technology. (n.d.). *Bluetooth SIG*. Retrieved January 28, 2016 from https://developer.bluetooth.org/TechnologyOverview/Pages/BLE.aspx

Boldrini, C., Conti, M., Iacopini, I., & Passarella, A. (2007). HiBOp: A history based routing protocol for opportunistic networks. *2007 IEEE International Symposium on a World of Wireless, Mobile and Multimedia Networks, WOWMOM.*

Boldrini, C., Conti, M., & Passarella, A. (2008). Exploiting users' social relations to forward data in opportunistic networks: The HiBOp solution. *Pervasive and Mobile Computing*, *4*(5), 633–657. doi:10.1016/j.pmcj.2008.04.003

Boonstra, A., & Broekhuis, M. (2010). Barriers to the acceptance of electronic medical records by physicians from systematic review to taxonomy and interventions. *BMC Health Services Research BMC Health Serv Res*, *10*(1), 231. doi:10.1186/1472-6963-10-231 PMID:20691097

Brouwers, N., & Langendoen, K. (2012). Pogo: a middleware for mobile phone sensing. In *Proceedings of the 13th International Middleware Conference*. doi:10.1007/978-3-642-35170-9_2

Brucker, A. D., & Petritsch, H. (2009, June). Extending access control models with break-glass. In *Proceedings of the 14th ACM symposium on Access control models and technologies* (pp. 197-206). ACM. doi:10.1145/1542207.1542239

Brunette, W., Sodt, R., Chaudhri, R., Goel, M., Falcone, M., Van Orden, J., & Borriello, G. (2012). Open data kit sensors: a sensor integration framework for android at the application-level. In *Proceedings of the 10th International Conference on Mobile Systems, Applications, and Services.* ACM.

Buecker, A., Forster, C., Muppidi, S., & Safabakhsh, B. (2009). *Flexible Policy Management for IT Security Services Using IBM Tivoli Security Policy Manager.* IBM Red Paper Publication REDP-451200. Retrieved December 10, 2015, from http://asmarterplanet.com/mobile-enterprise/blog/2014/12/mobile-infrastructure-analytics.html

Burke, J., Estrin, D., Hansen, M., Ramanathan, N., Reddy, S., & Srivastava, M. B. (2006). Participatory sensing. In *Workshop on World-Sensor-Web (WSW'06): Mobile Device Centric Sensor Networks and Applications*, (pp. 117-134).

Caine, K., & Hanania, R. (2013). Patients want granular privacy control over health information in electronic medical records. *Journal of the American Medical Informatics Association*, *20*(1), 7–15. doi:10.1136/amiajnl-2012-001023 PMID:23184192

Campbell, A. T., Eisenman, S. B., Lane, N. D., Miluzzo, E., & Peterson, R. A. (2006). People-centric urban sensing. In *Proceedings of 2nd Annual International Workshop on Wireless Internet*. doi:10.1145/1234161.1234179

Capzule. (2012). *Capzule PHR*. Retrieved from http://www.capzule.com/

Care360. (2014). *MyQuest*. Retrieved from https://myquest.questdiagnostics.com/web/home

Carroll, J. J., Dickinson, I., Dollin, C., Reynolds, D., Seaborne, A., & Wilkinson, K. (2004). Jena: Implementing the semantic web recommendations. In *Proceedings of the 13th international World Wide Web conference on alternate track papers & posters* (pp. 74–83). New York, NY: ACM. doi:10.1145/1013367.1013381

Chang, W., Streiff, B., & Lin, C. (2008). *Efficient and Extensible Security Enforcement Using Dynamic Data Flow Analysis*. CCS. doi:10.1145/1455770.1455778

Chen, D., & Zhao, H. (2012, March). Data security and privacy protection issues in cloud computing. In *Computer Science and Electronics Engineering (ICCSEE), 2012 International Conference on* (Vol. 1, pp. 647-651). IEEE. doi:10.1109/ICCSEE.2012.193

Chen, T., Chiu, C., & Tu, T. (2003). *Mixing and Combining with OAO and TOA for the Enhanced Accuracy of Mobile Location*. Retrieved January 28, 20166, from http://ieeexplore.ieee.org/stamp/stamp.jsp?tp=&arnumber=1350199

Cheng, B., & Hwu, W. W. (2000). Modular Interprocedural Pointer Analysis Using Access Paths: Design, Implementation, and Evaluation. In *Proceedings of the ACM SIGPLAN 2000 Conference on Programming language design and implementation*. doi:10.1145/349299.349311

Chen, H., Finin, T., & Joshi, A. (2003, September). An ontology for context-aware pervasive computing environments. *The Knowledge Engineering Review*, *18*(3), 197–207. doi:10.1017/S0269888904000025

Chen, K.-Y., Harniss, M., Lim, J., Han, Y., Johnson, K., & Patel, S. (2013). uLocate: A Ubiquitous Location Tracking System for People Aging with Disabilities. In *8th International Conference on Body Area Networks, BODYNETS 2013*. doi:10.4108/icst.bodynets.2013.253584

Chen, N., Lin, J., Hoi, S. C., Xiao, X., & Zhang, B. (2014, May). AR-Miner: mining informative reviews for developers from mobile app marketplace. In *Proceedings of the 36th International Conference on Software Engineering*(pp. 767-778). ACM doi:10.1145/2568225.2568263

Cisco. (2014). *Cisco Visual Networking Index: Global Mobile Data Traffic Forecast Update, 2014-2019*. Retrieved from http://www.cisco.com/c/en/us/solutions/collateral/service-provider/visual-networking-index-vni/mobile-white-paper-c11-520862.html

Cleary, J. G., & Trigg, L. E. (1995). *K^*: An Instance-based Learner Using an Entropic Distance Measure*. ICML.

CNN Money. (2015). *OPM hack's unprecedented haul: 1.1 million fingerprints*. Retrieved from http://money.cnn.com/2015/07/10/technology/opm-hack-fingerprints/

Cognos. (n.d.). Retrieved December 10, 2015, from http://www.ibm.com/software/analytics/cognos

Cohen, J. (2015, January 7). *11 Health And Fitness Apps That Achieve Top Results*. Retrieved from http://www.forbes.com/sites/jennifercohen/2015/01/07/the-11-top-health-fitness-apps-that-achieve-the-best-results/#11f5c21a1aca

CommCare Accredited Social Health Activist (ASHA). (2014). Retrieved from http://www.commcarehq.org/users/commcareasha/

Connecticut General Assembly. (2015). *Substitute for Raised H.B. No. 6722*. Retrieved from https://www.cga.ct.gov/asp/CGABillStatus/CGAbillstatus.asp?which_year=2015&selBillType=Bill&bill_num=HB6722

Conn, J. (2014). EHR makers' mobile medical apps grow in popularity. *Modern Healthcare*, *29*(November). Retrieved from http://www.modernhealthcare.com/article/20141129/MAGAZINE/311299981 PMID:25671868

Conti, M., Nguyen, V. T. N., & Crispo, B. (2011). Crepe: Context-related policy enforcement for android. In M. Burmester, G. Tsudik, S. Magliveras, & I. Ilic (Eds.), *Information security* (Vol. 6531, pp. 331–345). Springer Berlin Heidelberg. doi:10.1007/978-3-642-18178-8_29

Cornelius, C., Kapadia, A., Kotz, D., Peebles, D., Shin, M., & Triandopoulos, N. (2008). Anonysense: Privacy-aware people-centric sensing. In *Proceedings of the 6th International Conference on Mobile Systems, Applications, and Services, MobiSys '08*.

Couchbase. (n.d.). *Couchbase Mobile*. Retrieved from; http://www.couchbase.com/nosql-databases/couchbase-mobile

Coursaris, C. K., & Kim, D. J. (2011). A meta-analytical review of empirical mobile usability studies. *Journal of Usability Studies*, *6*(3), 117–171.

Cousot, P., & Cousot, R. (1977). Abstract Interpretation: A Unified Lattice Model for Static Analysis of Programs by Construction or Approximation of Fixpoints. In POPL. doi:10.1145/512950.512973

Coyne, E., & Weil, T. (2013). ABAC and RBAC: Scalable, Flexible, and Auditable Access Management. *IT Professional*, *15*(3), 14–16. doi:10.1109/MITP.2013.37

Cranor, L. F. (2003). P3P: Making privacy policies more useful. *IEEE Security and Privacy*, *1*(6), 50–55. doi:10.1109/MSECP.2003.1253568

Crittercism. (n.d.). Retrieved December 10, 2015, from http://www.crittercism.com

Cryderman, J. (2011). *Anywhere Computing: How Mobile Apps Are Changing the World*. Pipeline Publishing. Retrieved from http://www.pipelinepub.com/0111/Anywhere-Computing-Mobile-Apps1.html

Cuadrado, F., & Dueñas, J. C. (2012). Mobile application stores: Success factors, existing approaches, and future developments. *Communications Magazine, IEEE*, *50*(11), 160–167. doi:10.1109/MCOM.2012.6353696

Damianou, N., Dulay, N., Lupu, E., & Sloman, M. (2001). The ponder policy specification language. In *Policies for Distributed Systems and Networks* (pp. 18–38). Springer Berlin Heidelberg. doi:10.1007/3-540-44569-2_2

Das, T., Mohan, P., Padmanabhan, V. N., Ramjee, R., & Sharma, A. (2010). Prism: platform for remote sensing using smart phones. In *Proceedings of the 8th International Conference on Mobile Systems, Applications, and Services*. doi:10.1145/1814433.1814442

de Montjoye, Y. A., Shmueli, E., Wang, S. S., & Pentland, A. S. (2014). openpds: Protecting the privacy of metadata through safeanswers. *PLoS ONE*, *9*(7), e98790. doi:10.1371/journal.pone.0098790 PMID:25007320

Dean, J., Grove, D., & Chambers, C. (1995). Optimization of Object-oriented Programs Using Static Class Hierarchy Analysis. In *Proceedings of the 9th European Conference on Object-Oriented Programming, ECOOP '95*. doi:10.1007/3-540-49538-X_5

Denning, D. E. (1976). A Lattice Model of Secure Information Flow. *Communications of the ACM*, *19*(5), 236–243. doi:10.1145/360051.360056

Denning, D. E., & Denning, P. J. (1977). Certification of Programs for Secure Information Flow. *Communications of the ACM*, *20*(7), 504–513. doi:10.1145/359636.359712

Deo, M. G. (2013). Doctor population ratio for india-the reality. *The Indian Journal of Medical Research*, *137*(4), 632. PMID:23703329

Deutsch, A. (1992). *A Storeless Model of Aliasing and Its Abstractions Using Finite Representations of Right regular Equivalence Relations*. ICCL. doi:10.1109/ICCL.1992.185463

Developer Guidelines. (n.d.). Retrieved May 31, 2016, from http://www-03.ibm.com/able/guidelines/index.html

Developers. (n.d.a). *Saving Data*. Retrieved from: http://developer.android.com/training/basics/data-storage

Developers. (n.d.b). *Creating and Monitoring Geofences*. Retrieved from: http://developer.android.com/training/location/geofencing.html

Dey, A. K., & Abowd, G. D. (1999). Towards a better understanding of context and context-awareness. In *First int. symposium on handheld and ubiquitous computing (HUC)*.

Dierks, T., & Rescorla, E. (2008). *The Transport Layer Security (TLS) Protocol Version 1.2*. RFC 5246.

Duckham, M., & Kulik, L. (2005). A formal model of obfuscation and negotiation for location privacy. In *Pervasive Computing* (pp. 152–170). Springer Berlin Heidelberg. doi:10.1007/11428572_10

Dupont, C., & Lookingbill, A. (2012). *System and Method for Managing Indoor Geolocation Conversions*. US Patent Office, Patent # US9147203 B1, USPTO.

Dwivedi, H. (2010). *Mobile application security*. Tata McGraw-Hill Education.

Dwork, C., & Aaron R. (2014) The algorithmic foundations of differential privacy. *Foundations and Trends in Theoretical Computer Science, 9*(3-4).

Eagle, N., & Pentland, A. (2006). Reality mining: Sensing complex social systems. *Personal and Ubiquitous Computing*, *10*(4), 255–268. doi:10.1007/s00779-005-0046-3

Elkhodr, M., Shahrestani, S., & Cheung, H. (2011). *Enhancing the security of mobile health monitoring systems through trust negotiations. In Local Computer Networks (LCN), 2011 IEEE 36th Converence on* (pp. 754–757). Bonn: IEEE.

eMarketer. (2015, January 9). *Tablet Users to Surpass 1 Billion Worldwide in2015*. Retrieved from http://www.emarketer.com/Article/Tablet-Users-Surpass-1-Billion-Worldwide-2015/1011806

Enck, W., Gilbert, P., Chun, B.-G., Cox, Jung, J., & Sheth, A. N. (2010). Taintdroid: an information- flow tracking system for real-time privacy monitoring on smartphones. In *Proceedings of the 9th usenix conference on operating systems design and implementation* (pp. 1–6).

Enck, W., Gilbert, P., Chun, B.-G., Cox, L. P., Jung, J., McDaniel, P., & Sheth, A. N. (2010). TaintDroid: An Information-Flow Tracking System for Realtime Privacy Monitoring on Smartphones. *Osdi*, *10*(49), 1–6.

Enck, W., Gilbert, P., Han, S., Tendulkar, V., Chun, B.-G., Cox, L. P., & Jung, J. et al. (2014). TaintDroid. *ACM Transactions on Computer Systems*, *32*(2), 1–29.

Epic. (1979). *Epic EHR*. Retrieved from http://www.epic.com/about-index.php

Eriksson, J., Girod, L., Hull, B., Newton, R., Madden, A., & Balakrishnan, H. (2008). The pothole patrol: using a mobile sensor network for road surface monitoring. In *Proceedings of the 6th International Conference on Mobile Systems, Applications, and Services*. Doi:10.1145/1378600.1378605

Ester, M., Kriegel, H. P., Sander, J., & Xu, X. (1996). A density-based algorithm for discovering clusters in large spatial databases with noise. In KDD (Vol. 96, No. 34, pp. 226-231).

Estrin, D. (2010). Participatory sensing: Applications and architecture. *IEEE Internet Computing*, *14*(1), 12–42. doi:10.1109/MIC.2010.12

Eugster, P. T., Garbinato, B., & Holzer, A. (2005). Location-based publish/subscribe: Network Computing and Applications.*4th IEEE International Symposium*. doi:10.1109/NCA.2005.29

Farrell, S., & Housley, R. (2002, April). *An Internet Attribute Certificate Profile for Authorization*. Retrieved from The Internet Engineering Task Force (IETF®): https://www.ietf.org/rfc/rfc3281.txt

FDA. (2015, February 9). *FDA Guidelines for Mobile Medical Applications*. Retrieved from http://www.fda.gov/downloads/MedicalDevices/.../UCM263366.pdf

Fehnker, Huuck, Seefried, & Tapp. (2010). Fadetogrey: Tuning Static Program Analysis. *ENTCS*, 266.

Fernández-Alemán, J., Señor, I., Lozoya, P., & Toval, A. (2013). Security and privacy in electronic health records: A systematic literature review. *Journal of Biomedical Informatics*, *46*(3), 541–562. doi:10.1016/j.jbi.2012.12.003 PMID:23305810

Ferraiolo, D. F., & Kuhn, D. R. (2009). *Role-based access controls*. arXiv preprint arXiv:0903.2171

Ferraiolo, D. F., Barkley, J. F., & Kuhn, D. R. (1999). A role-based access control model and reference implementation within a corporate intranet. *ACM Transactions on Information and System Security*, *2*(1), 34–64. doi:10.1145/300830.300834

Ferraiolo, D. F., Sandhu, R., Gavrila, S., Kuhn, D. R., & Chandramou, R. (2001). Proposed NIST standard for role-based access control. *ACM Transactions on Information and System Security*, *4*(3), 224–274. doi:10.1145/501978.501980

Ferraiolo, D., Cugini, J., & Kuhn, D. R. (1995, December). Role-based access control (RBAC): Features and motivations. In *Proceedings of 11th annual computer security application conference* (pp. 241-48).

Ferraiolo, D., & Kuhn, R. (1992). Role-Based Access Control. In *Proceedings of the NIST-NSA National (USA) Computer Security Conference*.

Ferreira, A., Chadwick, D., & Antunes, L. (2007). Modelling access control for healthcare information systems. In *Proceedings of the Doctoral Consortium at the 9th International Conference on Enterprise Information Systems (ICEIS)*.

Flinn, J. (2012). Cyber Foraging: Bridging Mobile and Cloud Computing. *Synthesis Lectures on Mobile and Pervasive Computing*, *7*(2), 1–103. doi:10.2200/S00447ED1V01Y201209MPC010

Fluid UI. (n.d.). Retrieved December 10, 2015, from https://www.fluidui.com

Flurry. (n.d.). Retrieved April 18, 2016, from http://www.flurry.com

Fonseca, O. (2012, November). *Byod leads to data breaches in the workplace*. Retrieved from https://github.com/lencinhaus/androjena

Foursquare. (n.d.). Retrieved from: http://foursquare.com

Fu, Q., Lou, J. G., Wang, Y., & Li, J. (2009, December). Execution anomaly detection in distributed systems through unstructured log analysis. In *Data Mining, 2009. ICDM'09. Ninth IEEE International Conference on* (pp. 149-158). doi:10.1109/ICDM.2009.60

Fu, B., Lin, J., Li, L., Faloutsos, C., Hong, J., & Sadeh, N. (2013, August). Why people hate your app: Making sense of user feedback in a mobile app store. In *Proceedings of the 19th ACM SIGKDD international conference on Knowledge discovery and data mining* (pp. 1276-1284). ACM. doi:10.1145/2487575.2488202

Fuhrer, R., Tip, F., Kieżun, A., Dolby, J., & Keller, M. (2005). *Efficiently Refactoring Java Applications to Use Generic Libraries*. ECOOP. doi:10.1007/11531142_4

Furao, S., Ogura, T., & Hasegawa, O. (2007). An enhanced self-organizing incremental neural network for online unsupervised learning. *Neural Networks*, *20*(8), 893–903. doi:10.1016/j.neunet.2007.07.008 PMID:17826947

GAA-API. (n.d.). *Generic Authorization and Access-control API (GAA-API)*. Retrieved from http://gost.isi.edu/info/gaaapi/

Gafni, R. (2009). Usability issues in mobile-wireless information systems. *Issues in Informing Science and Information Technology*, *6*, 755–769.

Gajanayake, R., Iannella, R., & Sahama, T. (2014). Privacy oriented access control for electronic health records. *Electronic Journal of Health Informatics*, *8*(2), e15.

Ganti, R. K., Ye, F., & Lei, H. (2011). Mobile crowd sensing: Current state and future challenges. *Communications Magazine, IEEE*, *49*(11), 32–39. doi:10.1109/MCOM.2011.6069707

Garg, S., Singh, P., Ramanathan, P., & Sen, R. (2014). VividhaVahana: smartphone based vehicle classification and its applications in developing region.*Proceedings of the 11th International Conference on Mobile and Ubiquitous Systems: Computing, Networking and Services*. doi:10.4108/icst.mobiquitous.2014.257982

Gartner. (2012). *Mobile Device Management (MDM)*. Retrieved from http://www.gartner.com/it-glossary/mobile-device-management-mdm

Gartner. (2015). *Gartner Says Global Devices Shipments to Grow 2.8 Percent in 2015*. Retrieved from http://www.gartner.com/newsroom/id/3010017

Gartner. (2015, March 19). *Gartner Says Global Devices Shipments to Grow 2.8 Percent in2015*. Retrieved from http://www.gartner.com/newsroom/id/3010017

Ghosh, D. (2012). *Context based privacy and security in smartphones*. (Unpublished master's thesis). University of Maryland, Baltimore County.

Ghosh, D., Joshi, A., Finin, T., & Jagtap, P. (2012). Privacy control in smart phones using semantically rich reasoning and context modeling. In Security and privacy workshops (spw), 2012 IEEE symposium on (pp. 82-85). doi:10.1109/SPW.2012.27

Giannotti, F., Nanni, M., Pinelli, F., & Pedreschi, D. (2007). Trajectory pattern mining. In *Proceedings of the 13th ACM SIGKDD international conference on Knowledge discovery and data mining* (pp. 330-339). ACM. doi:10.1145/1281192.1281230

Gibler, C., Crussell, J., Erickson, J., & Chen, H. (2012). AndroidLeaks: Automatically detecting potential privacy leaks in Android applications on a large scale. Lecture Notes in Computer Science, 7344, 291-307.

Godik, S., Anderson, A., Parducci, B., Humenn, P., & Vajjhala, S. (2002). *OASIS eXtensible access control 2 markup language (XACML) 3. Tech. rep*. OASIS.

Goguen, J. A., & Meseguer, J. (1982). Security Policies and Security Models.S&P. doi:10.1109/SP.1982.10014

Gomez, L., Neamtiu, I., Azim, T., & Millstein, T. (2013, May). Reran: Timing-and touch-sensitive record and replay for android. In *Software Engineering (ICSE), 2013 35th International Conference on* (pp. 72-81). IEEE.

Google Analytics. (n.d.). Retrieved December 10, 2015, from https://www.google.com/analytics

Google Inc. (2016). *Lint*. Retrieved from http://developer.android.com/tools/help/lint.html

Google Inc. (2016a). *Google TalkBack*. Retrieved from https://play.google.com/store/apps/details?id=com.google.android.marvin.talkback&hl=en

Google Inc. (2016b). *Accessibility Scanner*. Retrieved from https://play.google.com/store/apps/details?id=com.google.android.apps.accessibility.auditor&hl=en

Google Location Services for Android. Making Your App Location-Aware. (2015). Retrieved January 28, 2016 from http://developer.android.com/training/location/index.html

Google Mobile Analytics. (n.d.). Retrieved December 10, 2015, from https://www.google.com/analytics/mobile

Google Now. (n.d.). *Landing Now*. Retrieved from: https://www.google.com/landing/now

Google Play. (2013). *Fitness Tracker*. Retrieved from https://play.google.com/store/apps/details?id=com.realitinc.fitnesstracker

Google Play. (n.d.). *Maps*. Retrieved from: http://play.google.com/store/apps/details?id=com.google.android.apps.maps

Google Translate. (n.d.). Retrieved from: http://play.google.com/store/apps/details?id=com.google.android.apps.translate

Guarnieri, S., Pistoia, M., Tripp, O., Dolby, J., Teilhet, S., & Berg, R. (2011). *Saving the World Wide Web from Vulnerable JavaScript*. ISSTA. doi:10.1145/2001420.2001442

Guha, A., Krishnamurthi, S., & Jim, T. (2009). *Using Static Analysis for Ajax Intrusion Detection*. WWW. doi:10.1145/1526709.1526785

Guo, B., Yu, Z., Zhou, X., & Zhang, D. (2014). From participatory sensing to Mobile Crowd Sensing. *2014 IEEE International Conference on Pervasive Computing and Communication Workshops, PERCOM WORKSHOPS 2014* (pp. 593-598).

Gupta, A., Thapar, J., Singh, A., Singh, P., Srinivasan, V., & Vardhan, V. (2013). Simplifying and Improving Mobile Based Data Collection. In *Proceedings of the Sixth International Conference on Information and Communications Technologies and Development Notes*. ACM Press. doi:10.1145/2517899.2517929

Gupta, A., Thapar, J., Singh, A., Singh, P., Srinivasan, V., & Vardhan, V. (n.d.). Simplifying and improving mobile based data collection. In *Proceedings of the 6th International Conference on Information and Communications Technologies and Development*.

Gu, T., Wang, X. H., Pung, H. K., & Zhang, D. Q. (2004, January). An ontology-based context model in intelligent environments. In *Proceedings of communication networks and distributed systems modeling and simulation conference* (pp. 270-275).

Haberle, C. (2014, August 27). Changing Healthcare Through Mobile Technology. *phx a cost management company*. Retrieved from http://www.phx-online.com/ecudednews/changing-healthcare-through-mobile-technology/

Hachem, S., Pathak, A., & Issarny, V. (2013). Probabilistic registration for large-scale mobile participatory sensing. In *Pervasive Computing and Communications (PerCom), IEEE International Conference*, (pp. 132-140).

Haderer, N., Rouvoy, R., & Seinturier, L. (2013). A preliminary investigation of user incentives to leverage crowd sensing activities. *Pervasive Computing and Communications Workshops (PERCOM Workshops),2013 IEEE International Conference*.

Hadoop. (n.d.). Retrieved December 10, 2015, from https://hadoop.apache.org

Hammer, C., Krinke, J., & Snelting, G. (2006). Information Flow Control for Java Based on Path Conditions in Dependence Graphs.S&P.

Handheld Tele-ECG Instrument for Rural Health Care. (2014). Retrieved from http://www.barc.gov.in/technologies/ecg/ecg br.html

Hardt, D. (Ed.). (2012). *The OAuth 2.0 Authorization Framework*. Internet Engineering Task Force (IETF). Retrieved from https://tools.ietf.org/html/rfc6749

Harrison, R., Flood, D., & Duce, D. (2013). Usability of mobile applications: Literature review and rationale for a new usability model. *Journal of Interaction Science*, *1*(1), 1–16. doi:10.1186/2194-0827-1-1

Hartung, C., Lerer, A., Anokwa, Y., Tseng, C., Brunette, W., & Borriello, G. (2010). Open data kit: Tools to build information services for developing regions. In *Proceedings of the 4th ACM/IEEE International Conference on Information and Communication Technologies and Development*. doi:10.1145/2369220.2369236

Hastie, T., Tibshirani, R., & Friedman, J. (2009). *The Elements of Statistical Learning* (Vol. 2). Springer. doi:10.1007/978-0-387-84858-7

Health Information Privacy. (2002). *Minimum Necessary Requirement*. Retrieved from http://www.hhs.gov/hipaa/for-professionals/privacy/guidance/minimum-necessary-requirement/index.html

Health Level Seven International. (2010). *Introduction to HL7 Standards*. Retrieved from http://www.hl7.org/implement/standards/

Healthcare, G. E. (2002). *GE Healthcare Centricity EMR*. Retrieved from http://www3.gehealthcare.com/en/products/categories/healthcare_it/electronic_medical_records/centricity_emr

Heckerman, D., Geiger, D., & Chickering, D. M. (1995). Learning Bayesian Networks: The Combination of Knowledge and Statistical Data. *Machine Learning*, *20*(3), 197–243. doi:10.1007/BF00994016

Heckle, R., Lutters, W., & Gurzick, D. (2008). Network Authentication Using Single Sign-on: The Challenge of Aligning Mental Models.*Proceedings of the 2nd ACM Symposium on Computer Human Interaction for Management of Information Technology* (pp. 6:1-6:10). San Diego, CA: ACM. doi:10.1145/1477973.1477982

Heintze, N., & Tardieu, O. (2001). *Demand-Driven Pointer Analysis*. PLDI. doi:10.1145/378795.378802

Herzberg, A. (2003, May). Payments and Banking with Mobile Personal Devices. *Communications of the ACM*, *46*(5), 53–58. doi:10.1145/769800.769801

Herzig, K., Just, S., & Zeller, A. (2013, May). It's not a bug, it's a feature: how misclassification impacts bug prediction. In *Proceedings of the 2013 International Conference on Software Engineering* (pp. 392-401). IEEE Press. doi:10.1109/ICSE.2013.6606585

HHS. (1996). *Health Insurance Portability and Accountability Act of 1996*. U.S. Department of health & Human Services. Retrieved from http://www.hhs.gov/hipaa/index.html

HHS.gov. (2013). *Health Information Privacy*. Retrieved from http://www.hhs.gov/hipaa/index.html

Himiss. (2014). *How mHealth is Changing Health and Healthcare*. Retrieved from http://www.himss.org/ResourceLibrary/mHimssRoadmapLanding.aspx?ItemNumber=30562

Himss. (2014, June 17). How #mHealth is Changing Health and Healthcare. *Himss transforming health through IT*. Retrieved from http://www.himss.org/how-mhealth-changing-health-and-healthcare

HL7 Security Technical Committee. (2007, September). *HL7 Role-Based Access Control (RBAC) Role Engineering Process*. Retrieved from http://csrc.nist.gov/groups/SNS/rbac/documents/hl7_role-based_access_control_(rbac).pdf

Housley, R., Polk, W., Ford, W., & Solo, D. (2002, April). *Internet X.509 Public Key Infrastructure Certificate and CRL Profile*. Retrieved from The Internet Engineering Task Force: http://www.ietf.org/rfc/rfc3280.txt

Hsu, W., & Pan, J. (2013). The Secure Authorization Model for Healthcare Information System. *Journal of Medical Systems*, *37*(5), 1–5. doi:10.1007/s10916-013-9974-z PMID:24061706

Huang, C. M., Lan, K. C., & Tsai, C. Z. (2008). A survey of opportunistic networks. *Proceedings - International Conference on Advanced Information Networking and Applications, AINA* (pp. 1672-1677). doi:10.1109/WAINA.2008.292

Huang, K. L., Kanhere, S. S., & Hu, W. (2010). Are You Contributing Trustworthy Data? The Case for a Reputation System in Participatory Sensing. In ACM Modeling, Analysis and Simulation of Wireless and Mobile Systems (MSWiM).

Huang, J., Xu, Q., Tiwana, B., Mao, Z. M., Zhang, M., & Bahl, P. (2010, June). Anatomizing application performance differences on smartphones. In *Proceedings of the 8th international conference on Mobile systems, applications, and services* (pp. 165-178). ACM.

Hubert, R. (2006). Accessibility and usability guidelines for mobile devices in home health monitoring. *ACM Sigaccess Accessibility and Computing*, (84), 26-29.

Hu, C., & Neamtiu, I. (2011, May). Automating GUI testing for Android applications. In *Proceedings of the 6th International Workshop on Automation of Software Test* (pp. 77-83). ACM. doi:10.1145/1982595.1982612

Hussain, A., & Kutar, M. (2009). *Usability metric framework for mobile phone application*. PostGraduate Symposium on The Convergence of Telecommunications, Networking and Broadcasting.

Hussain, A., & Ferneley, E. (2008, November). Usability metric for mobile application: a goal question metric (GQM) approach. In *Proceedings of the 10th International Conference on Information Integration and Web-based Applications & Services* (pp. 567-570). ACM. doi:10.1145/1497308.1497412

Hwang, S., Cho, S., & Park, S. (2009). Keystroke dynamics-based authentication for mobile devices. *Computers & Security*, *28*(1-2), 85–93. doi:10.1016/j.cose.2008.10.002

iBeacon for Developers, Apple Inc. (2016). Retrieved January 28, 2016 from https://developer.apple.com/ibeacon/

IBM Corp. (2015a, June 24). *IBM's commitment to people with disabilities*. Retrieved from http://www.ibm.com/able/product_accessibility/ibmcommitment.html

IBM Corp. (2015b). *Automated Accessibility Tester*. Retrieved from http://ibm.biz/bluemix-aat

IBM Corp. (2015c). *Digital Content Checker*. Retrieved from http://ibm.biz/bluemix-dcc

IBM MobileFirst. (n.d.). Retrieved December 10, 2015, from http://www.ibm.com/mobilefirst

IBM TeaLeaf. (n.d.). Retrieved December 10, 2015, from http:// www.ibm.com/software/info/tealeaf

IChangeMyCity. (n.d.). Retrieved from the Janaagraha Wiki: http://www.ichangemycity.com/ichangemystreet

IEEE Standard 802.15-2005. (2011). *Part 15.1: Wireless Medium Access Control (MAC) and Physical Layer (PHY) Specifications for Wireless Personal Area Networks (WPANs)*. Author.

IEEE Standard for Information Technology. (2012). *Telecommunications and information exchange between systems Local and metropolitan area networks*. IEEE Standard 802.11.

IHS. (n.d.). *Health Information Exchange and Master Patient Index.* Retrieved from https://www.ihs.gov/hie/index.cfm?module=dsp_hie_mpi

Informatica. (n.d.). Retrieved December 10, 2015, from https://www.informatica.com

Instruments User Guide. (n.d.). *About Instruments.* Retrieved May 31, 2016, from https://developer.apple.com/library/prerelease/mac/documentation/DeveloperTools/Conceptual/InstrumentsUserGuide/index.html

InVision. (n.d.). Retrieved December 10, 2015, from http://www.invisionapp.com

iOS 9. (2014). *Health: An innovative new way to use your health and fitness information.* Retrieved from https://www.apple.com/ios/health/

iTunes. (n.d.). *iTunes App Store Medical Apps.* Retrieved from https://itunes.apple.com/us/genre/ios-medical/id6020?mt=8

Jagtap, P., Joshi, A., Finin, T., & Zavala, L. (2011a, Sept). Preserving privacy in context-aware systems. In *Semantic computing (ICSC), 2011 fifth IEEE international conference on* (p. 149-153). doi:10.1109/ICSC.2011.87

Jagtap, P., Joshi, A., Finin, T., & Zavala, L. (2011b). Privacy preservation in context aware geosocial networking applications. *Organization.*

Jaimes, L., Vergara-Laurens, I., & Labrador, M. (2012). A location-based incentive mechanism for participatory sensing systems with budget constraints. In IEEE PerCom. doi:10.1109/PerCom.2012.6199855

Jain, S., Narayanaswamy, B., & Narahari, Y. (2014). *A multiarmed bandit incentive mechanism for crowdsourcing demand response in smart grids.* AAAI.

Jain, V., Sharma, A., & Subramanian, L. (2012). Road traffic congestion in the developing world. In *Proceedings of the 2nd ACM Symposium on Computing for Development.* doi:10.1145/2160601.2160616

Jamieson, K., Malloy, M., Nowak, R., & Bubeck, S. (2013). *UCB: An optimal exploration algorithm for multi-armed bandits.* arXiv preprint arXiv:1312.7308

Jaramillo, D., Smart, R., Furht, B., & Agarwal, A. (2013, April). A secure extensible container for hybrid mobile applications. In Southeastcon, 2013 Proceedings of IEEE (pp. 1-5). doi:10.1109/SECON.2013.6567439

Jar-The Java Archive Tool. (n.d.). Retrieved May 31, 2016, from http://docs.oracle.com/javase/7/docs/technotes/tools/windows/jar.html

Johnson, B., Song, Y., Murphy-Hill, E., & Bowdidge, R. (2013). Why Don't Software Developers Use Static Analysis Tools to Find Bugs? ICSE.

Josang, A., & Ismail, R. (2002). The Beta Reputation System. In *15th Bled Electronic Commerce Conference.*

Joseph, S. R. H., & Uther, M. (2009). Mobile devices for language learning: Multimedia approaches. *Research and Practice in Technology Enhanced Learning, 4*(1), 1–26. doi:10.1142/S179320680900060X

Junker, M., Huuck, R., Fehnker, A., & Knapp, A. (2012). *SMT-based False Positive Elimination in Static Program Analysis.* ICFEM. doi:10.1007/978-3-642-34281-3_23

Kaasila, J., Ferreira, D., Kostakos, V., & Ojala, T. (2012, December). Testdroid: automated remote UI testing on Android. In *Proceedings of the 11th International Conference on Mobile and Ubiquitous Multimedia* (p. 28). ACM. doi:10.1145/2406367.2406402

Kagal, L., & Abelson, H. (2010). Access control is an inadequate framework for privacy protection. In *W3C Privacy Workshop.*

Kagal, L., & Berners-Lee, T. (2005). *Rein: Where policies meet rules in the semantic web.* Cambridge, MA: Computer Science and Artificial Intelligence Laboratory, Massachusetts Institute of Technology.

Kagal, L., Finin, T., & Joshi, A. (2003, October). A policy based approach to security for the semantic web. In *International Semantic Web Conference* (Vol. 2870, pp. 402-418). doi:10.1007/978-3-540-39718-2_26

Karamshuk, D., Boldrini, C., Conti, M., & Passarella, A. (2011). Human mobility models for opportunistic networks. *IEEE Communications Magazine*, *49*(12), 157–165. doi:10.1109/MCOM.2011.6094021

Karnin, Z., Koren, T., & Somekh, O. (2013). *Almost optimal exploration in multi-armed bandits*. ICML.

Kelly, S. M. (2014, June 27). *In Google Fit vs. Apple HealthKit, Fitness Apps Stay Neutral.* Retrieved from http://mashable.com/2014/06/27/healthkit-google-fit-apps/#nf_r0U7flmqC

Khonji, M., Iraqi, Y., & Jones, A. (2013). Phishing detection: A literature survey. *IEEE Communications Surveys and Tutorials*, *15*(4), 2091–2121. doi:10.1109/SURV.2013.032213.00009

Kiehn, A., Raj, P., & Singh, P. (2014). A Causal Checkpointing Algorithm for Mobile Computing Environments. *LNCS*, *8314*, 134–148.

Kim, S., Park, S., & Jeong, Y. (2013, December). RFID-based indoor location tracking to ensure the safety of the elderly in smart home environments. *Personal and Ubiquitous Computing*, *17*(8), 1699–1707. doi:10.1007/s00779-012-0604-4

Kirkpatrick, M. S., Damiani, M. L., & Bertino, E. (2011, November). Prox-RBAC: a proximity-based spatially aware RBAC. In *Proceedings of the 19th ACM SIGSPATIAL International Conference on Advances in Geographic Information Systems* (pp. 339-348). ACM.

Kiukkonen, N., Blom, J., Dousse, O., Gatica-Perez, D., & Laurila, J. (2010). Towards rich mobile phone datasets: Lausanne data collection campaign.*Proc. ICPS.*

Kotz, D., Fu, K., Gunter, C., & Rubin, A. (2015). Privacy and Security for Mobile and Cloud Frontiers in Healthcare. *Communications of the ACM*, *58*(8), 21–23. doi:10.1145/2790830

Koutsopoulos, I. (2013). *Optimal incentive-driven design of participatory sensing systems.* INFOCOM. doi:10.1109/INFCOM.2013.6566934

Krause, A., Horvitz, E., Kansal, A., & Zhao, F. (2008). Toward Community Sensing. In *Proceedings of the 7th International Conference on Information processing in sensor networks*, (pp. 481-492).

Krontiris, I., & Albers, A. (2012). Monetary incentives in participatory sensing using multi-attributive auctions. *International Journal of Parallel Emerg. Distrib. Syst.*, *27*(4), 317–336. doi:10.1080/17445760.2012.686170

Kuhn, D. R., Coyne, E. J., & Weil, T. R. (2010). Adding attributes to role-based access control. *IEEE Computer*, *43*(6), 79–81. doi:10.1109/MC.2010.155

Lane, N. D., Chon, Y., Zhou, L., Zhang, Y., Li, F., & Kim, D., … Cha, H. (2013). \Piggyback crowd sensing (pcs): energy efficient crowd sourcing of mobile sensor data by exploiting smartphone app opportunities. In *Proceedings of the 11th ACM Conference on Embedded Networked Sensor Systems.* doi:10.1145/2517351.2517372

Lane, N. D., Miluzzo, E., Lu, H., Peebles, D., Choudhury, T., & Campbell, A. T. (2010). A survey of mobile phone sensing. *Communications Magazine, IEEE*, *48*(9), 140–150. doi:10.1109/MCOM.2010.5560598

Lars Ole Andersen. (1994). *Program Analysis and Specialization for the C Programming Language.* (PhD thesis). University of Copenhagen, Copenhagen, Denmark.

Laurila, J. K., Gatica-Perez, D., Aad, I., Bornet, O., Do, T. M. T., Dousse, O., . . . Miettinen, M. (2012). The mobile data challenge: Big data for mobile computing research. In Pervasive Computing (No. EPFL-CONF-192489).

Lee Ventola, C. (2014, May). *Mobile Devices and Apps for Health Care Professionals: Uses and Benefits.* PMC US National Library of Medicine National Institutes of Health. Retrieved from http://www.ncbi.nlm.nih.gov/pmc/articles/PMC4029126/

Lewis, G. A., Echeverría, S., Simanta, S., Bradshaw, B., & Root, J. (2014). Cloudlet-Based Cyber-Foraging for Mobile Systems in Resource-Constrained Edge Environments. *ICSE, 14*, 412–415.

Lewis, N. (2011). 80% Of Doctors Use Mobile Devices At Work. *Information Week, 21*(October). Retrieved from http://www.informationweek.com/mobile/80--of-doctors-use-mobile-devices-at-work/d/d-id/1100880

Le, X. H., Doll, T., Barbosu, M., Luque, A., & Wang, D. (2012). An enhancement of the role-based access control model to facilitate information access management in context of team collaboration and workflow. *Journal of Biomedical Informatics, 45*(6), 1084–1107. doi:10.1016/j.jbi.2012.06.001 PMID:22732236

Lhoták, O., & Hendren, L. J. (2006). Context-Sensitive Points-to Analysis: Is It Worth It? CC.

Liang, H., Song, H., Fu, Y., Cai, X., & Zhang, Z. (2011, June). A remote usability testing platform for mobile phones. In *Computer Science and Automation Engineering (CSAE), 2011 IEEE International Conference on* (Vol. 2, pp. 312-316). IEEE.

Liang, Q., Cheng, X., & Chen, D. (2011). Opportunistic Sensing in Wireless Sensor Networks: Theory and Application. *2011 IEEE Global Telecommunications Conference - GLOBECOM 2011, 63*(8), 1-5.

Liang, C. J. M., Lane, N. D., Brouwers, N., Zhang, L., Karlsson, B. F., & Liu, H. et al. (2014, September). Caiipa: automated large-scale mobile app testing through contextual fuzzing. In *Proceedings of the 20th annual international conference on Mobile computing and networking* (pp. 519-530). ACM. doi:10.1145/2639108.2639131

Liang, C. J. M., Lane, N., Brouwers, N., Zhang, L., Karlsson, B., Chandra, R., & Zhao, F. (2013). *Contextual fuzzing: automated mobile app testing under dynamic device and environment conditions*. Microsoft.

Liebowitz, M. (2011). *Developer sneaks fake apps into android market.* Retrieved from http://www.nbcnews.com/id/45641853/ns/technology_and_science-security/t/developer-sneaks-fake-apps-android-market/

Lin, J. (2013). *Understanding and capturing people's mobile app privacy preferences* (Unpublished doctoral dissertation). Carnegie Mellon University, Pittsburgh, PA, USA. (AAI3577905)

Lin, J., Liu, B., Sadeh, N., & Hong, J. I. (2014, July). Modeling users' mobile app privacy preferences: Restoring usability in a sea of permission settings. In Symposium on usable privacy and security (soups 2014) (pp. 199–212). Menlo Park, CA: USENIX Association. Retrieved from https://www.usenix.org/conference/soups2014/proceedings/presentation/lin

Lin, W. (2015). *Guidercare Makes Smart Watch Ideal Solution for Elderly Care and Activity Tracking*. SMA100. Retrieved January 28, 2016 from http://www.mysmahome.com/COMPANY/4459/guidercare-makes-smart-watch-ideal-solution-for-elderly-care-and-activity-tracking.aspx

Lindorfer, M., Volanis, S., Sisto, A., Neugschwandtner, M., Athanasopoulos, E., Maggi, F., & Ioannidis, S. (2014). Andradar: Fast discovery of android applications in alternative markets. In S. Dietrich (Ed.), Detection of intrusions and malware, and vulnerability assessment (Vol. 8550, pp. 51-71). Springer International Publishing. doi:4 doi:10.1007/978-3-319-08509-8

Lint. (n.d.). Retrieved May 31, 2016, from https://en.wikipedia.org/wiki/Lint_(software)

Liu, L. (2007). *From Data Privacy to Location Privacy: Models & Algorithms.* Retrieved March 6, 2016 from http://web.calstatela.edu/faculty/hpguo/Research/database/liu07.pdf

Livshits, V. B., & Lam, M. S. (2005). Finding Security Vulnerabilities in Java Applications with Static Analysis. USENIX Security.

Localytics. (n.d.). Retrieved April 18, 2016, from https://www.localytics.com

Location and Maps Programming Guide. Apple Inc. (2015). Retrieved January 28, 2016 from https://developer.apple.com/library/ios/documentation/UserExperience/Conceptual/LocationAwarenessPG/UsingGeocoders/UsingGeocoders.html#//apple_ref/doc/uid/TP40009497-CH4-SW1

Lodderstedt, T., Basin, D. A., & Doser, J. SecureUML: A UML-Based Modeling Language for Model-Driven Security. In *Proceeding UML '02 Proceedings of the 5th International Conference on The Unified Modeling Language* (pp. 426-441). Springer-Verlag. doi:10.1007/3-540-45800-X_33

Lorecarra. (2009). *Androjena: Jena android porting.* Retrieved from http://www.experian.com/blogs/data-breach/2012/05/02/medical-and-mobile-convenience-trumps-security/

Loreto, S., Mecklin, T., Opsenica, M., & Rissanen, H. M. (2009). Service broker architecture: Location business case and mashups. *Communications Magazine, IEEE, 47*(4), 97–103. doi:10.1109/MCOM.2009.4907414

Lu, X., Li, D., Xu, B., Chen, W., & Ding, Z. (2013). Minimum cost collaborative sensing network with mobile phones. In Communications (ICC), 2013 IEEE International Conference.

Lu, H., Pan, W., Lane, N. D., Choudhury, T., & Campbell, A. T. (2009). Soundsense: scalable sound sensing for people-centric applications on mobile phones. In *Proceedings of the 7th International Conference on Mobile Systems, Applications, and Services.* doi:10.1145/1555816.1555834

Lu, K. et al.. (2015). *Checking More and Alerting Less: Detecting Privacy Leakages via Enhanced Data-flow Analysis and Peer Voting.* NDSS; doi:10.1145/1899475.1899487

Machiry, A., Tahiliani, R., & Naik, M. (2013, August). Dynodroid: An input generation system for Android apps. In *Proceedings of the 2013 9th Joint Meeting on Foundations of Software Engineering* (pp. 224-234). ACM. doi:10.1145/2491411.2491450

Mallare, I. J. G., & Pancho-Festin, S. (2013, December). Combining Task-and Role-Based Access Control with Multi-Constraints for a Medical Workflow System. In *IT Convergence and Security (ICITCS), 2013 International Conference on* (pp. 1-4). IEEE. doi:10.1109/ICITCS.2013.6717814

Mantyjarvi, J., Lindholm, M., Vildjiounaite, E., Makela, S.-M., & Ailisto, H. (2005). Identifying users of portable devices from gait pattern with accelerometers. *Acoustics, Speech, and Signal Processing, 2005. Proceedings. (ICASSP '05). IEEE International Conference on.* IEEE. doi:10.1109/ICASSP.2005.1415569

Mare, S., Sorber, J., Shin, M., Cornelius, C., & Kotz, D. (2014). Hide-n-sense: Preserving privacy efficiently in wireless mhealth. *Mobile Networks and Applications, 19*(3), 331–344.

Mathur, S., Jin, T., Kasturirangan, N., Chandrasekaran, J., Xue, W., Gruteser, M., & Trappe, W. (2010). ParkNet: drive-by sensing of road-side parking statistics. In *Proceedings of the 8th international conference on Mobile systems, applications, and services (MobiSys '10).* doi:10.1145/1814433.1814448

Mauro, C., Sunyaev, A., Leimeister, J., Schweiger, A., & Krcmar, H. (2008). A Proposed Solution for Managing Doctor's Smart Cards in Hospitals Using a Single Sign-On Central Architecture.*Hawaii International Conference on System Sciences, Proceedings of the 41st Annual* (pp. 2565-266). Waikoloa, HI: IEEE. doi:10.1109/HICSS.2008.33

Mavridis, I., Georgiadis, C., Pangalos, G., & Khair, M. (2001, January-March). Access Control based on Attribute Certificates for Medical Intranet Applications. *Journal of Medical Internet Research*, *3*(1), e9. doi:10.2196/jmir.3.1.e9 PMID:11720951

McCamant, S., & Ernst, M. D. (2008). *Quantitative Information Flow as Network Flow Capacity*. PLDI. doi:10.1145/1375581.1375606

MedWatcher. (2012). *MedWatcher*. Retrieved from https://medwatcher.org/

Method Swizzling. (2014). Retrieved May 31, 2016, from http://nshipster.com/method-swizzling/

Microsoft. (2007). *Microsoft HealthVault*. Retrieved from https://www.healthvault.com/us/en

Minamide, Y. (2005). *Static Approximation of Dynamically Generated Web Pages*. WWW. doi:10.1145/1060745.1060809

Mirzaei, N., Malek, S., Păsăreanu, C. S., Esfahani, N., & Mahmood, R. (2012). Testing android apps through symbolic execution. *Software Engineering Notes*, *37*(6), 1–5. doi:10.1145/2382756.2382798

Mittal, R., Kansal, A., & Chandra, R. (2012, August). Empowering developers to estimate app energy consumption. In *Proceedings of the 18th annual international conference on Mobile computing and networking* (pp. 317-328). ACM. doi:10.1145/2348543.2348583

MixPanel. (n.d.). Retrieved April 18, 2016, from https://mixpanel.com

Mobile Analytics: Why You Should Care. (n.d.). Retrieved December 10, 2015, from http://asmarterplanet.com/mobile-enterprise/blog/2013/10/mobile-analytics-why-you-should-care.html

MobileIron Solutions EMM. (2015). *Enable business transformation MobileIron's Enterprise Mobility Management (EMM) Platform*. Retrieved from https://www.mobileiron.com/en/solutions/enterprise-mobile-management-emm

MobileIron Solutions MAM. (2015). *Do more with secure mobile application management (MAM)*. Retrieved from https://www.mobileiron.com/en/solutions/mobile-application-management-mam

MobileIron Solutions MCM. (2015). *Secure mobile content management (MCM) keeps data safe and business moving*. Retrieved from https://www.mobileiron.com/en/solutions/mobile-content-management-mcm

MobileIron Solutions MDM. (2015). *Mobile Device Management (MDM): The foundation for a secure mobile enterprise*. Retrieved from https://www.mobileiron.com/en/solutions/mobile-device-management-mdm

MobileIron. (2015). *MobileIron*. Retrieved from https://www.mobileiron.com/en

Mohan, P., Padmanabhan, V. N., & Ramjee, R. (2008). Nericell: rich monitoring of road and traffic conditions using mobile smartphones. In *Proceedings of the 6th ACM conference on Embedded network sensor systems (SenSys '08)*. doi:10.1145/1460412.1460444

Montoliu, R., & Gatica-Perez, D. (2010). Discovering human places of interest from multimodal mobile phone data. In *Proceedings of the 9th International Conference on Mobile and Ubiquitous Multimedia* (p. 12). ACM.

Montopoli, B. (2013). *For criminals, smartphones becoming prime targets*. Retrieved from http://www.cbsnews.com/news/for-criminals-smartphones-becoming-prime-targets/

Mossakowski, T., Drouineaud, M., & Sohr, K. (2003, July). A temporal-logic extension of role-based access control covering dynamic separation of duties. In *Temporal Representation and Reasoning, 2003 and Fourth International Conference on Temporal Logic. Proceedings. 10th International Symposium on* (pp. 83-90). IEEE. doi:10.1109/TIME.2003.1214883

Motta, G., & Furuie, S. (2003). A contextual role-based access control authorization model for electronic patient record. *Proceedings of the IEEE Transactions on Information Technology in Biomedicine*, *7*(3), 202–207. doi:10.1109/TITB.2003.816562 PMID:14518734

MTBC PHR. (2011). *MTBC PHR*. Retrieved from https://phr.mtbc.com/

Mun, M., Reddy, S., Shilton, K., & Yau, N. (2009). PEIR, the personal environmental impact report, as a platform for participatory sensing systems research. *MobiSys*, (pp. 55-68).

Mun, M., Reddy, S., Shilton, K., Yau, N., Burke, J., Estrin, D., Hansen, M.… & Boda, P. (2009). PEIR, the personal environmental impact report, as a platform for participatory sensing systems research. In *ACM MobiSys*.

Muske, T. B., Baid, A., & Sanas, T. (2013). *Review Efforts Reduction by Partitioning of Static Analysis Warnings*. SCAM. doi:10.1109/SCAM.2013.6648191

Muske, T. B., Datar, A., Khanzode, M., & Madhukar, K. (2013). *Efficient Elimination of False Positives Using Bounded Model Checking*. VALID.

Musolesi, M., Piraccini, M., Fodor, K., Corradi, A., & Campbell, A. T. (2010). Supporting energy-efficient uploading strategies for continuous sensing applications on mobile phones. In *Pervasive Computing* (pp. 355–372). Springer. doi:10.1007/978-3-642-12654-3_21

My Imaging Records App. (2013). *My Imaging Records App*. Retrieved from http://myimagingrecords.com/index.html

Myers, A. C. (1999). *JFlow: Practical Mostly-static Information Flow Control*. POPL. doi:10.1145/292540.292561

Myers, A. C., & Liskov, B. (1997). *A Decentralized Model for Information Flow Control*. SOSP. doi:10.1145/268998.266669

Nadler, S., Soroka, V., Fuchs, O., Korenshtein, R., & Sonsino, E. (2008). Presence Zones for Contextual Location Based Services. In Innovations in Clouds, Internet and Networks, 2008. ICIN.

Nandakumar, V., Ekambaram, V., & Sharma, V. (2013). Appstrument-A Unified App Instrumentation and Automated Playback Framework for Testing Mobile Applications. In Mobile and Ubiquitous Systems: Computing, Networking, and Services (pp. 474-486). Springer International Publishing.

Na, S., & Cheon, S. (2000). Role Delegation in Role-based Access Control.*Proceedings of the Fifth ACM Workshop on Role-based Access Control* (pp. 39-44). Berlin, Germany: ACM. doi:10.1145/344287.344300

Necula, G. C. (2002). *Proof-carrying code. Design and Implementation*. Springer Netherlands.

Network Emulators from iTrinegy. (n.d.). Retrieved May 31, 2016, from http://www.itrinegy.com/index.php/products/network-emulators

Network Virtualization. (n.d.). Retrieved May 31, 2016, from http://www8.hp.com/in/en/software-solutions/network-virtualization/

Neuman, B., & Ts'o, T. (1994, September). Kerberos: An Authentication. *IEEE Communications Magazine*, *32*(9), 33–38. doi:10.1109/35.312841

Nevill-Manning, C. G., Holmes, G., & Witten, I. H. (1995). The Development of Holte's 1R Classifier. In *Proceedings of the Second New Zealand International Two-Stream Conference on Artificial Neural Networks and Expert Systems*. IEEE. doi:10.1109/ANNES.1995.499480

Newsome, J., & Song, D. (2005). *Dynamic Taint Analysis for Automatic Detection, Analysis, and Signature Generation of Exploits on Commodity Software*. NDSS.

NEXTGEN Healthcare. (2009). *Nextgen EHR*. Retrieved from https://www.nextgen.com/Products-and-Services/Ambulatory/Electronic-Health-Records-EHR

Nexus. (n.d.). *Fingerprint security on Nexus devices*. Retrieved from https://support.google.com/nexus/answer/6300638?hl=en

Ng, A. Y., & Jordan, M. I. (2002). On Discriminative vs. Generative Classifiers: A Comparison of Logistic Regression and Naive Bayes. *Advances in Neural Information Processing Systems*, 2.

Nguyen, N., Kleinrock, L., & Reiher, P. (2012). Debugging Ubiquitous Computing Applications With the Interaction Analyzer. *International Journal on Advances in Software*, *5*(3 & 4), 2012.

NIST Computer Security Division. (2013). *Attribute-Based Access Control*. Retrieved from http://csrc.nist.gov/projects/abac/

NIST. (2015, January 15). *The NIST RBAC Standards*. NIST. Retrieved from http://csrc.nist.gov/groups/SNS/rbac/

OASIS. (2013, January 22). *eXtensible Access Control Markup Language (XACML)*. OASIS. Retrieved from http://docs.oasis-open.org/xacml/3.0/xacml-3.0-core-spec-os-en.html

OpenEMR. (2012). *Open Source EMR*. Retrieved from http://www.open-emr.org/

Ostrovsky, D., & Rodenski, Y. (2014). Couchbase Lite on Android. In *Pro Couchbase Server* (pp. 283–292). Apress.

Paek, J., Kim, J., & Govindan, R. (2010). Energy-efficient rate-adaptive GPS-based positioning for smartphones. In *Proceedings of the 8th international conference on Mobile systems, applications, and services* (pp. 299-314). ACM.

Paganini, P. (2015, June 17). *Mobile App Security: Threats and Best Practices*. Retrieved from https://www.veracode.com/blog/2015/05/mobile-app-security-threats-and-best-practices-sw

Park, J., & Sandhu, R. (2004). The UCON ABC usage control model. *ACM Transactions on Information and System Security*, *7*(1), 128–174. doi:10.1145/984334.984339

Parse. (n.d.). *Parse Server*. Retrieved from: http://parse.com

Pashalidis, A., & Mitchell, C. (2003). A Taxonomy of Single Sign-On Systems. *8th Australasian Conference, ACISP. 2727*. Wollongong, Australia: Springer-Verlag Berlin Heidelberg.

Pathak, A., Hu, Y. C., & Zhang, M. (2012, April). Where is the energy spent inside my app?: fine grained energy accounting on smartphones with eprof. In *Proceedings of the 7th ACM european conference on Computer Systems* (pp. 29-42). ACM. doi:10.1145/2168836.2168841

Patwardhan, A., Korolev, V., Kagal, L., & Joshi, A. (2004, August). Enforcing policies in pervasive environments. In *Mobile and Ubiquitous Systems: Networking and Services, 2004. MOBIQUITOUS 2004. The First Annual International Conference on* (pp. 299-308). IEEE. doi:10.1109/MOBIQ.2004.1331736

Peleg, M., Beimel, D., Dori, D., & Denekamp, Y. (2008). Situation-Based Access Control: Privacy management via modeling of patient data access scenarios. *Journal of Biomedical Informatics*, *41*(6), 1028–1040. doi:10.1016/j.jbi.2008.03.014 PMID:18511349

Pelusi, L., Passarella, A., & Conti, M. (2006). Opportunistic networking: Data forwarding in disconnected mobile ad hoc networks. *IEEE Communications Magazine*, *44*(11), 134–141. doi:10.1109/MCOM.2006.248176

Pepe, M. S. (2003). *The Statistical Evaluation of Medical Tests for Classification and Prediction*. Oxford University Press.

Pew Research Center. (2012, February 23). *Mobile Technology Fact Sheet*. Retrieved from http://www.pewinternet.org/fact-sheets/mobile-technology-fact-sheet/

Pham, H. N., Sim, B. S., & Youn, H. Y. (2011). A novel approach for selecting the participants to collect data in participatory sensing: in Applications and the Internet (SAINT).*11th International Symposium IEEE/IPSJ*. doi:10.1109/SAINT.2011.17

Pharmacy, C. V. S. (2015). *myCVS On the Go*. Retrieved from http://www.cvs.com/mobile-cvs

Pillai, J., Patel, V., Chellappa, R., & Ratha, N. (2010). Sectored Random Projections for Cancelable Iris Biometrics. *Acoustics Speech and Signal Processing (ICASSP), 2010 IEEE International Conference on* (pp. 1838-1841). Dallas, TX: IEEE. doi:10.1109/ICASSP.2010.5495383

Pistoia, M., Flynn, R. J., Koved, L., & Sreedhar, V. C. (2005). *Interprocedural Analysis for Privileged Code Placement and Tainted Variable Detection*. ECOOP. doi:10.1007/11531142_16

Pontil, M., & Verri, A. (1998). Properties of Support Vector Machines. *Neural Computation*, 10. PMID:9573414

Posada, M. (2014, June 23). *The Evolving Landscape of Medical Apps in Healthcare*. HIT Consultant. Retrieved from http://hitconsultant.net/2014/06/23/the-evolving-landscape-of-medical-apps-in-healthcare/

PrimeLife. (2011). Retrieved January 28, 2016 from http://primelife.ercim.eu/

Profiling with Traceview and dmtracedump. (n.d.). Retrieved May 31, 2016, from http://developer.android.com/tools/debugging/debugging-tracing.html

Progno C. I. S. (2010). *The PrognoCIS EHR*. Retrieved from http://prognocis.com/

Quinlan, J. R. (1993). *C4.5: Programs for Machine Learning* (Vol. 1). Morgan Kaufmann.

Radicati, S. (2014). *Mobile Statistics Report, 2014-2018*. Retrieved from http://www.radicati.com/wp/wp-content/uploads/2014/01/Mobile-Statistics-Report-2014-2018-Executive-Summary.pdf

Ra, M. R., Liu, B., La Porta, T. F., & Govindan, R. (2012). Medusa: A programming framework for crowd-sensing applications. In *Proceedings of the 10th International Conferenceon Mobile Systems, Applications, and Services, MobiSys '12*. doi:10.1145/2307636.2307668

Ramey, K. (2013, December 14). Mobile Technology – 7 Location Based Apps Changing The Mobile Industry. *Use of Technology*. Retrieved from http://www.useoftechnology.com/7-location-based-apps/

Ramos, J. (2003). Using tf-idf to determine word relevance in document queries. In *Proceedings of the first instructional conference on machine learning*.

Rana, R. K., Chou, C. T., Kanhere, S. S., Bulusu, N., & Hu, W. (2010). Ear-phone: an end-to-end Participatory urban noise mapping system. In *Proceedings of the 9th ACM/IEEE International Conference on Information Processing in Sensor Networks*. doi:10.1145/1791212.1791226

Ranvier, J. E., Catasta, M., Vasirani, M., & Aberer, K. (2015). RoutineSense: A Mobile Sensing Framework for the Reconstruction of User Routines. In *2th International Conference on Mobile and Ubiquitous Systems: Computing, Networking and Services* (No. EPFL-CONF-208793). doi:10.4108/eai.22-7-2015.2260055

Ratha, N., Connell, J., Bolle, R., & Chikkerur, S. (2006). Cancelable Biometrics: A Case Study in Fingerprints. *Pattern Recognition, 2006. ICPR 2006. 18th International Conference on, 4*. doi:10.1109/ICPR.2006.353

Ratha, N., Connell, J., & Bolle, R. (2001). Enhancing security and privacy in biometrics-based authentication systems. *IBM Systems Journal*, *40*(3), 614–634. doi:10.1147/sj.403.0614

Ravindranath, L., Padhye, J., Agarwal, S., Mahajan, R., Obermiller, I., & Shayandeh, S. (2012). AppInsight: mobile app performance monitoring in the wild. In *Presented as part of the 10th USENIX Symposium on Operating Systems Design and Implementation (OSDI 12)* (pp. 107-120).

Ravindranath, L., Nath, S., Padhye, J., & Balakrishnan, H. (2014, June). Automatic and scalable fault detection for mobile applications. In *Proceedings of the 12th annual international conference on Mobile systems, applications, and services* (pp. 190-203). ACM. doi:10.1145/2594368.2594377

Ray, I., & Toahchoodee, M. (2007). A spatio-temporal role-based access control model. In *Data and Applications Security XXI* (pp. 211–226). Springer Berlin Heidelberg. doi:10.1007/978-3-540-73538-0_16

Ray, I., & Toahchoodee, M. (2007). A spatio-temporal role-based access control model.*Proceedings of the 21st annual IFIP WG 11.3 working conference on Data and applications security* (pp. 211-226). Redondo Beach, CA: Springer-Verlag.

Realm. (n.d.). Retrieved from: http://realm.io

Reddy, S., Estrin, D., & Srivastava, M. (2010). Recruitment framework for participatory sensing data collections. *Pervasive Computing,* 138-155.

Reps, T., Horwitz, S., & Sagiv, M. (1995). *Precise Interprocedural Dataflow Analysis via Graph Reachability.* POPL. doi:10.1145/199448.199462

Rindfleisch, T. C. (1997). Privacy, information technology, and health care. *Communications of the ACM, 40*(8), 92–100. doi:10.1145/257874.257896

Russello, G., Dong, C., & Dulay, N. (2008). A Workflow-Based Access Control Framework for e-Health Applications.*22nd International Conference on Advanced Information Networking and Applications - Workshops (AINAW 2008).* doi:10.1109/WAINA.2008.131

Ryutov, T., Zhou, L., Neuman, C., Leithead, T., & Seamons, K. (2005). Adaptive Trust Negotiation and Access Control. *SACMAT '05 Proceedings of the tenth ACM symposium on Access control models and technologies* (pp. 139-146). New York: ACM.

Sabater, J., & Sierra, C. (2005, September). Review on computational trust and reputation models. *Artificial Intelligence Review, 24*(1), 33–60. doi:10.1007/s10462-004-0041-5

Sabelfeld, A., & Myers, A. C. (2003). Language-based Information-flow Security. *IEEE Journal on Selected Areas in Communications, 21*(1), 5–19. doi:10.1109/JSAC.2002.806121

Sadeh, N. M., Chan, T.-C., Van, L., Kwon, O. B., & Takizawa, K. (2003). A semantic web environment for context-aware m-commerce. In *Proceedings of the 4th ACM conference on electronic commerce* (pp. 268–269). New York, NY: ACM. doi:10.1145/779928.779992

Saitoh, S. (1988). *Theory of Reproducing Kernels and Its Applications.* Longman.

Sana Technology Platform. (2014). Retrieved from http://sana.mit.edu/platform/

Sandhu, R., Ferraiolo, D. F., & Kuhn, R. (2000). The NIST Model for Role Based Access Control: Toward a Unified Standard. In *Proceedings of the Fifth ACM Workshop on Role-based Access Control (RBAC '00).* doi:10.1145/344287.344301

Sandhu, R., & Samarati, P. (1996, March). Authentication, access control, and audit. *ACM Computing Surveys, 28*(1), 241–243. doi:10.1145/234313.234412

Santos-Pereira, C., Augusto, A. B., Correia, M. E., Ferreira, A., & Cruz-Correia, R. (2012). A Mobile Based Authorization Mechanism for Patient Managed Role Based Access Control. LNCS, 7451, 54-68.

SAS. (n.d.). Retrieved December 10, 2015, from https://www.sas.com

Satyanarayanan, M. (2010). Mobile Computing: the Next Decade. *Scenario, 15*, 1-6. Retrieved from http://dl.acm.org/citation.cfm?id=1810936

Satyanarayanan, M., Bahl, P., Cáceres, R., & Davies, N. (2009). The case for VM-based cloudlets in mobile computing. *IEEE Pervasive Computing / IEEE Computer Society [and] IEEE Communications Society*, *8*(4), 14–23. doi:10.1109/MPRV.2009.82

Satyanarayanan, M., Kistler, J. J., Kumar, P., Okasaki, M. E., Siegel, E. H., & Steere, D. C. (1990). Coda: A Highly Available File System for a Distributed Workstation Environment. *IEEE Transactions on Computers*, *39*(4), 447–459. doi:10.1109/12.54838

Savitz, E. (2012, June 4). 5 Ways Mobile Apps Will transform Healthcare. *Forbes*. Retrieved from http://www.forbes.com/sites/ciocentral/2012/06/04/5-ways-mobile-apps-will-transform-healthcare/#7490182d6509

Schefer-Wenzl, S., & Strembeck, M. (2013). Modelling Context-Aware RBAC Models for Mobile Business Processes. *International Journal of Wireless and Mobile Computing*, *6*(5), 448. doi:10.1504/IJWMC.2013.057387

Scholkopf, B., Smola, A. J., Williamson, R. C., & Bartlett, P. L. (2000). New Support Vector Algorithms. *Neural Computation*, 12.

Scholl, M. A., Stine, K. M., Hash, J., Bowen, P., Johnson, L. A., Smith, C. D., & Steinberg, D. I. (2008). *An Introductory Resource Guide for Implementing the Health Insurance Portability and Accountability Act (HIPAA) Security Rule*. Gaithersburg, MD: National Institute of Standards & Technology. doi:10.6028/NIST.SP.800-66r1

Schulz, T., Gladhorn, F., & Sæther, J. A. (2015). *Best Practices for Creating Accessible Mobile Applications*. Report Norsk Regnesentral–Norwegian Computing Center.

Schwartz, A. (2015). Moovit Crowdsources Public Transit Data, So You'll Never Get Stuck Waiting For The Bus Again. *Fast Company*. Retrieved from http://www.fastcoexist.com/3041915/moovit-crowdsources-public-transit-data-so-youll-never-get-stuck-waiting-for-the-bus-again

Science Application International Corporation (SAIC). (2004, May 11). *Role-Based Access Control (RBAC) Role Engineering Process*. Retrieved from http://csrc.nist.gov/groups/SNS/rbac/documents/HealthcareRBACTFRoleEngineeringProcessv3.0.pdf

Seabrook, H., Stromer, J. N., Shevkenek, C., Bharwani, A., Grood, J., & Ghali, W. A. (2014). Medical applications: A database and characterization of apps in Applie iOS and Android Platforms. *BMC Research Notes*, *7*(1), 573. doi:10.1186/1756-0500-7-573 PMID:25167765

Security, L. M. (2015). *App vetting API*. Retrieved from https://www.mylookout.com/app-vetting-api

Sen, R., Maurya, A., Raman, B., Mehta, R., Kalyanaraman, R., Vankadhara, N., … Sharma, P. (2012). Kyun queue: a sensor network system to monitor road traffic queues. In *Proceedings of the 10th ACM Conference on Embedded Network Sensor Systems*. doi:10.1145/2426656.2426670

Sen, R., Cross, A., Vashistha, A., Padmanabhan, V. N., Cutrell, E., & Thies, W. (2013). Accurate speed and density measurement for road traffic in India. In *Proceedings of the 3rd ACM Symposium on Computing for Development*. doi:10.1145/2442882.2442901

Sen, R., Raman, B., & Sharma, P. (2010). Horn-ok-please. In *Proceedings of the 8th International Conference on Mobile Systems, Applications, and Services*. Doi:10.1145/1814433.1814449

Shankar, U., Talwar, K., Foster, J. S., & Wagner, D. (2001). Detecting Format String Vulnerabilities with Type Qualifiers. USENIX Security.

Sheng, X., Tang, J., & Zhang, W. (2012). Energy-efficient collaborative sensing with mobile Phones: in INFOCOM. Proceedings IEEE, 1916-1924.

Shen, H., Fang, J., & Zhao, J. (2011). *EFindBugs: Effective Error Ranking for FindBugs*. ICST.

Shin, M., Cornelius, C., Kapadia, A., Triandopoulos, N., & Kotz, D. (2015, June). Location Privacy for Mobile Crowd Sensing through Population Mapping. *Sensors (Basel, Switzerland)*, *15*(7), 15285–15310. doi:10.3390/s150715285 PMID:26131676

Signal, O. (2013). *Android fragmentation visualized.* Retrieved from opensignal. com: http://opensignal.com/reports/fragmentation-2013

Simon, R. T., & Zurko, M. E. (1997, June). Separation of duty in role-based environments. In *Computer Security Foundations Workshop, 1997. Proceedings., 10th* (pp. 183-194). IEEE.

Singh, A., Naik, V., Lal, S., Sengupta, R., Saxena, D., Singh, P., & Puri, A. (2011). Improving the efficiency of healthcare delivery system in underdeveloped rural areas. *2011 3rd International Conference on Communication Systems and Networks, COMSNETS 2011.*

Singh, P., Juneja, N., & Kapoor, S. (2013). Using mobile phone sensors to detect driving behavior. *Proceedings of the 3rd ACM Symposium on Computing for Development - ACM DEV '13.* doi:10.1145/2442882.2442941

Slevin, L. A., & Macfie, A. (2007). Role Based Access Control for a Medical Database. In *Proceedings of Software Engineering and Applications* (pp. 195–199). SEA.

SlideShare. (2012). *Constrained RBAC diagram.* Retrieved from http://image.slidesharecdn.com/rbac6576-121205031439-phpapp01/95/rbac-18-638.jpg?cb=1354677352

SlimFramework. (n.d.). *Slim framework.* Retrieved from http://www.slimframework.com/

Smith, A. (2015, April 1). *U.S. Smartphone Use in2015.* Retrieved from http://www.pewinternet.org/2015/04/01/us-smartphone-use-in-2015/

Snelting, Robschink, & Krinke. (2006). Efficent Path Conditions in Dependence Graphs for Software Safety Analysis. *TOSEM*, *15*(4).

Splunk. (n.d.). Retrieved December 10, 2015, from http:// www.splunk.com

SPSS. (n.d.). Retrieved December 10, 2015, from http://www.ibm.com/software/analytics/spss

Sridharan, M., Artzi, S., Pistoia, M., Guarnieri, S., Tripp, O., & Berg, R. (2011). *F4F: Taint Analysis of Framework-based Web Applications.* OOPSLA. doi:10.1145/2048066.2048145

Sridharan, M., & Bodík, R. (2006). Refinement-based Context-sensitive Points-to Analysis for Java. In *ACM SIGPLAN Conference on Programming Language Design and Implementation (PLDI 2006).* doi:10.1145/1133981.1134027

Sridharan, M., Fink, S. J., & Bodík, R. (2007). *Thin Slicing.* PLDI. doi:10.1145/1250734.1250748

Srikant, R., & Agrawal, R. (1996). *Mining sequential patterns: Generalizations and performance improvements.* Springer Berlin Heidelberg.

Srinivasan, V., Vardhan, V., Kar, S., Asthana, S., Narayanan, R., Singh, P., … Seth, A. (n.d.). Airavat: An automated system to increase transparency and accountability in social welfare schemes in India. In *Proceedings of the 6th International Conference on Information and Communications Technologies and Development.*

Srinivasan, V., Vardhan, V., Kar, S., Asthana, S., Narayanan, R., & Singh, P. (2013). Airavat: An Automated System to Increase Transparency and Accountability in Social Welfare Schemes in India. In *Proceedings of the Sixth International Conference on Information and Communications Technologies and Development Notes*. ACM Press.

SSB Bart Group. (2015, Mar 6). *AMP for Mobile*. Retrieved from http://info.ssbbartgroup.com/AMPforMobile.html

Starov, O., Vilkomir, S., & Kharchenko, V. (2013). Cloud Testing for Mobile Software Systems Concept and Prototyping.*8th International Conference on Software Engineering and Applications (ICSOFT-EA)*

State of Connecticut. (n.d.). *An Act Concerning Young Athletics and Concussions*. Retrieved from http://www.cga.ct.gov/2014/act/pa/pdf/2014PA-00066-R00HB-05113-PA.pdf

Statista. (2015). *Number of apps available in leading app stores as of July 2015.* Retrieved from http://www.statista.com/statistics/

Stonebraker, M., Bruckner, D., Ilyas, I. F., Beskales, G., Cherniack, M., Zdonik, S. B., & Xu, S. (2013, January). Data Curation at Scale: The Data Tamer System. In *Proceedings of the Conference on Innovative Data Systems Research. CIDR 2013.*

Stormpath. (2011). *Stormpath*. Retrieved from https://stormpath.com/

Stormpath. (2015). *User Authorization Management*. Retrieved from https://stormpath.com/product/authorization

Sujansky, W. V., Faus, S. A., Stone, E., & Brennan, P. F. (2010). A method to implement fine-grained access control for personal health records through standard relational database queries. *Journal of Biomedical Informatics*, *43*(5), 46–50. doi:10.1016/j.jbi.2010.08.001 PMID:20696276

Sundelin, T. L. (2003). *Surrogate Trust Negotiation: Solving Authentication and Authorization Issues in Dynamic Mobile Networks*. Brigham Young University.

Sweeney, L. (2002). k-anonymity: A model for protecting privacy. *International Journal of Uncertainty, Fuzziness and Knowledge-based Systems*, *10*(05), 557–570. doi:10.1142/S0218488502001648

Symantec. (2014, October 7). *Securing Mobile App Data - Comparing Containers and App Wrappers*. Retrieved from https://www.symantec.com/content/en/us/enterprise/white_papers/b-securing-mobile-app-data-comparing-containers-wp-21333969.pdf

Tang, J., Song, Y., Miller, H. J., & Zhou, X. (2015). Estimating the most likely space-time paths, dwell times and path uncertainties from vehicle trajectory data: A time geographic method. *Transportation Research Part C, Emerging Technologies*. doi:10.1016/j.trc.2015.08.014

Tateishi, T., Pistoia, M., & Tripp, O. (2011). *Path- and Index-sensitive String Analysis Based on Monadic Second-order Logic*. ISSTA. doi:10.1145/2001420.2001441

The Johns Hopkins University Applied Physics Laboratory. (2014, December 19). *Digital Policy Management Framework for Attribute-Based Access Control*. Retrieved from https://www.ise.gov/sites/default/files/DigitalPolicyFramework-ABAC.pdf

Thepvilojanapong, N., Zhang, K., Tsujimori, T., Ohta, Y., Zhao, Y., & Tobe, Y. (2013). Participation-Aware Incentive for Active Crowd Sensing. In *Proceedings of the 11th IEEE/IFIP International Conference on Embedded and Ubiquitous Computing (EUC).*

Thiagarajan, A., Ravindranath, L., Balakrishnan, H., Madden, S., & Girod, L. (2011). Accurate, low-energy trajectory mapping for mobile devices. In *Proceedings of the 8th USENIX Conference on Networked Systems Design and Implementation.*

Thiagarajan, A., Ravindranath, L., LaCurts, K., Madden, S., Balakrishnan, H., Toledo, S., & Eriksson, J. (2009). Vtrack: Accurate, energy-aware road traffic delay estimation using mobile phones. In *Proceedings of the 7th ACM Conference on Embedded Networked Sensor Systems, SenSys '09.* doi:10.1145/1644038.1644048

Torgan, C. (2009, November 6). The mHealth Summit: Local & Global Converge. *Kinetics: From Lab Bench to Park Bench.* Retrieved from http://caroltorgan.com/mhealth-summit/

Torres, M. H. C., Haesevoets, R., & Holvoet, T. (2013). Coos: coordination support for mobile collaborative applications. In *Mobile and Ubiquitous Systems: Computing, Networking, and Services* (pp. 152–163). Springer. doi:10.1007/978-3-642-40238-8_13

Tripp, O., Pistoia, M., Cousot, P., Cousot, R., & Guarnieri, S. (2013). Andromeda: Accurate and Scalable Security Analysis of Web Applications. FASE.

Tripp, O., Pistoia, M., Fink, S. J., Sridharan, M., & Weisman, O. (2009). TAJ: Effective Taint Analysis of Web Applications. PLDI.

Tripp, O., Ferrara, P., & Pistoia, M. (2014). *Hybrid Security Analysis of Web JavaScript Code via Dynamic Partial Evaluation.* ISSTA. doi:10.1145/2610384.2610385

Tsai, P. S., Lee, C. C., & Chen, A. L. (1999). An efficient approach for incremental association rule mining. In *Methodologies for Knowledge Discovery and Data Mining* (pp. 74–83). Springer Berlin Heidelberg. doi:10.1007/3-540-48912-6_10

Ubuntu Manuals. (n.d.). Retrieved May 31, 2016, from http://manpages.ubuntu.com/manpages/wily/man1/aapt.1.html

US Access Board. (2000, December 21). *Section 508 Standards for Electronic and Information Technology.* Retrieved from https://www.access-board.gov/guidelines-and-standards/communications-and-it/about-the-section-508-standards/section-508-standards

US Census Bureau. (2015, July 25). *Nearly 1 in 5 People Have a Disability in the U.S., Census Bureau Reports.* Retrieved from https://www.census.gov/newsroom/releases/archives/miscellaneous/cb12-134.html

Ushahidi. (n.d.). Retrieved from the Ushahidi Wiki: https://wiki.ushahidi.com/pages/viewpage.action?pageId=13598724

Uszok, A., Bradshaw, J., Jeffers, R., Suri, N., Hayes, P., Breedy, M., & Lott, J. (2003, June). KAoS policy and domain services: Toward a description-logic approach to policy representation, deconfliction, and enforcement. In *Policies for Distributed Systems and Networks, 2003. Proceedings. POLICY 2003. IEEE 4th International Workshop on* (pp. 93-96). IEEE.

Uther, M., Singh, P., & Uther, J. (2005). Mobile Adaptive CALL (MAC): An adaptive s/w for computer assisted language learning. *Proceedings - IEEE International Conference on Pervasive Services, ICPS 2005* (Vol. 2005, pp. 413-416).

Uther, M., Uther, J., Athanasopoulos, P., Singh, P., & Akahane-Yamada, R. (2007). Mobile adaptive CALL (MAC). A lightweight speech-based intervention for mobile language learners. *Proceedings of the Annual Conference of the International Speech Communication Association, INTERSPEECH* (Vol. 2, pp. 1445-1448).

Uther, M., Uther, J., Athanasopoulos, P., Singh, P., & Reiko, A. Y. (2007). *Mobile Adaptive CALL (MAC): A lightweight speech-based intervention for mobile language learners*. Academic Press.

Uther, M., Zipitria, I., Uther, J., & Singh, P. (2005). Mobile Adaptive CALL (MAC): A case-study in developing a mobile learning application for speech/audio language training. *Proceedings - IEEE International Workshop on Wireless and Mobile Technologies in Education, WMTE 2005* (Vol. 2005, pp. 187-191).

van der Horst, T. W., Sundelin, T., Seamons, K. E., & Knutson, C. D. (2004). Mobile Trust Negotiation: Authentication and Authorization in Dynamic Mobile Networks. *Proc. of the Eighth IFIP Conference on Communications and Multimedia Security*.

van der Horst, T. W., Sundelin, T., Seamons, K. E., & Knutson, C. D. (2005). Mobile Trust Negotiation. In D. a. Chadwick (Ed.), *Communications and Multimedia Security* (Vol. 175, pp. 97–109). Springer. doi:10.1007/0-387-24486-7_7

Vawdrey, D. K., Sundelin, T. L., Seamons, K. E., & Knutson, C. D. (2003). Trust Negotiation for Authentication and Authorization in Healthcare Information Systems. *Engineering in Medicine and Biology Society, 2003.Proceedings of the 25th Annual International Conference of the IEEE* (pp. 1406-1409). IEEE.

Ventola, C. L. (2014, May). Mobile Devices and Apps for Health Care Professionals: Uses and Benefits. *Pharmacy and Therapeutics, 39*(5), 356-364. Retrieved from http://www.ncbi.nlm.nih.gov/pmc/articles/PMC4029126

Venturebeat Staff. (2014, September 29). Mobile Apps 2015: 5 Predictions of How Apps Will Change. *VentureBeat*. Retrieved from http://venturebeat.com/2014/09/29/mobile-apps-2015-5-predictions-of-how-apps-will-change/

Vilkomir, S., Marszalkowski, K., Perry, C., & Mahendrakar, S. (2015). Effectiveness of multi-device testing mobile applications. In *Mobile Software Engineering and Systems (MOBILESoft), 2015 2nd ACM International Conference on* (pp. 44-47). IEEE. doi:10.1109/MobileSoft.2015.12

Volpano, Irvine, & Smith. (1996). A Sound Type System for Secure Flow Analysis. *JCS, 4*(2-3).

Vosloo & Kourie. (2008). Server-centric Web Frameworks: An Overview. *ACM Comput. Surv., 40*(2), 4:1–4:33.

W3C. (2008, December 11). *Web Content Accessibility Guidelines (WCAG) 2.0*. Retrieved from https://www.w3.org/TR/WCAG20/

Wadhwa, R., Mehra, A., Singh, P., & Singh, M. (2015). A pub/sub based architecture to support public healthcare data exchange. *Communication Systems and Networks (COMSNETS).7th International Conference*. doi:10.1109/COMSNETS.2015.7098706

Wadhwa, R., Singh, P., Singh, M., & Kumar, S. (2015). An EMR-enabled medical sensor data collection framework. *Communication Systems and Networks (COMSNETS),7th International Conference*.

WAI-ARIA Overview. (n.d.). Retrieved May 31, 2016, from https://www.w3.org/WAI/intro/aria

Wainwright, A. (2012, June 21). *7 Benefits of BYOD on Enterprise Wireless Networks*. Retrieved from http://www.securedgenetworks.com/blog/7-Benefits-of-BYOD-on-Enterprise-Wireless-Networks

Wang, D., Kaplan, L., Abdelzaher, T., & Aggarwal, C. (2013). On Credibility Tradeoffs in Assured Social Sensing. *JSAC, 31*(6), 1026 – 1037.

Wang, S., Su, L., Li, S., & Hu, S. (2015). Scalable Social Sensing of Interdependent Phenomena. In *The 14th ACM/IEEE Conference on Information Processing in Sensor Networks (IPSN)*. doi:10.1145/2737095.2737114

Wang, S., Wang, D., Su, L., & Kaplan, L. (2014). Towards Cyber-physical Systems in Social Spaces: The Data Reliability Challenge. In *IEEE 35th Real-Time Systems Symposium (RTSS)*.

Wang, Y., & Uzum, A. (2012). Tracommender – Exploiting Continuous Background Tracking Information on Smartphones for Location-Based Recommendations. In *Proceedings of the 5th International Conference, Mobile Wireless Middleware, Operation Systems, and Applications*. Berlin, Germany: Springer.

Wang, D., Kaplan, L., & Abdelzaher, T. (2014). On Truth Discovery in Social Sensing with Conflicting Observations: A Maximum Likelihood Estimation Approach. *ACM Transactions on Sensor Networks*, *10*(2), 30. doi:10.1145/2530289

Wang, D., Kaplan, L., Le, H., & Abdelzaher, T. (2012). On Truth Discovery in Social Sensing: A Maximum Likelihood Estimation Approach. In *11th ACM/IEEE Conference on Information Processing in Sensor Networks (IPSN)*. doi:10.1145/2185677.2185737

Wassermann, G., & Su, Z. (2007). *Sound and Precise Analysis of Web Applications for Injection Vulnerabilities*. PLDI. doi:10.1145/1250734.1250739

Wassermann, G., & Su, Z. (2008). *Static Detection of Cross-site Scripting Vulnerabilities*. ICSE. doi:10.1145/1368088.1368112

Weiss, N. E., & Miller, R. S. (2015, February). The Target and Other Financial Data Breaches: Frequently Asked Questions. In *Congressional Research Service, Prepared forMembers and Committees of Congress February* (*Vol. 4*).

Weitzner, D. J., Abelson, H., Berners-Lee, T., Feigenbaum, J., Hendler, J., & Sussman, G. J. (2008). Information accountability. *Communications of the ACM*, *51*(6), 82–87. doi:10.1145/1349026.1349043

West, D. (2012) How Mobile Devices are Transforming Healthcare. *Issues in Technology Innovation*, *19*. Retrieved from http://www.brookings.edu/~/media/research/files/papers/2012/5/22-mobile-health-west/22-mobile-health-west.pdf

West, D., & Miller, E. A. (2009). Digital Medicine: Health Care in the Internet Era. Brooking Institution Press.

Westin, A. F. (1970). *Privacy and freedom*. Academic Press.

Whaley, J., & Lam, M. S. (2004). *Cloning Based Context-Sensitive Pointer Alias Analysis Using Binary Decision Diagrams*. PLDI. doi:10.1145/996841.996859

Wheeler, D. A. (2004). *SLOC count user's guide*. Retrieved from http://www.dwheeler.com/sloccount/sloccount.html

White, K. (2015). Determining Accessibility for iOS Applications: Piloting a Checklist for Practitioners. Theses and Dissertations. University of Wisconsin-Milwaukee.

Wiech, D. (2013, April 17). *Role-Based Access Control for Healthcare Data Security*. Retrieved from http://healthcare-executive-insight.advanceweb.com/Features/Articles/Role-based-Access-Control-for-Healthcare-Data-Security.aspx

Winsborough, W. H., Seamons, K. E., & Jones, V. E. (2000). *Automated trust negotiation. In DARPA Information Survivability Conference and Exposition, 2000. DISCEX '00. Proceedings* (pp. 88–102). Hilton Head, SC: IEEE; doi:10.1109/DISCEX.2000.824965

Witten, I. H., & Frank, E. (2005). *Data Mining: Practical Machine Learning Tools and Techniques*. Morgan Kaufmann.

World Health Organization and The World Bank. (2011). *World Report on Disability*. Retrieved from http://www.who.int/disabilities/world_report/2011/en/

X.509. (n.d.). *Standard*. Retrieved from https://tools.ietf.org/html/rfc5280

Xiao, Z., & Xiao, Y. (2013). Security and privacy in cloud computing. *IEEE Communications Surveys and Tutorials*, *15*(2), 843–859. doi:10.1109/SURV.2012.060912.00182

Xiong, H., Zhang, D., Wang, L., Gibson, J. P., & Zhu, J. (2015). \Eemc: Enabling energy-e_cient mobile crowd sensing with anonymous participants. *ACM Transactions on Intelligent Systems and Technology*, *6*(3), 39. doi:10.1145/2644827

Xu, H., Zhou, Y., & Lyu, M. R. (2014). Towards Continuous and Passive Authentication via Touch Biometrics: An Experimental Study on Smartphones.*Symposium On Usable Privacy and Security (SOUPS 2014)* (pp. 187-198). Menlo Park, CA: USENIX Association.

Yadav, K., Naik, V., Singh, A., Singh, P., & Chandra, U. (2012). Low energy and sufficiently accurate localization for non-smartphones. *Proceedings - 2012 IEEE 13th International Conference on Mobile Data Management, MDM 2012* (pp. 212-221). doi:10.1109/MDM.2012.32

Yadav, K., Naik, V., Singh, A., Singh, P., Kumaraguru, P., & Chandra, U. (2010). *Challenges and Novelties While Using Mobile Phones As ICT Devices for Indian Masses: Short Paper*. Academic Press.

Yaeli, A., & Bak, P. (2014). Understanding Customer Behavior Using Indoor Location Analysis and Visualization. IBM Systems Journal, 58(5-6).

Yan, D., Xu, G., & Rountev, A. (2011). Demand-driven context-sensitive alias analysis for java. In *Proceedings of the 2011 International Symposium on Software Testing and Analysis*. doi:10.1145/2001420.2001440

Yarmand, M. H., Sartipi, K., & Down, D. G. (2008, June). Behavior-based access control for distributed healthcare environment. In *Computer-Based Medical Systems, 2008. CBMS'08. 21st IEEE International Symposium on* (pp. 126-131). IEEE. doi:10.1109/CBMS.2008.14

Ye, Y., Zheng, Y., Chen, Y., Feng, J., & Xie, X. (2009, May). Mining individual life pattern based on location history. In *Mobile Data Management: Systems, Services and Middleware, 2009. MDM'09. Tenth International Conference on* (pp. 1-10). IEEE. doi:10.1109/MDM.2009.11

Yoon, J., Noble, B., & Liu, M. (2007). Surface street traffic estimation. In *Proceedings of the 5th International Conference on Mobile Systems, Applications and Services*. Doi:10.1145/1247660.1247686

You, C. W., Lane, N. D., Chen, F., Wang, R., Chen, Z., Bao, T. J., & Campbell, A. T. (2013). CarSafe app: alerting drowsy and distracted drivers using dual cameras on smartphones. In *Proceeding of the 11th Annual International Conference on Mobile Systems, Applications, and Services*. Doi:10.1145/2462456.2465428

Yu, J., Wang, G., & Mu, Y. (2012). Provably Secure Single Sign-on Scheme in Distributed Systems and Networks. *TRUSTCOM '12 Proceedings of the 2012 IEEE 11th International Conference on Trust, Security and Privacy in Computing and Communications* (pp. 271-278). Washington, DC: IEEE.

Zargar, S. T., Joshi, J., & Tipper, D. (2013). A survey of defense mechanisms against distributed denial of service (DDoS) flooding attacks. *IEEE Communications Surveys and Tutorials*, *15*(4), 2046–2069. doi:10.1109/SURV.2013.031413.00127

Zhang, X., Parisi-Presicce, F., Sandhu, R., & Park, J. (2005). Formal model and policy specification of usage control. *ACM Transactions on Information and System Security*, *8*(4), 351–387. doi:10.1145/1108906.1108908

Zhao, D., Li, X. Y., & Ma, H. (2014). How to crowdsource tasks truthfully without sacrificing utility: online incentive mechanisms with budget constraint. In Proceedings of IEEE INFOCOM. doi:10.1109/INFOCOM.2014.6848053

Zheng, X., & Rugina, R. (2008). Demand-driven alias analysis for c. In *Proceedings of the 35th annual ACM SIGPLAN-SIGACT Symposium on Principles of Programming Languages*. doi:10.1145/1328438.1328464

Zuo, J., Ratha, N., & Connell, J. (2008). Cancelable Iris Biometric. *Pattern Recognition, 2008. ICPR 2008. 19th International Conference on* (pp. 1-4). Tampa, FL: IEEE. doi:10.1109/ICPR.2008.4761886

About the Contributors

Sougata Mukherjea is currently the Program Director of Hybrid Integration Solutions in IBM Global Technology Services. Prior to that he was the head of the Telecom & Mobile Research department in IBM India Research Lab. His research interests include Middleware technologies and its applications to Mobile, Analytics and Visualization. He has several patents as well as multiple publications in reputed Computer Science conferences and journals in these research areas. His group has developed many innovative technologies in the Telecom and Mobile domain some of which have been integrated into IBM products or licensed to customers. Sougata has been awarded an IBM Outstanding Technical Achievement award for his work in the area of Network Analytics. Sougata received his Ph.D. from Georgia Institute of Technology, USA in Computer Science. Prior to joining IBM Sougata worked in Research & Development for several companies in the Silicon Valley (California) including NEC Research, Inktomi and BEA Systems.

* * *

Karl Aberer is a full professor for Distributed Information Systems at EPFL Lausanne, Switzerland, since 2000. His research interests are on decentralization and self-organization in information systems with applications in peer-to-peer search, semantic web, trust management and social, mobile and sensor networks. Karl Aberer received his Ph.D. in mathematics in 1991 from the ETH Zürich. From 1991 to 1992 he was postdoctoral fellow at the International Computer Science Institute (ICSI) at the University of California, Berkeley. In 1992 he joined the Integrated Publication and Information Systems institute (IPSI) of GMD in Germany, where he was leading the research division Open Adaptive Information Management Systems. From 2005 to 2012 he was the director of the Swiss National Research Center for Mobile Information and Communication Systems (NCCR-MICS, www.mics.ch). Since September 2012 he is Vice-President of EPFL responsible for information systems. He is member of the editorial boards of VLDB Journal, ACM Transaction on Autonomous and Adaptive Systems and World Wide Web Journal.

Thomas P. Agresta is a Full Professor and Director of Medical Informatics in the Department of Family Medicine at the University of Connecticut Health Center, Director of Clinical Informatics for the Biomedical Informatics Division of the Center for Quantitative Medicine, and is the Section Leader for Informatics at the Connecticut Institute for Primary Care Innovation. He is a family physician with more than 25 years of experience caring for patients at Family Medicine Center in Hartford, CT. He is Board Certified in Clinical Informatics and is nationally known for his innovative approach to teaching students, residents and physicians to use technology at point of care to provide patient-centered, evidence-based care.

Tom Brunet is a Senior Software Engineer for IBM Accessibility, part of IBM research. Tom's work in accessibility began in 2000, building content extraction components for IBM Homepage Reader, the first Javascript enabled browser for the visually impaired. His recent work has focused on automated compliance verification and tracking systems to support rich internet applications, EPUB documents, integration into continuous integration processes, and other scenarios to support IBM's commitment to accessibility. Tom received a BS in Computer Science from the University of Texas at Austin, and an MS in Computer Science with a focus in Biomedical Imaging from the University of Wisconsin-Madison.

Michele Catasta is a research scientist and lecturer in Data Science at EPFL, Switzerland. During his PhD (EPFL, 2015), he let human memories and information systems have their first dance. To make this debut happen, he added new bells and whistles (human computation, machine learning, psychology) to his original researcher hat (big data analytics, information retrieval, semantic technologies). Michele was in the founding team of Sindice.com, the largest Semantic Web search engine (now SIREn Solutions). He also worked for MIT Media Lab, Google and Yahoo Labs. In the past years, he received several awards and recognitions - among them, a focused grant from Samsung Research USA.

Deepthi Chander is a Research Scientist in the Distributed and Mobile Computing Group, XRCI.

George Christodoulou has received his Bachelor in computer science from the University of Cyprus in 2012. He was accepted for a Master's program from Ecole Polytechnique Fédérale de Lausanne and was awarded the academic excellence scholarship. He completed his MS with a specialization in Internet Computing in 2015. Since graduating he has been working for Cisco Systems in the areas of networking, wireless and mobility.

Joanne Conover is an Associate Professor in the Department of Physiology and Neurobiology at the University of Connecticut, with research interests of: developmental neurobiology, stem cell biology and neurodegenerative diseases. Dr. Conover chairs the Goldwater Nomination Committee and is involved with SURF, University Scholar and McNair fellowship programs, and is the Director of UConn's Women in Math, Science, and Engineering (WiMSE) Learning Community from 2013-2015. Dr. Conover has over 32 archival publications, in the following categories: 32 peer-reviewed journal articles and 2 book chapters.

Prajit Kumar Das is a PhD candidate in the Ebiquity Research Group at UMBC. His research focuses on the primary goal of achieving efficient, context-driven access control on mobile platforms. He has previously worked in the industry as a developer for the largest software exporter in the world, Tata Consultancy Services Ltd., India. He has also worked as an intern at Samsung, Apple and Symantec Corporation. He is an experienced backend developer with years of experience in data management and analytics. He is also an expert of mobile app development and web development. He did his undergraduate degree in Engineering with Information Technology major, from Jadavpur University in India.

Koustuv Dasgupta manages the Distributed and Mobile Computing group at Xerox Research Centre India (XRCI). In his current role, Koustuv leads research topics that leverage the cloud infrastructure to create large-scale, optimized and sustainable service offerings. His work primarily focuses on the use of novel technologies for improved collaboration, communication, and process optimization in agile enterprises. He is also responsible for growing a competency around mobile and pervasive systems that can be leveraged for solving societal-scale problems and impact multiple domains ranging from smarter cities, to better health and education, as well as increased productivity and user engagement. Prior to XRCI, Koustuv was a Research Scientist at the IBM India Research Lab - where he contributed to multiple innovations in the areas of Telecom Service Delivery Platforms, contextual applications and architectures for next-generation networks; and predictive analytics for business intelligence. His work has led to ten US patents (numerous pending grant) and 40+ research publications in premier journals and conferences; while receiving several awards including an IBM Research Division award for contributions to the IBM Telecom Service Delivery Platform. Outside XRCI, Koustuv continues to collaborate with faculty and students in leading academic institutions, in the areas of mobile and pervasive computing. He serves on the committees of premiere conferences and workshops, including the role of Industry Chair for the ACM 2014 conference on Distributed Event Based Systems (DEBS). Koustuv holds a Ph.D. in Computer Science from the University of Maryland Baltimore County.

Steven A. Demurjian is a Full Professor in Computer Science & Engineering at the University of Connecticut, and co-Director of Research Informatics for the Biomedical Informatics Division, with research interests of: secure-software engineering, security for biomedical applications, and security-web architectures. Dr. Demurjian has over 150 archival publications, in the following categories: 1 book, 2 edited collections, 58 journal articles and book chapters, and 98 refereed conference/workshop articles.

Kuntal Dey is a Senior Research Software Engineer in IBM Research India. An alumnus of the Department of Computer Science and Engineering in IIT Bombay, Dey has has worked across multiple research disciplines at IBM Research India. This includes social network analysis and mobile analytics. Prior to IBM Research India, Dey was a part of the R&D division at Microsoft IDC. Dey has a number of research publications and patents to his credit.

Michael Diamond is an undergraduate student in Computer Science at Pomona College and spent the summer funded on a National Science Foundation Research Experience for Undergraduates (REU) in Trustable Computing Systems.

Vijay Ekambaram is a seasoned Mobile & Wearable Technologist currently working at IBM Research focusing on product ideation, innovation and development starting from conception to launch. In 3+ years of overall industrial experience at IBM Research Labs and Intel R&D, he has invented over 90+ US Patents (file-rated applications) in mobile/wearable/IoT domain and he is one of the youngest IBM Master Inventor. He holds a Masters degree in Computer Science from IIT Madras. He is specialized in Mobile/Wearable Computing, Software Engineering/Security/Analytics for Mobile/Wearable Apps/ Frameworks, Android Internals, Wireless Communications and Content Centric Networking.

Sharanya Eswaran is currently a Research Scientist at Xerox Research Center India. Her core research interests lie in the area of mobile communications, crowd sourcing and IoT. She holds a PhD degree in Computer Science and Engineering from Pennsylvania State University.

Tim Finin is a Professor of Computer Science and Electrical Engineering at the University of Maryland, Baltimore County (UMBC). He has over 30 years of experience in applications of Artificial Intelligence to problems in information systems and language understanding. His current research is focused on the Semantic Web, mobile computing, analyzing and extracting information from text and online social media, and on enhancing security and privacy in information systems. He is AAAI Fellow, received an IEEE Technical Achievement award in 2009 and was selected as the UMBC Presidential Research Professor in 2012. Finin received an S.B. degree in Electrical Engineering from MIT and a Ph.D. degree in Computer Science from the University of Illinois at Urbana-Champaign. He has held full-time positions at UMBC, Unisys, the University of Pennsylvania, and the MIT AI Laboratory. He is the author of over 300 refereed publications and has received research grants and contracts from a variety of sources. He participated in the DARPA/NSF Knowledge Sharing Effort and helped lead the development of the KQML agent communication language and was a member of the W3C Web Ontology Working Group that standardized the OWL Semantic Web language. Finin has chaired of the UMBC Computer Science Department, served on the board of directors of the Computing Research Association, been a AAAI councilor, and chaired several major research conferences. He is currently an editor-in-chief of the Elsevier Journal of Web Semantics and a co-editor of the Viewpoints section of the Communications of the ACM.

Ivan Gavrilovic obtained Bachelor's degree at University of Belgrade (2012), followed by Master's degree in Computer Science at Ecole Polytechnique Fédérale de Lausanne (2015). In 2016 he was hired by Google where he currently works as a software engineer. Following his affiliation with the mobile applications development, currently his interest is in the Android build system, a tool used by Android developers to build their applications.

Dibyajyoti Ghosh is currently a software engineer in Adobe Systems Inc for Acrobat document services. Did graduate research on Android privacy, semantic web based context modeling and NASA GEOSv5 weather model computational improvement. Previously, a backend developer in network security startup iViZ Security (acquired by Cigital Inc). Interned with Symantec Corporation intrusion detection systems group.

Pramod Jagtap is a software development engineer with extensive experience in "Web Application Security" domain. Currently working with Amazon to build Trade-In program that allows customers to receive an Amazon Gift Card in exchange for hundreds of thousands of eligible items including phones, electronics, video games, books, DVDs, and CDs.

Anupam Joshi is the Oros Family Professor and Chair of Computer Science and Electrical Engineering Department at the University of Maryland, Baltimore County(UMBC). He is the Director of UMBC's Center for Cybersecurity, and the Co-Technical Director of the newly announced National Cybersecurity FFRDC. He is a Fellow of IEEE. Dr. Joshi obtained a B.Tech degree from IIT Delhi in 1989, and a Masters and Ph.D. from Purdue University in 1991 and 1993 respectively. His research interests are in the broad area of networked computing and intelligent systems. His primary focus has been on data management and security/privacy in mobile/pervasive computing environments, and policy driven approaches to security and privacy. He is also interested in Semantic Web and Data/Text/Web Analytics, especially their applications to (cyber) security. He has published over 200 technical papers with an h-index of 71 and over 17000 citations (per Google scholar), filed and been granted several patents, and has obtained research support from National Science Foundation (NSF), NASA, Defense Advanced Research Projects Agency (DARPA), US Dept of Defense (DoD), NIST, IBM, Microsoft, Qualcom, Northrop Grumman, and Lockheed Martin amongst others.

David Lubensky is a Senior Manager of Collaborative Technologies and Analytics and the Director of the Center for Mobile Enterprise Research (CMER) at the IBM T.J. Watson Research Center. He received his B.S. in Computer Science and M.S. in Electrical Engineering from Drexel University in 1984 and 1987, respectively. He joined Siemens Research in Princeton NJ in 1984 through 1989, where he was a Senior Researcher responsible for R&D of speech recognition algorithms and multimodal solutions targeted towards eyes/hands busy applications in healthcare and manufacturing. In 1989-1995, he joined Verizon Science & Technology Center in White Plains NY, where he was leading projects in the area of large-scale network-based speech solutions, and was a core contributor to the first commercial deployment of Voice-Dialing solution. Since 1995, he has been with the IBM T.J. Watson Research Center. His research topics at IBM have been in speech recognition, machine translation, social and cognitive analytics and mobile enterprise software. He led the first deployment of the IBM telephony-based speech solution with a major commercial client. He also managed a large World Wide IBM team and was PI on the IBM commercialization of Real Time Translation solution which resulted in the first pervasive deployment of machine translation inside the global enterprise (a.k.a. n.Fluent). Recently, he developed strategic direction and responsible for execution plan for the multidisciplinary team in two major focus areas: Social and Cognitive Analytics, and Mobile Enterprise Software. Mr. Lubensky's team is currently leading delivery on 3 DARPA contracts: Anomaly Detection at Multiple Scales, Social Media in Strategic Communication, and Social and Cognitive Network Academic Research Center. In the Mobile area, Mr. Lubensky's team is developing technologies and solutions in support of an emerging enterprise trend of Bring Your Own Device (BYOD), addressing issues such as Application Scanning, Management, and Security. He is a Member of the IBM Academy of Technology, and the recipient of numerous IBM Invention and Outstanding Technical Achievement Awards. David Lubensky has over 20 publications and holds more than 40 issued and filed patents.

Tridib Mukherjee is a Senior Research Scientist at the Xerox Research Center India (XRCI). He works in the broad areas of Distributed Computing, Cloud Computing, Large-scale Systems, Green Computing, Sensor Networks, Mobile Computing, and Services Computing. He joined XRCI in September 2011 and has since been involved in exploring research and business opportunities in a sustainable marketplace consisting of cloud-based services. Tridib's work on CloudAdvisor, a cloud configuration recommendation system for enterprise application, won the prestigious IEEE Cloud Cup in 2013. He is currently working on platform optimizations in enterprise clouds for efficient service delivery as well as for large scale mobile, participatory, & pervasive sensing, data integration and analytics (especially for smarter city applications). Prior to joining Xerox, Tridib was a Postdoctoral Research Scholar at the Arizona State University where he worked on data center optimization and body sensor networks. Tridib has co-authored a book, multiple book chapters, and published nearly 50 papers in reputed journals and conferences. Tridib also has 3 granted US patents in addition to more than 30 US patents filed.

Amanda Murphy is an undergraduate student in Computer Science at Canisius College and spent the summer funded on a National Science Foundation Research Experience for Undergraduates (REU) in Trustable Computing Systems.

Sima Nadler is a Senior Program Manager in IBM's Research Division. She has 25 years of experience serving clients in multiple industries and has worked with customers worldwide on providing unique and innovative solutions. In her current global role Sima is the global Research leader for Retail and acts as the liaison between IBM Research and the Retail sales, services and development arms of IBM and its customers. Her personal Research focus is in the area of privacy, mobile, in-store location identification, and making the "Blurring of the Virtual and Physical" a reality. Sima invented and lead the development of the initial version of the Presence Insights product, IBM's solution for indoor location tracking and analytics. Recently she is focused on the privacy implications of this and other technologies. In 2015 she lead the Retail standards body (ARTS) committee focused on privacy that defined best practices and recommended changes to standards to increase privacy compliance.

Marco Pistoia, Ph.D., has worked for IBM Corporation since January 1996 and is currently an IBM Distinguished Researcher and Senior Manager at the IBM Thomas J. Watson Research Center in New York, where he manages the Mobile Enterprise Software research group. In January 2010, he was one of 38 IBM employees worldwide to be bestowed the title of IBM Master Inventor. He is the inventor of 119 patents issued by the United States Patent and Trademark Office, and 175 patent applications. Dr. Pistoia has designed and implemented numerous analysis components and contributed large amounts of code to IBM's two main products for static quality analysis: IBM Rational Software Analyzer and IBM Security AppScan Source. He has also contributed code and technology to the main IBM products in the area of mobile computing: IBM Fiberlink MaaS360, the IBM MobileFirst Platform Development Foundation, IBM Tealeaf CX Mobile, and IBM Rational Test Workbench. Dr. Pistoia has written ten books and published numerous papers and journal articles on various aspects of Program Analysis and Language-Based Security. Most recently, he has published his Ph.D. thesis, and has been the lead author of the books Enterprise Java Security, published by Addison-Wesley in 2004 (and available in Chinese since 2006), and Java 2 Network Security, published by Prentice Hall in 1999, both used as textbooks in many universities worldwide. Along with his colleague Omer Tripp, he is now writing a new book, Usable Program Security Analysis, to be published by MIT Press. He has published and presented at numerous conferences worldwide, including OOPSLA, ECOOP, PLDI, ICSE, ACSAC, ISSTA, CCS, PLAS and the IEEE Symposium on Security and Privacy. He has also been invited to lecture at several research institutions worldwide, including Harvard University, New York University, University of Maryland, Rutgers University, Virginia Tech, Stony Brook University, University of Texas at Austin and Stevens Institute of Technology in the United States, Tohoku University and the National Institute of Informatics in Japan, École Normale Supérieure in France, Dagstuhl School of Informatics and Saarland University in Germany, Eidgenössische Technische Hochschule (ETH) Zürich in Switzerland, La Sapienza University and Tor Vergata University in Italy, Tel Aviv University, Israel Institute of Technology (Technion) and Ben Gurion University of the Negev in Israel, University of Porto in Portugal, Chalmers University of Technology in Sweden, and The Royal Society in the United Kingdom. He has been an Adjunct Professor of Computer Science at New York University, Polytechnic School of Engineering since 2000 and at Fordham University since 2015. He was the General and Program Co-chair of PLAS 2008, and the Program Chair of the ACM Student Research Competition at PLDI 2009. Furthermore, he has served as Program Committee member on several conferences, including ICSE 2012 and 2017, ICST 2012, ISSTA 2011, PLAS 2007, 2009, 2010, 2011 and 2012, 2014 and 2015, NDSS 2009, IEEE SSIRI 2009, 2010 and 2011, IEEE SERE 2012, ACSAC 2008 and 2009, CISIM 2012, PLDI 2016 and 2017 and CCS 2016. Dr. Pistoia received his Ph.D. in Mathematics from the New York University, Polytechnic School of Engineering in May 2005 with a thesis entitled A Unified Mathematical Model for Stack- and Role-Based Authorization Systems, and his Master of Science and Bachelor of Science degrees in Mathematics summa cum laude from the University of Rome, Italy in July 1995, with a thesis entitled Theory of Reductive Algebraic Groups and Their Representations. His mathematical interests include lattices and invariant theory. His computer interests include mobile-code security, program analysis and secure language design. Dr. Pistoia has been the recipient of several awards, including three ACM SIGSOFT Distinguished Paper Awards (2007, 2011 and 2014), an IBM Research Pat Goldberg Memorial Best Paper Award (3 papers selected our of 130), an IBM Research Outstanding Technical Achievement Award, two IBM Research Outstanding Innovation Awards, four IBM Research Division Awards, and a European Community Erasmus Fellowship Award. In September 2007, the Italian Ministry of Education, University and Research, the National Committee of the Italian Presidents of Faculties of Sciences and Technologies, and Confindustria, Italy's leading organization representing all the Italian manufacturing and service companies, presented Pistoia as one of the 70 most successful Italian mathematicians who graduated from an Italian university between the years 1980 and 2000. His biography was published in the book Matematici al Lavoro. Dr. Pistoia became a citizen of the United States of America in March 2014. On a personal level, Marco is the father of two beautiful children, Juliet Alexandra, 6, and Charles William, 4. He is also a passionate body builder. His photograph was featured on page 28 ot the Winter 2015 issue of MuscleSport Magazine.

Horia Radu received his MSc in Computer Science from Ecole polytechnique fédérale de Lausanne (EPFL) in 2014. His Master Thesis was done inside the Distributed Information Systems laboratory, on MEmoIt and afterwards, he continued to work on the project as an intern. Currently he works as a Software Engineer for a start-up in Romania and as a part time Teaching Assistant for the "Politehnica" University of Timisoara (2016, Timisoara).

Nitendra Rajput defines and leads the MobileFirst agenda for IBM Research in India. His group invests in developing research technologies that can add value and help differentiate the IBM Mobile portfolio, be it a product, a service or an industry specific solution. Nitendra also drives the global IBM research strategy for MobileFirst by collaborating with research labs across Latin America, Asia and Europe. On the technical front, he is broadly interested in mobile interactions, HCI, statistical signal processing, recommender systems and multimodal search. Nitendra is an IBM Master Inventor and an IBM Academy of Technology member. In 2012, he co-authored a book titled "Speech in Mobile and Pervasive Environments" that was published by John Wiley & Sons. Nitendra is an ACM Distinguished Scientist and a Senior Member of the IEEE.

P. G. Ramachandran is the Program Director of Advanced Tech. team within IBM Accessibility. He has extensive experience in applied research and software development. He holds a number of patents, has been leading the research of accessibility related technologies and incorporating those into IBM products and services. Here is Ram's LinkedIn profile - https://www.linkedin.com/in/reachout2ram.

Venkatraman Ramakrishna is a researcher in the Blockchain and Smart Contracts Group in IBM Research, India. Until the end of 2015, he was a member of the Analytics and Mobile Enabled Solutions (AMES) group in the IBM Research Lab. His research interests lie in the areas of mobile and ubiquitous computing, blockchain technology and smart contracts, security, analytics, mobile commerce, service-oriented computing, cloud computing, and human-computer interaction. Prior to joining IBM Research, he was a member of the Infrastructure team of Microsoft's Bing search engine in Redmond, WA. Ramakrishna graduated with a B.Tech. in Computer Science and Engineering from IIT Kharagpur in 2001. Subsequently he obtained an M.S. and Ph.D. in Computer Science from UCLA, where his advisor was Professor Peter Reiher. His doctoral dissertation is titled "Policy Management and Interoperation Through Negotiation in Ubiquitous Computing."

Jean-Eudes Ranvier obtained his engineer's degree from UTT (France) in 2012 with a specialisation in information systems and telecommunication. Since then, he is a PhD candidate at EPFL (Switzerland) in the distributed information systems laboratory of Professor Aberer. His research focuses on the inference of complex user's states based on wearable sensors. More concretely, using smartphone sensors and off-the-shelf sensing devices, he endeavours to capture certain physical and physiological features of the user.

Yaira K. Rivera Sánchez is a Ph.D. student at the Computer Science & Engineering department at the University of Connecticut. Her research interests include security for mobile applications, specifically in the authentication/authorization process of them (allow mobile applications to access, share, and exchange information from different sources/applications) and emphasizing on solutions that are suitable for applications in the biomedical and healthcare domains.

Eugene Sanzi is a Ph.D. student at the Computer Science & Engineering department at the University of Connecticut. His research interests include the usage of credentials to define user profiles that can be utilized to provide authentication to a system that a user has never been authorized to utilize for time-critical situations.

Xian Shao is a Ph.D. student at the Computer Science & Engineering department at the University of Connecticut. Her research interests include security for mobile applications, specifically to allow mobile applications on mobile devices to dynamically adjust permissions based on both the system being utilized and the actual location of the user/device.

Tiziano Signo received his MS in Computer Science from the École polytechnique fédérale de Lausanne in 2016, with a thesis on satellite navigation in orbit. In the same year he was hired as a Software Engineer by the Boeing Company in Köln, where he currently works. He is the Webmaster and one of the co-founders of BuddyNS, a growing startup that operates since 2010 providing secondary DNS service.

Pushpendra Singh is an Associate Professor at Indraprastha Institute of Information Technology (IIIT-Delhi). His research interests are in the areas of mobile computing, ICT for development, and Human-Computer Interaction.

Omer Tripp is a Research Staff Member and Techincal Lead, leading research on mobile security and privacy in the mobile enterprise software group under Dr Marco Pistoia. Dr. Tripp's team researches topics under the broad umbrella of mobile reputation analysis, including e.g. malware detection, static and dynamic detection of integrity vulnerabilities in mobile systems, detection and quantification of unauthorized information sharing (aka data leaks), usability aspects of mobile security and privacy, identification and characterization of library dependencies in obfuscated mobile code, run-time application self protection (RASP), and developer tools to address security and privacy threats in mobile software. Dr. Tripp's formal education is in the area of program analysis. He completed his Ph.D. at Tel Aviv University under the supervision of Prof Mooly Sagiv. Beyond his work on analysis and transformation of mobile software, he is also interested in concurrent programming and applications of program analysis therein, and in ways of integrating machine learning into program analysis.

Index

A

B

C

D

E

F

G

H

I

L

M

N

P

R

S

T

U

W

www.ingramcontent.com/pod-product-compliance
Ingram Content Group UK Ltd.
Pitfield, Milton Keynes, MK11 3LW, UK
UKHW050951290926
15470UKWH00005B/1762

9 781522 509455